# Applications of Artificial Intelligence in New Energy Technology Systems

# Applications of Artificial Intelligence in New Energy Technology Systems

Editors

**Mehdi Seyedmahmoudian**
**Alex Stojcevski**
**Ben Horan**
**Saad Mekhilef**

MDPI • Basel • Beijing • Wuhan • Barcelona • Belgrade • Manchester • Tokyo • Cluj • Tianjin

*Editors*
Mehdi Seyedmahmoudian
Swinburne University of Technology
Melbourne, Australia

Alex Stojcevski
Swinburne University of Technology
Melbourne, Australia

Ben Horan
Deakin University
Geelong, Australia

Saad Mekhilef
Swinburne University of Technology
Melbourne, Australia

*Editorial Office*
MDPI
St. Alban-Anlage 66
4052 Basel, Switzerland

This is a reprint of articles from the Special Issue published online in the open access journal *Sustainability* (ISSN 2071-1050) (available at: https://www.mdpi.com/journal/sustainability/special_issues/ai_sus).

For citation purposes, cite each article independently as indicated on the article page online and as indicated below:

LastName, A.A.; LastName, B.B.; LastName, C.C. Article Title. *Journal Name* **Year**, *Volume Number*, Page Range.

**ISBN 978-3-0365-8416-4 (Hbk)**
**ISBN 978-3-0365-8417-1 (PDF)**

# Contents

**About the Editors** . . . . . . . . . . vii

**Preface to "Applications of Artificial Intelligence in New Energy Technology Systems"** . . . . ix

**Muhammad Usman, Haris Hussain, Fahid Riaz, Muneeb Irshad, Rehmat Bashir, Muhammad Haris Shah, et al.**
Artificial N eural N etwork L ed O ptimization o f O xyhydrogen H ybridized D iesel Operated Engine
Reprinted from: *Sustainability* **2021**, *13*, 9373, doi:10.3390/su13169373 . . . . . . . . . . 1

**Alamera Nouran Alquennah, Mohamed Trabelsi, Khaled Rayane, Hani Vahedi and Haitham Abu-Rub**
Real-Time Implementation of an Optimized Model Predictive Control for a 9-Level CSC Inverter in Grid-Connected Mode
Reprinted from: *Sustainability* **2021**, *13*, 8119, doi:10.3390/su13158119 . . . . . . . . . . 25

**Ahmad Zuhairi Muzakir, Eng Hwa Yap and Teuku Meurah Indra Mahlia**
The Way towards an Energy Efficient Transportation by Implementation of Fuel Economy Standards: Fuel Savings and Emissions Mitigation
Reprinted from: *Sustainability* **2021**, *13*, 7348, doi:10.3390/su13137348 . . . . . . . . . . 41

**Farhood Sarrafzadeh Javadi and Rahman Saidur**
Thermodynamic and Energy Efficiency Analysis of a Domestic Refrigerator Using $Al_2O_3$ Nano-Refrigerant
Reprinted from: *Sustainability* **2021**, *13*, 5659, doi:10.3390/su13105659 . . . . . . . . . . 59

**Jafar Tavoosi, Amir Abolfazl Suratgar, Mohammad Bagher Menhaj, Amir Mosavi, Ardashir Mohammadzadeh and Ehsan Ranjbar**
Modeling Renewable Energy Systems by a Self-Evolving Nonlinear Consequent Part Recurrent Type-2 Fuzzy System for Power Prediction
Reprinted from: *Sustainability* **2021**, *13*, 3301, doi:10.3390/su13063301 . . . . . . . . . . 75

**Hossein Moayedi and Amir Mosavi**
Synthesizing Multi-Layer Perceptron Network with Ant Lion Biogeography-Based Dragonfly Algorithm Evolutionary Strategy Invasive Weed and League Champion Optimization Hybrid Algorithms in Predicting Heating Load in Residential Buildings
Reprinted from: *Sustainability* **2021**, *13*, 3198, doi:10.3390/su13063198 . . . . . . . . . . 97

**Rostislav Krč, Martina Kratochvílová, Jan Podroužek, Tomáš Apeltauer, Václav Stupka and Tomáš Pitner**
Machine Learning-Based Node Characterization for Smart Grid Demand Response Flexibility Assessment
Reprinted from: *Sustainability* **2020**, *13*, 2954, doi:10.3390/su13052954 . . . . . . . . . . 117

**Hossein Moayedi and Amir Mosavi**
Electrical Power Prediction through a Combination of Multilayer Perceptron with Water Cycle Ant Lion and Satin Bowerbird Searching Optimizers
Reprinted from: *Sustainability* **2021**, *13*, 2336, doi:10.3390/su13042336 . . . . . . . . . . 135

**Younes Zahraoui, Ibrahim Alhamrouni, Saad Mekhilef, M. Reyasudin Basir Khan, Mehdi Seyedmahmoudian, Alex Stojcevski and Ben Horan**
Energy Management System in Microgrids: A Comprehensive Review
Reprinted from: *Sustainability* **2021**, *13*, 10492, doi:10.3390/su131910492 . . . . . . . . . . 153

# About the Editors

**Mehdi Seyedmahmoudian**

Associate Professor Mehdi Seyedmahmoudian is an esteemed expert in renewable energy integration and AI-based inverter control systems. He holds the position of Associate Professor and Discipline Leader for Electrical and Electronics at Swinburne University of Technology (SUT). With extensive experience and expertise, he serves as the Director of Siemens Swinburne Energy Transition Hub and is a founding member and co-leader of the New Energy Technologies (NET) research group at SUT. A/Prof Mehdi has completed complex energy projects and secured over AUD 8.2 million in research funding in the last five years. His research contributions are evident in his 180+ published research articles, covering diverse topics in renewable energy system integration and new energy technology development. Recognized for his scholarly achievements, A/Prof Mehdi ranks among the top 2% of researchers worldwide according to SCOPUS.

**Alex Stojcevski**

Professor Alex Stojcevski is a renowned expert in power system analysis and the integration of renewable energy systems, the inclusion of cutting-edge advancements in sustainable energy solutions, power system stability and electric vehicle charging stations. As the Dean of the School of Science, Computing, and Engineering Technologies at Swinburne University of Technology (SUT), he has made significant contributions to the field. His research focuses on developing efficient and cost-effective strategies for integrating renewable energy and electric vehicles into the electrical grid. With over AUD 80M in project funding and 300+ publications, including book chapters, journal articles, and conference papers, Professor Stojcevski's expertise and research findings have garnered international recognition.

**Ben Horan**

Professor Ben Horan is an esteemed expert in Virtual Reality (VR), advanced Human–Computer Interaction (HCI), and AI in energy applications. Apart from his research contributions, Professor Horan holds a prominent leadership role as the Head of School, or Dean, for the School of Engineering at Deakin University and he is also the founder and leader of the CADET Virtual Reality Training and Simulation Research Lab, known for its work in VR training and AI in energy and simulation applications. His prime focus lies in exploring the intersection of AI and energy systems. His research aims to unlock the potential of AI to optimize energy systems, improve sustainability, and foster a transition towards cleaner and more efficient energy solutions by enhancing energy efficiency, grid management, and renewable energy integration.

**Saad Mekhilef**

Professor Saad Mekhilef is an accomplished researcher in power electronics and renewable energy. He holds the position of Professor at Swinburne University of Technology (SUT) and is widely recognized for his exceptional research contributions. With over 20 years of experience, he is the founder of PEARL and has secured AUD 15M in research grants, developed 12 patents, and published 550+ scientific papers. His work has an H-index of 95 and >40,000 citations. As a leader, he provides scientific guidance, mentors researchers, and promotes the commercialization of smart solar modular multilevel converters.

# Preface to "Applications of Artificial Intelligence in New Energy Technology Systems"

Dear Readers,

It is our great pleasure to present this book, titled "Advances in Energy Technologies: Exploring Innovations for a Sustainable Future." This collection of research and insights brings together a diverse range of topics and expertise, all centered around the advancement and application of energy technologies for a greener and more sustainable world.

In recent years, the integration of distributed power generation and emerging energy technologies has gained significant attention. However, the successful integration of these technologies into existing power systems requires overcoming various challenges, such as efficiency limitations, unreliable control strategies, inaccurate prediction methods, and high operating costs. To tackle these obstacles and maximize the potential of new energy technologies, researchers have turned to the power of AI and soft computing techniques.

This book, which is a compilation of the articles from the Special Issue, serves as a platform for researchers to share their original work on AI-enabled solutions that improve, develop, and manage new energy technologies. The selected papers cover a wide range of topics, including AI-enabled control systems for renewable energy systems, advanced energy management systems, innovative energy prediction techniques, AI-enabled energy planning strategies, grid integration of new energy systems, smart grid communication systems empowered by AI, power electronics and industrial electronics applications, electric vehicles and storage systems, virtual reality visualization and simulation for new energy technologies, and virtual power plants.

By showcasing the latest advancements in AI for new energy technologies, we aim to foster collaboration, spark new ideas, and contribute to the development of sustainable and efficient energy systems. We believe that the research presented in this Special Issue will pave the way for a future where new energy technologies seamlessly integrate into our daily lives and contribute to a greener and more sustainable world.

We would like to express our sincere gratitude to all the authors who contributed their valuable research to this Special Issue book. Their dedication and expertise have ensured the high quality of the papers included. We would also like to extend our appreciation to the reviewers for their meticulous evaluation and constructive feedback, which have helped shape this collection.

We invite all readers to explore the articles in this book and delve into the cutting-edge research at the intersection of AI and new energy technologies. Together, let us accelerate the progress toward a cleaner, smarter, and more sustainable future.

**Mehdi Seyedmahmoudian, Alex Stojcevski, Ben Horan, and Saad Mekhilef**

*Editors*

*sustainability*

MDPI

*Article*

# Artificial Neural Network Led Optimization of Oxyhydrogen Hybridized Diesel Operated Engine

**Muhammad Usman [1,*], Haris Hussain [1], Fahid Riaz [2], Muneeb Irshad [3], Rehmat Bashir [1], Muhammad Haris Shah [1], Adeel Ahmad Zafar [1], Usman Bashir [1], M. A. Kalam [4,*], M. A. Mujtaba [5,*] and Manzoore Elahi M. Soudagar [6]**

1 Department of Mechanical Engineering, University of Engineering and Technology, Lahore 54890, Pakistan; haris.hussain@uet.edu.pk (H.H.); rehmatbshir@uet.edu.pk (R.B.); haris1049shah@outlook.com (M.H.S.); adeelzafar966@gmail.com (A.A.Z.); usman.bashir1002@gmail.com (U.B.)

2 Department of Mechanical Engineering, National University of Singapore, Singapore 117575, Singapore; fahid.riaz@u.nus.edu

3 Department of Physics, University of Engineering and Technology Lahore, Lahore 54890, Pakistan; muneeb_irshad@uet.edu.pk

4 Center for Energy Science, Department of Mechanical Engineering, University of Malaya, Kuala Lumpur 50603, Malaysia

5 Department of Mechanical Engineering, New Campus, University of Engineering and Technology, Lahore 54890, Pakistan

6 Department of Mechanical Engineering, School of Technology, Glocal University, Delhi-Yamunotri Marg, SH-57, Mirzapur Pole, Saharanpur 247121, Uttar Pradesh, India; me.soudagar@gmail.com

* Correspondence: muhammadusman@uet.edu.pk (M.U.); kalam@um.edu.my (M.A.K.); m.mujtaba@uet.edu.pk (M.A.M.)

**Citation:** Usman, M.; Hussain, H.; Riaz, F.; Irshad, M.; Bashir, R.; Haris Shah, M.; Ahmad Zafar, A.; Bashir, U.; Kalam, M.A.; Mujtaba, M.A.; et al. Artificial Neural Network Led Optimization of Oxyhydrogen Hybridized Diesel Operated Engine. *Sustainability* **2021**, *13*, 9373. https://doi.org/10.3390/su13169373

Academic Editor: Mehdi Seyedmahmoudian

Received: 9 July 2021
Accepted: 17 August 2021
Published: 20 August 2021



**Abstract:** The prevailing massive exploitation of conventional fuels has staked the energy accessibility to future generations. The gloomy peril of inflated demand and depleting fuel reservoirs in the energy sector has supposedly instigated the urgent need for reliable alternative fuels. These very issues have been addressed by introducing oxyhydrogen gas (HHO) in compression ignition (CI) engines in various flow rates with diesel for assessing brake-specific fuel consumption (BSFC) and brake thermal efficiency (BTE). The enrichment of neat diesel fuel with 10 $dm^3/min$ of HHO resulted in the most substantial decrease in BSFC and improved BTE at all test speeds in the range of 1000–2200 rpm. Moreover, an Artificial Intelligence (AI) approach was employed for designing an ANN performance-predicting model with an engine operating on HHO. The correlation coefficients (R) of BSFC and BTE given by the ANN predicting model were 0.99764 and 0.99902, respectively. The mean root errors (MRE) of both parameters (BSFC and BTE) were within the range of 1–3% while the root mean square errors (RMSE) were 0.0122 kg/kWh and 0.2768% for BSFC and BTE, respectively. In addition, ANN was coupled with the response surface methodology (RSM) technique for comprehending the individual impact of design parameters and their statistical interactions governing the output parameters. The $R^2$ values of RSM responses (BSFC and BTE) were near to 1 and MRE values were within the designated range. The comparative evaluation of ANN and RSM predicting models revealed that MRE and RMSE of RSM models are also well within the desired range but to be outrightly accurate and precise, the choice of ANN should be potentially endorsed. Thus, the combined use of ANN and RSM could be used effectively for reliable predictions and effective study of statistical interactions.

**Keywords:** diesel; oxyhydrogen; artificial neural network; response surface methodology; prediction; desirability

## 1. Introduction

Ever-growing industrialization and unprecedented use of non-renewable fuels have brought us to a very feeble junction where we have to be a bit vigilant or we may run out of fossil fuels [1]. Hydrocarbon reserves found in nature are going to become extinct in the future if we continue using them without any restrictions because these reserves are being consumed at a rate faster than their formation [2,3]. The world energy demand is excessively soaring and is likely to be 28% higher in 2040 if consumption continues at the current pace. The intensifying pressure of depleting energy resources and fear of damage to the environment has consequently made scientists to look for alternative or green fuels. The use of oxygenated, alcoholic, and hydrogen fuels in CI engines has been the locus of interest of engineers for the past few years [4–10].

Compression ignition (CI) engines have long been the power generation source for heavy machinery in energy and aquatic transport owing to their high efficiency, torque, and feasibility of operation on a lean mixture of air and fuel [11]. Diesel is a commonly used fuel in CI engines. However, due to the incessant usage of fossil fuels to generate diesel, the focus of the researchers, engineers, and scientists made a paradigm shift towards the study of more efficient, promising, and greener fuels [12–16].

Hydrogen itself cannot be used in CI engines due to its high auto-ignition temperatures, which requires a very high compression ratio, but it can be mixed with fuel with low autoignition temperatures. Hydrogen gas is a good blending agent and could be effectively used in engines because of its low ignition temperature and high flammability [17]. The use of hydrogen as a mixing fuel is a concept with novelty and therefore much work has been reported. The earliest studies were conducted by T. Litzinger et al. on the operations of the IC Engines with multi-blended fuels. They validated the role of $H_2$ inside the IC engines and found it as a potential replacement of fossil fuels [18]. Moreover, with scientific and technological development, many researchers have discovered that $H_2$ can be used as a blend with other gases to reduce its combustibility and increase its ignition energy. H.K Abdel Aal carried out one such study to generate a safe method of $H_2$ enrichment by using methane as a blending agent with $H_2$. He used Le Chatelier's principle for predicting flammability and determining a safe ratio [19].

Similarly, among many fuels available for mixing, the use of diesel has also been an area of interest of researchers [20,21]. Kadir Aydin et al. conducted experimentation on a Mitsubishi 4 stroke CI engine using HHO gas $dm^3/min$ as a blended fuel with diesel. They observed a 19.1% increase in brake power (BP) and a 14% decrease in brake-specific fuel consumption (BSFC) with the addition of HHO as compared with simple diesel. Their study identified 1750 rpm as a critical speed, below which the HHO addition was not favorable for engine performance [22]. Similarly, Alfredas et al. used an Audi 1.9 TDI (IZ type) CI engine to investigate the effect of HHO addition at three liters per minute (LPM) with diesel on the performance of the engine. Within the rpm range of 1900–3700, HHO proved to be favorable for engine performance [23]. Ali Yimilaz et al. also studied the outcome of HHO addition on engine performance and reported that engine torque was amplified by 19.1%, and an average gain of 14% was achieved using HHO [24]. The effect of HHO gas on the performance of a Mitsubishi Canter brand, four-stroke, water-cooled diesel engine was conducted by Raif et al. They varied the flow rate of HHO from 3 LPM to 7 LPM and observed that with HHO enrichment, the torque and brake power increased, whereas fuel consumption decreased [25]. HHO has also the potential to be used as a blending agent with other fuels. In this regard, Usman et al. used HHO with LPG and CNG and reported improved performance and reduced emissions with the addition of HHO for both cases [26].

Owing to the cost and time of simply random experiments, researchers nowadays are utilizing computer systems to attain similar efficiency by performing the least number of experimental runs. Artificial neural network (ANN) and response surface methodology (RSM) techniques are currently used to solve problems in science and engineering, especially where classical modeling methods have pathetically failed. The predictive capability of the ANN model is based on the training of experimental data values and followed by validation. If new data values are not desired, the ANN model may re-learn to enhance the performance [27,28]. Raif Kenanoglu et al. used an artificial neural network for performance and emission analysis of a hydroxy gas-enriched CI engine for odd flow rates of 3.5 and 7 L/min. They used the Levenberg-Marquardt (LM) training function and found a 95.82% accuracy for torque [29]. Similarly, Yildrium et al. studied the effect of HHO enrichment on three different biodiesels using ANN. They used artificial intelligence for fixing optimum hydrogen enrichment and found that developed models had a coefficient of determination close to 1 [30].

Similarly, the RSM has the statistical regression technique for prediction [31,32]. Over the last few years, the combined application of ANN and RSM methods has been hailed with significant success in the power industry. Ghobadian et al. utilized diesel and biodiesel fuel blends for ANN-based prediction of performance and emission. The developed ANN model was viable with correlation coefficients (R-values) of 0.999 and 0.9487 for BSFC and torque, respectively [33]. The ANN coupled RSM-based optimization of SI engine was carried by Samet Uslu et al., which rendered the use of the ANN-supported RSM model as an effective tool for performance prediction [34].

Considering the literature cited, the use of ANN for predicting the performance of engines fueled with diesel HHO blends has already been studied [30]. However, the optimization of the engine with the same blend has not been reported so far. In the current study, ANN was used to predict the performance (BSFC and BTE) of a CI engine operating on diesel with HHO in flow rates of 2–10 LPM. Moreover, the ANN-assisted RSM optimization was applied to identify the optimized working conditions. The obtained optimum conditions were validated using experimentation. Thus, the combined use of artificial intelligence and RSM proved valuable in estimating and optimizing the performance of a CI engine.

## 2. Materials and Methods

### 2.1. HHO Generator

The HHO (hydroxy gas) was produced using an Ironside HHO Generator, shown in Figure 1. The features of the used HHO generator are itemized in Table 1. The power to the electrical unit was supplied using an AC source. Water was ionized using potassium hydroxide as a catalyst that generated hydrogen (positive charge) and hydroxide (negative charge) ions. On supplying the potential across ionized water, the generated HHO traversed first from a bubbler and later from a flow meter. Prior to injection in the engine's intake manifold, the flow rate of the gas was monitored using the flow meter. The potentiometer was connected with the electrical box for regulating the flow through the cell.

| 1. Control switch | 2. Ammeter | 5. Transformer |
|---|---|---|
| 3. Bubbler | 4. Bridge rectifier | |

**Figure 1.** HHO generator.

**Table 1.** HHO generator specifications.

| Feature | Description |
|---|---|
| Material of Plate | Stainless steel (316-l) |
| Dimensions of Plate | 16.5 cm by 16.5 cm by 0.1 cm |
| Electrode configuration | Center anodes, end cathodes |
| Plate spacing | 2 mm |
| HHO flow rate | up to 10 LPM |
| Maximum Voltage | 35 V |
| Maximum Current | 40 A |
| Relation between current and LPM | Direct relation up to 10 LPM |

### 2.2. Experimental Methodology and Test Fuels

A direct injection, four-stroke, three-cylinder diesel engine was used for performance tests whose specifications are shown in Table 2. The speed, load, and fuel flow measuring system were equipped with the engine, as comprehensively demonstrated in the experimental setup (see Figure 2). The engine was attached with a 3-phase AC generator having 85% efficiency. Five breakers with equal loading capacity were utilized from the control unit, as shown in the schematic of the engine testbed (Figure 3). The loads were applied to the test engine utilizing the generator.

**Table 2.** Engine specifications.

| Features | Description |
|---|---|
| Engine type | Perkin/AD 3.152 |
| Bore | 91.4 mm |
| Stroke | 127.0 mm |
| Number of holes of nozzles | 4 |
| Brake mean effective pressure | 7.1570 bars |
| Injection timing | $17^{0}$ BTDC |
| Displacement | 2.5 Liters |
| Compression ratio | 18.5 |
| Maximum speed | 2200 rpm |
| Maximum power | 36.8 kW at 1500 rpm |
| Maximum torque | 243 N.m at 1400 rpm |

The fuels used for conducting the experimental runs were pure diesel and HHO-mixed diesel with 2, 4, 6, 8, and 10 LPM enrichment. The physicochemical properties of liquid and gaseous fuels are shown in Table 3.

**Table 3.** Properties of fuels.

| Properties | Diesel | Hydrogen |
|---|---|---|
| Research octane number | 30 | >130 |
| Density at 20 °C | 833.1 kg/m$^3$ | 0.0827 kg/m$^3$ |
| Net heating value | 42.5 MJ/kg | 119.93 MJ/kg |
| Flame velocity | 30 cm/s | 265–325 cm/s |
| Autoignition temperature | 530 K | 858 K |
| Chemical composition | $C_{12}H_{23}$ | $H_2$ |

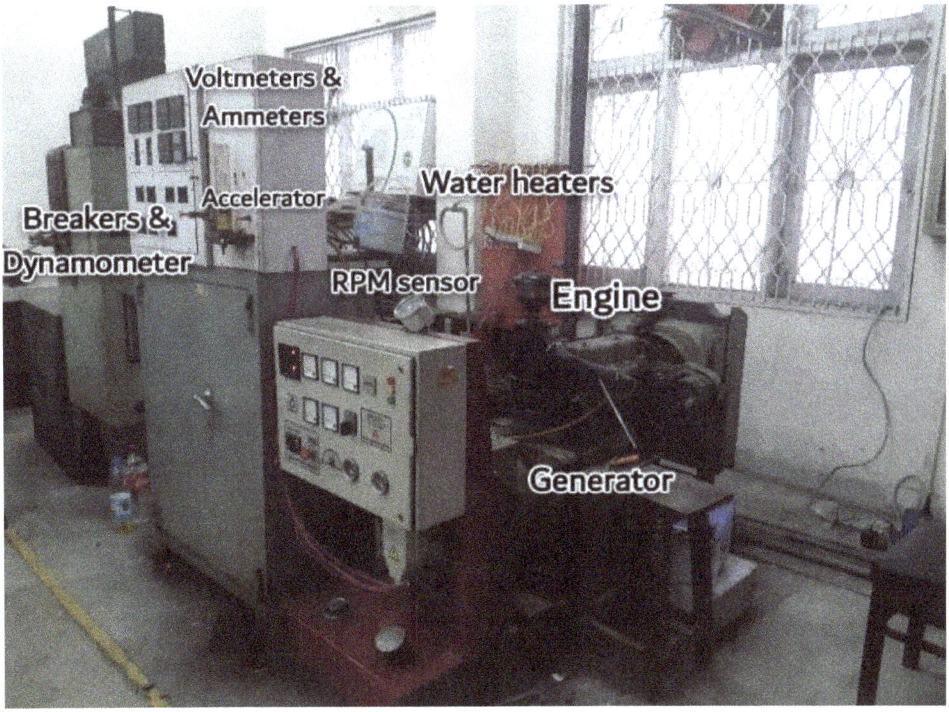


**Figure 2.** Experimental setup.

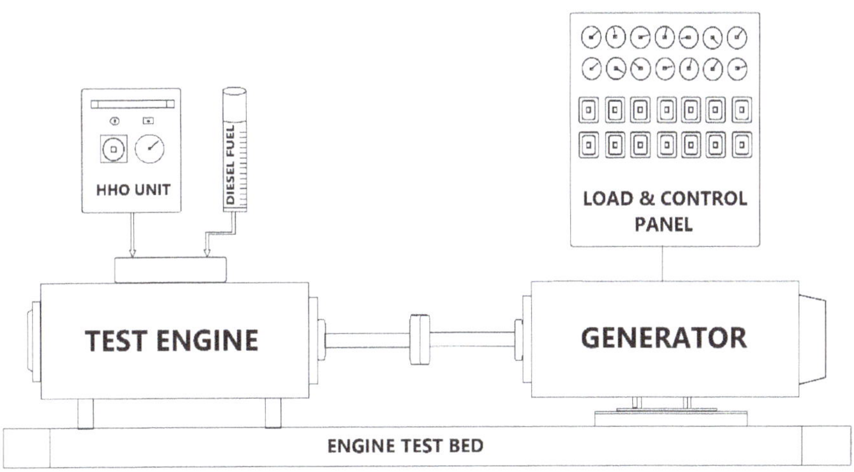


**Figure 3.** Schematic of engine test bed.

First, the diesel engine was allowed to warm up for 10 min ahead of recording the experimental observations. The tests were started at an engine speed of 1000 rpm and ceased at 2200 rpm, with equal increments of 200 rpm at each stage. The performance parameters, BSFC and BTE, were calculated at each constant strategic engine speed with varying loads (9%, 18%, 27%, 36%, 45%) and HHO concentration of 2, 4, 6, 8, and 10 LPM by using the following mathematical modelling equations:

$$\text{BTE} = \frac{\text{Break Power} \times 3600}{\text{Fuel Consumption} \times \text{Calorific Value}} \quad (1)$$

$$\text{BSFC} = \frac{\text{Fuel Consumption}}{\text{Break Power}} \quad (2)$$

Considering all the possible combinations of continuously varying factors, 210 experimental observations were documented. The experimental work was restricted to lower half partial loading conditions considering the safety aspect associated with HHO utilization in engine. Moreover, using hydrogen accompanies better combustion and higher flame speed and temperature inside the engine, which may lead to thermal degradation if operated at higher loading conditions. Later, the ANN technique was implemented for designing the predictable model of engine performance. Finally, the performance was optimized using response surface methodology (RSM) and the desirability aspect was investigated.

## 3. Experimental Results and Discussion

Non-renewable fuels are normally associated with enhanced performance and reduced exhaust emissions [35]. Hydroxy gas is believed to facilitate cleaner and smoother combustion compared with conventional fuels. The experimental deliverables significantly demonstrated the decrease in BSFC and increase in BTE by virtue of HHO addition to diesel. The detailed effect of HHO on performance parameters is presented in the following sections.

### 3.1. Brake Specific Fuel Consumption

The patterns of BSFC variation with varying flow rate and engine load are shown in Figure 4a–g. The BSFC of all operating conditions of HHO enriched fuel was lower than for neat diesel. The addition of gaseous fuel to diesel evidenced promising fuel economy. At the speed of 2200 rpm and a load of 9%, the parameter (BSFC) differed by 12% for neat diesel and the one having 10 LPM of HHO, with the latter being more fuel-efficient. Similarly, the speed of 1600 rpm rendered un-blended diesel less efficient on the account of an average of 8.44% higher fuel consumption when juxtaposed with 10 LPM HHO-enriched diesel. Moreover, the higher engine speeds seem to have a greater decrement in BSFC compared with low speeds. Equated at loads of 9% and 18% for speeds 1800 and 2200 rpm, the BSFC values diverged by 0.33 and 0.64 kg/kWh for low and high speed, respectively. The higher flammability at higher speeds is the reason for augmented variations (decline) in BSFC for increased revolutions of the power-producing shaft. The improved fuel economy of the engine functioned with HHO could be apprehended by the high calorific value of hydrogen and efficient combustion due to the availability of oxygen atoms in the HHO structure [19,36–38].

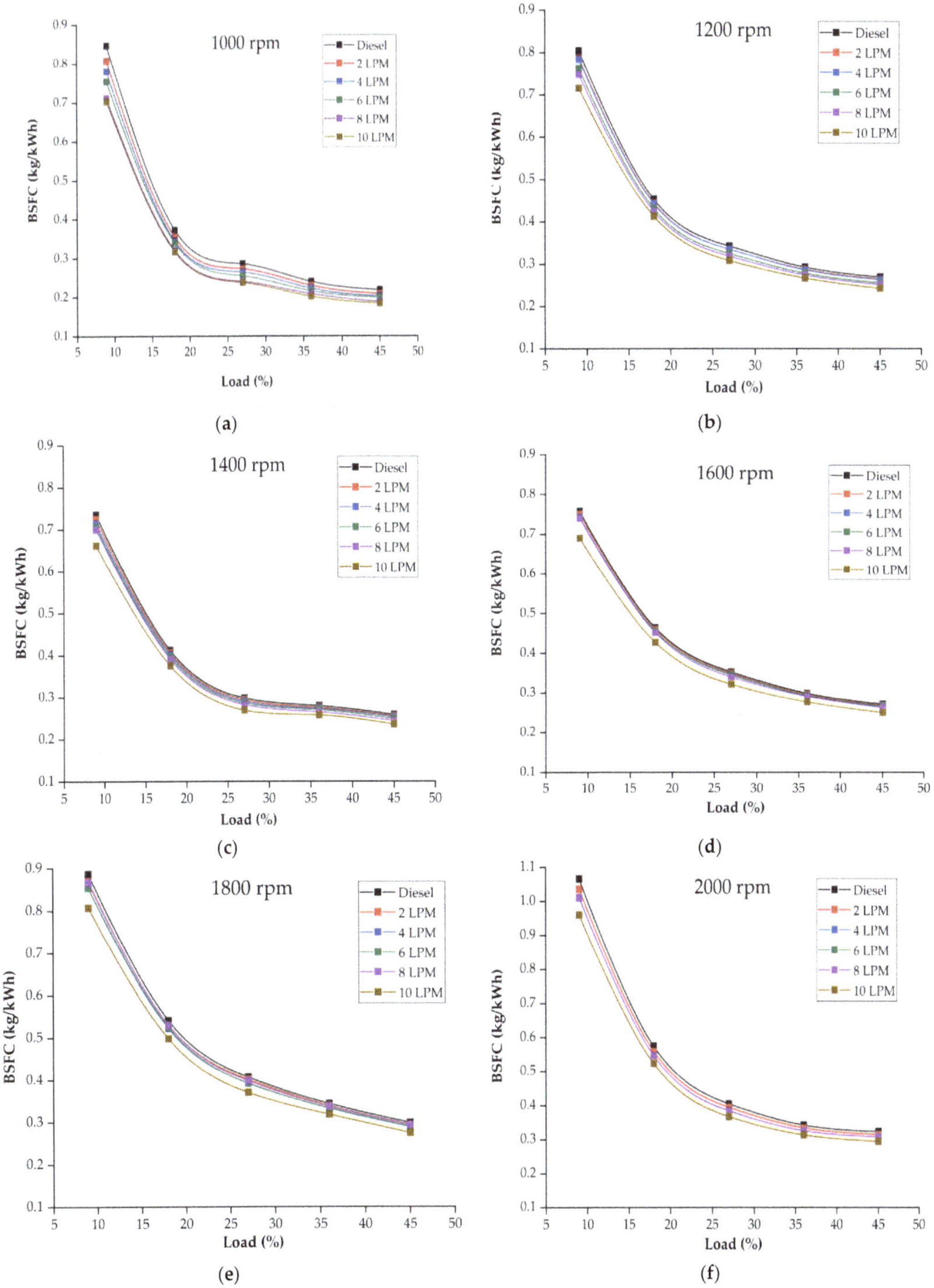


**Figure 4.** *Cont.*

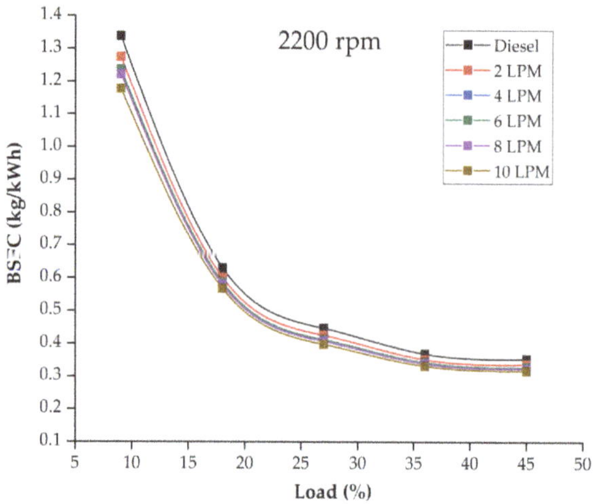


**(g)**

**Figure 4.** (**a**–**g**) Comparison of BSFC for diesel—HHO blends in speed range (1000–2200).

### 3.2. Brake Thermal Efficiency

Figure 5a–g shows the brake thermal efficiencies of all operating conditions at different HHO flow rates and loads. Identical to BSFC, the engine exhibited boosted thermal efficiencies for all HHO-enriched fuels compared with pure diesel. The test speed of 1400 rpm and flow rates of 2, 4, 6, 8, and 10 HHO at the constant load of 18% returned BTE values of 11.11%, 19.80%, 27.37%, 29.16%, and 31.55%, respectively. Thus, with the successive addition of HHO to the pure diesel, the engine exhibited a more efficient behavior. Figure 5c is seen to be following a different pattern compared with other test speeds. The manufacturer provided the maximum torque of 243 Nm at 1400 rpm (see Table 2). Therefore, at 36% loading condition for 1400 rpm, the fuel consumption was observed as higher, which resulted in a lower rate of increase in BTE, as made evident by the part of curve after the 25% load. Of all the experimental runs, the maximum recorded efficiency was 42.39% at the following conditions: 1000 rpm, 10 LPM HHO, and 45% load. Compared at a speed of 2000 rpm, diesel with 10 LPM HHO presented an average of 9.07% better performance than neat fuel. The chemical structure of diesel reveals the presence of 23 hydrogen atoms. The addition of hydroxy gas augments the number of hydrogen atoms and hydrogen to carbon ratio increases, which could be held accountable for significantly improved efficiency [39]. Moreover, enhanced combustion, the high calorific value of fuel mixture, diffusivity, and fast flame propagation speed of hydrogen are the phenomena governing the better performance of the engine [29,30,40,41].

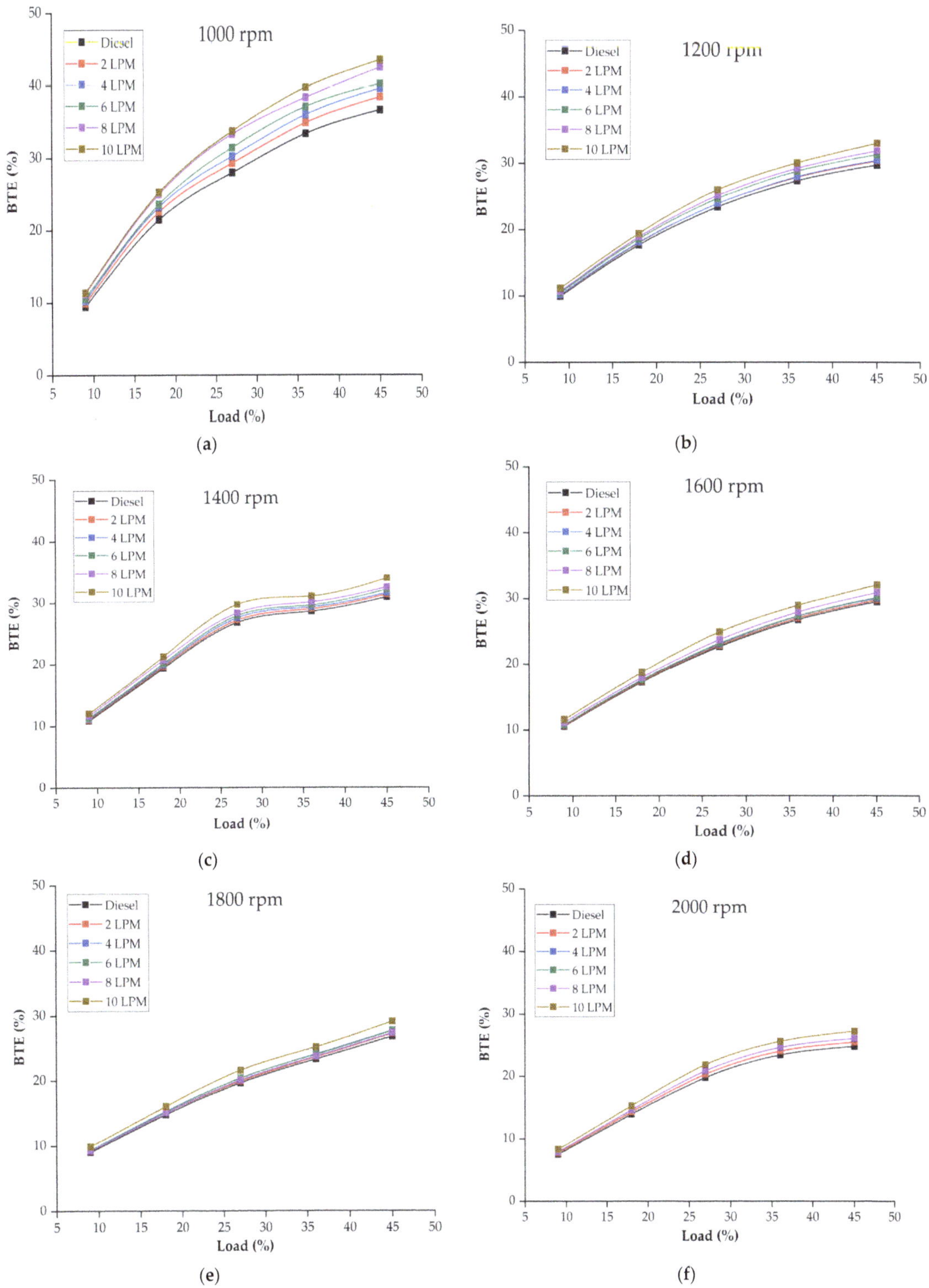


**Figure 5.** *Cont.*

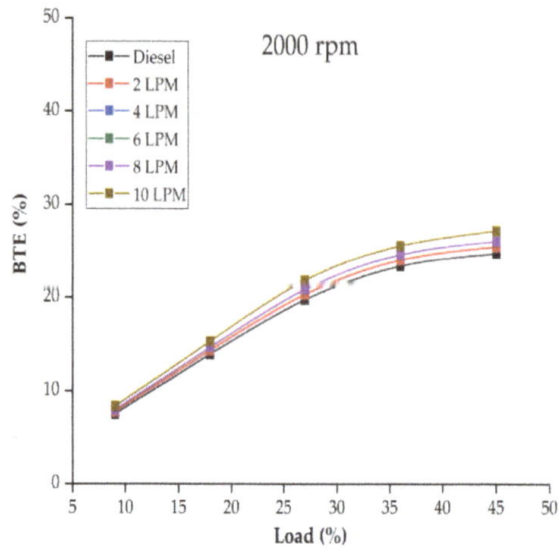


**(g)**

**Figure 5.** (**a–g**) Comparison of BTE for diesel—HHO blends in the speed range (1000–2200).

## 4. ANN Application

### *4.1. Data Preprocessing*

Prior to the development of the model, the input data (speed, load, and flow rate) and output data (BSFC and BTE) were normalized between the (0, 1) range using Equations (3) and (4). All the data were within 0 to 1 and no faulty data were detected.

$$\text{Input}^{\text{N}} = \frac{\text{input} - \min(\text{input})}{\max(\text{input}) - \min(\text{input})} \tag{3}$$

$$\text{Output}^{\text{N}} = \frac{\text{Output} - \min(\text{output})}{\max(\text{output}) - \min(\text{output})} \tag{4}$$

### *4.2. ANN Model*

ANN is chiefly a statistical model that stems from the very idea of the information processing system of the human brain [42]. Over the few decades, ANN models have been growing exceptionally more common, owing to their widespread use for analyzing, processing, system controls, and optimization applications. With similar popularity in other fields, it has also been expansively used in the automotive sector as performance parameters could be correctly estimated using it. Depending upon the complexity of data, the ANN could have many layers, but generically it is reported with three stages: input layer, hidden layer, and output layer [43,44]. Neurons are information carriers that act as a connecting medium between the three layers. The neurons are interlinked through communication links which are in turn connected with connection weights. The signals are transmitted to the neurons by connection weights.

In the current study, engine speed, HHO flow rate, and engine load were designated as input parameters for input layers while performance parameters (BSFC and BTE) were dedicated to the output layer. The number of experimental observations recorded were 210, which served as a dataset to the input layer of ANN. MATLAB NN Toolbox was used for developing the model, which randomly divided the input into three groups as training (70%), validation (15%), and testing (15%). The network used in the hidden and output layers of the ANN model was a feedforward backpropagation network because of its valuable uses in the modelling of the system, signal processing of data with non-linearities, and accuracy [43,45]. The Trainlm training function and mean square error (MSE) performance function were employed, which is generally a preferable combination for faultless predictions [46]. Owing to the magnitude of the data set and higher reported efficiency rates, the tansig transfer function was selected. For diminishing the errors, LEARNGDM learning function was used. The quantity of neurons in the hidden layer is central to the efficient prediction of the ANN model, as with too low a number the connection between input data and output predicted results could be feeble and the resultant model will be considered inappropriate [43,47]. Moreover, the criteria of ceasing the training on the escalation of validation error was used. Usually, the use of 10 neurons is widely reported in the literature and the same could also be considered for the present case. However, with such a rough guess, the results may often be misleading. Therefore, the optimum neurons were identified by iterations 5 points above and near 10 below which identified the use of 10 neurons for achieving desirable results. The trained ANN model may have an error in form overfitting, which is a considerable difference of error between the training and testing. Figure 6 shows the performance of ANN training for 1000 epochs, from which it could be arguably concluded that the test error is comparable to the training error and are converging at one value. Therefore, the ANN model is not over fitted. The attributes of the ANN models are epitomized in Table 4 and the detailed network structure of the ANN model is shown in Figure 7. The working of ANN for the current case is clarified by the process diagram shown in Figure 8. It encompasses three stages. Input parameters were introduced in the first stage, which were repetitively trained in the second stage for minimizing disparity, and checked for the desired results in the third stage.

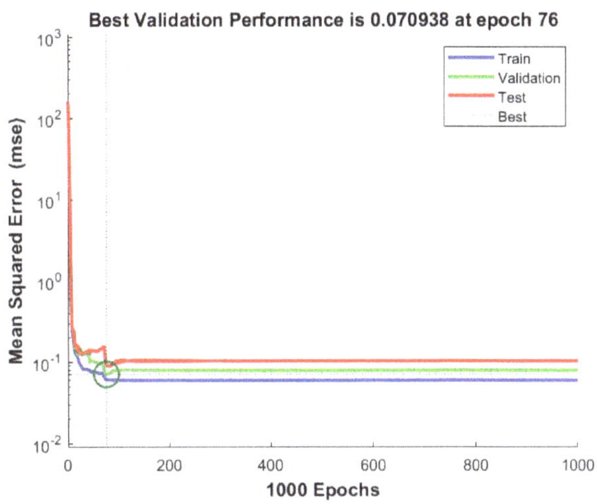


**Figure 6.** Performance validation of ANN.

**Table 4.** Attributes of the ANN model.

| Attributes | Description |
|---|---|
| Parameters | Three Inputs, Two Outputs, One hidden layer |
| Network Type | Feedforward backpropagation |
| Total number of data sets | 210 |
| Number of data sets for ANN training | 147 |
| Neuron in hidden layer | 10 |
| Data Division | 15% for validation, 15% for testing and 70% for training |

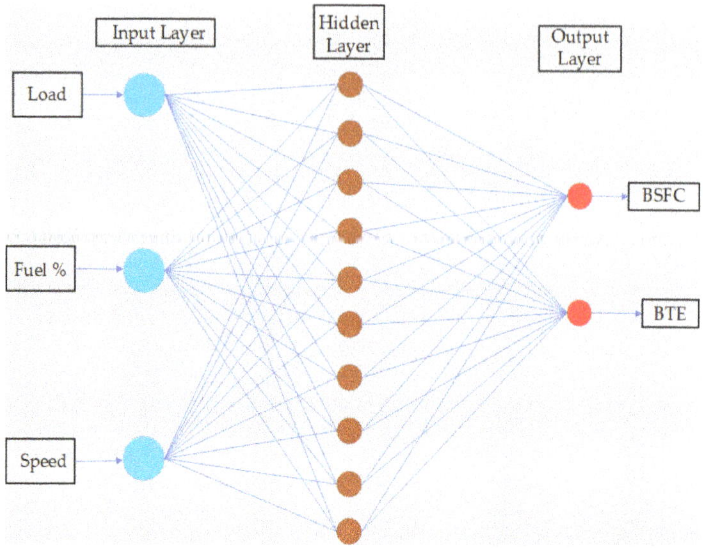


**Figure 7.** The ANN model.

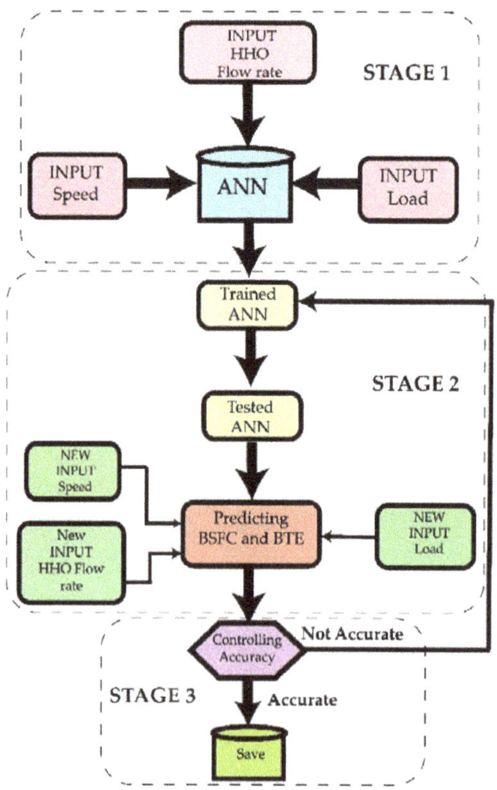


**Figure 8.** Working of ANN.

The output results of the ANN model were tested using the statistical measures of mean relative error (MRE), Root mean square error (RMSE), and correlation coefficient ($R^2$), as defined in Equations (5)–(7):

$$\mathrm{MRE}(\%) = \frac{1}{n}\sum_{i=1}^{n}\left|100\frac{t_i - p_i}{p_i}\right| \tag{5}$$

$$\mathrm{RMSE} = \sqrt{\frac{1}{n}\sum_{i=1}^{n}(p_i - o_i)^2} \tag{6}$$

$$R^2 = 1 - \left(\frac{\sum_{i=1}^{n}(p_i - o_i)^2}{\sum_{i=1}^{n}(o_i)^2}\right) \tag{7}$$

The best output results could be adjudicated by ensuing certain statistical ranges associated with the above formulas. In the current scenario, the ANN predicted outputs rendered appropriate based on two statistical indicators: (a) correlation coefficient (R) close to positive unity and (b) the MRE of input and output within the defined range of 1–3%. In the case the predicted results failed to meet the demarcated criterion, the ANN model learning rate was varied.

### *4.3. ANN Prediction Comparison and Discussion*

The prediction of performance characteristics of test engine fueled with diesel-HHO blends using the artificial intelligence approach proved exceptionally valuable. The overall regression graphs yielded by the ANN application are shown in Figure 9a–d. The results generated by the model were in line with the statistical criterion defined in the preceding sections. The correlation coefficients for the three stages of the developed neural network were found qua 0.99998 for training, 0.99988 for validation. 0.99978 for testing, and 0.99994 for training, testing, and validation as a whole. The correlation coefficients for all stages were precisely near to +1, which demonstrates the well-matching of the experimental and ANN-predicted results.

The further analysis was initiated by evaluating the predicted and experiment results of BSFC and BTE on an individual basis as shown in Figure 10a,b. The correlation coefficient for BSFC returned a value of 0.99764. The MRE and RMSE accuracy-defining equations proved solid testimonies of BSFC model-generated results with values of 2.64% and 0.0122 kg/kWh. The statistical parameters showed that the prediction of the BSFC of a diesel engine operating on blended fuel using ANN has enough competence and efficiency. Similarly, the BTE-guessed values were significantly close to experimental values shown with R, MRE, and RMSE values of 0.99902, 1.91%, and 0.2768%. The BTE ANN model proved remarkable in the prediction of performance parameters.

The comprehensive comparison of two data sets, experimental and predicted, for each observation of parameters (BSFC and BTE) is shown in Figure 11a,b. The plotted line graphs depict the overlapping data points for most of the test runs, signaling the negligible deviations. From 210 observations, there are only a few sets for which the predicted values were seen escalating on either extreme, but collectively, they could merely be ignored due to an inconsequential effect. The in-depth analysis of the obtained statistical parameters unequivocally advocates that the ANN prediction model is suitable for performance parameters.

(**a**) (**b**)

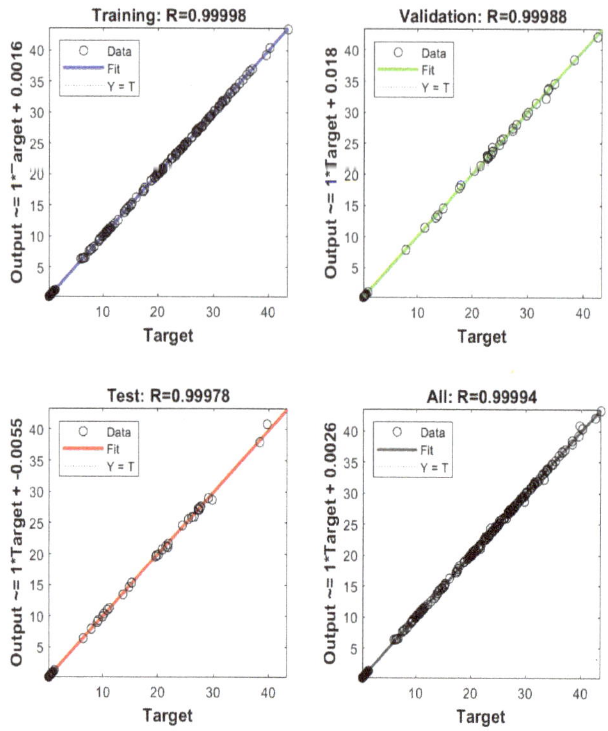


(**c**) (**d**)

**Figure 9.** (**a**) Training, (**b**) validation, (**c**) testing, and (**d**) overall correlation coefficients.

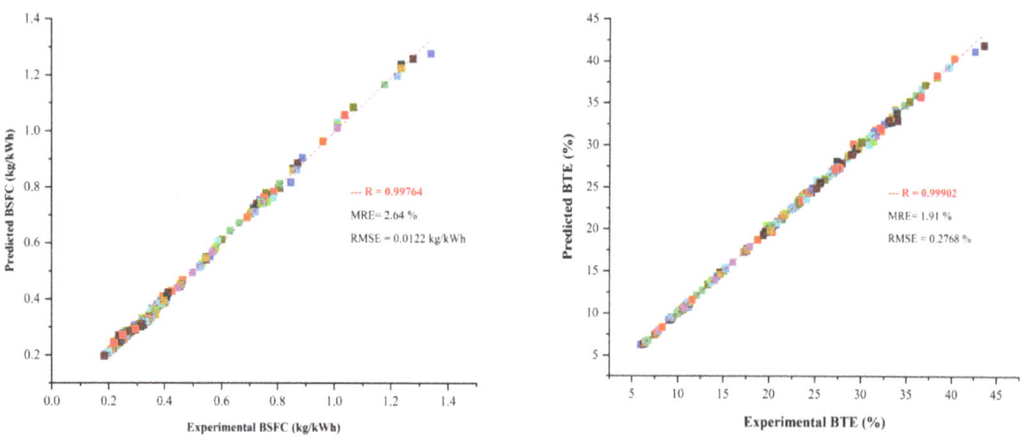


(**a**) (**b**)

**Figure 10.** Comparison of predicted results with experimental results for (**a**) BSFC and (**b**) BTE.

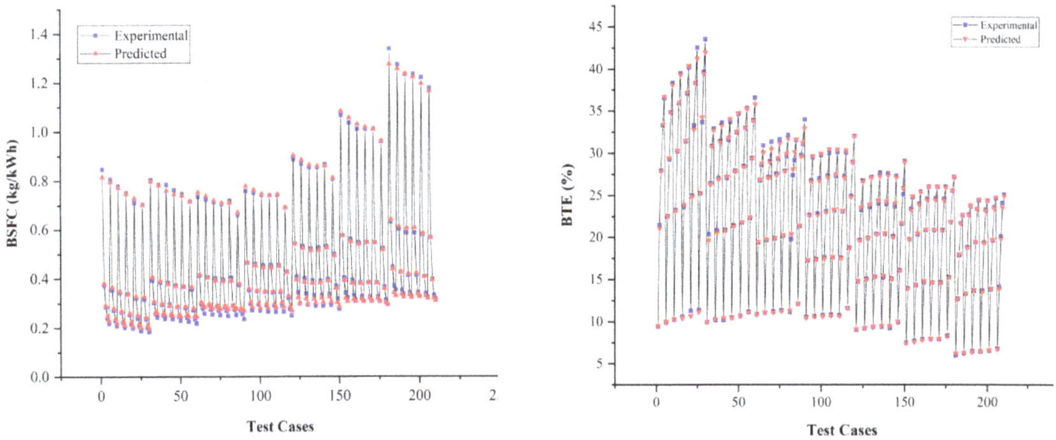


(a) (b)

**Figure 11.** Comparison of experimental and predicted values for (**a**) BSFC and (**b**) BTE.

## 5. RSM-Based Optimization

Optimization of any process aims at achieving the maximum output by manipulating the controlled variables. In any optimization technique, numerical constraints are typically introduced for either maximizing or minimizing the response variables. Several available optimization techniques could admirably define the optimized parameters within the provided range. Response surface methodology (RSM) is a well-known statistical technique employed for the optimization of involved parameters using experimentally extracted data for solving multiple simultaneous equations. Over the years, the RSM has been seen extensively used in the engineering sector owing to its accurate prediction of response(s) influenced by multiple discrete factors. In the current optimization study, the BSFC and BTE of the test engine were nominated as response variables. The goal was to maximize BTE and minimize BSFC. The RSM design factors considered for optimization of diesel engine performance attributes were engine speed (rpm), HHO flow rate (LPM), and load of the engine (%). Design Expert 11 was used for creating the model and response surfaces. A multilevel design for a pre-defined experimental strategy was developed using historical data feature. The model defining parameters, listed in Table 5, were three numeric factors, seven levels of speed, six levels of HHO blend, and five levels of engine load.

**Table 5.** Factors and levels.

| Factors | Units | Levels | L [1] | L [2] | L [3] | L [4] | L [5] | L [6] | L [7] |
|---|---|---|---|---|---|---|---|---|---|
| Speed | Rpm | 7 | 1000 | 1200 | 1400 | 1600 | 1800 | 2000 | 2200 |
| Flow rate | LPM | 6 | 0 | 2 | 4 | 6 | 8 | 10 | — |
| Load | % | 5 | 9 | 18 | 27 | 36 | 45 | — | — |

### 5.1. Selection of an Empirical Model

The fit summaries of BSFC and BTE are listed in Tables 6 and 7. Generally, the selection of the appropriate model is governed by (a) *p*-value (b) predicted $R^2$ and (c) reasonable agreement between predicted and adjusted $R^2$ [48]. Based on the mentioned assessing parameters, the first two models (linear and 2FI) had small values of $R^2$. However, the quadratic model had the signs of best fit, owing to $p < 0.0001$ and $R^2$ significantly close to 1. Recent studies have also shown that the engine combustion process is complex and therefore could be aptly described with a quadratic model [49]. Thus, referring to the deduced observations and published literature, the quadratic model was designated for optimization purposes.

**Table 6.** BSFC fit summary.

| Source | *p*-Value | Adjusted $R^2$ | Predicted $R^2$ |
|---|---|---|---|
| Linear | <0.05 | 0.7224 | 0.7135 |
| 2FI | <0.05 | 0.7482 | 0.7316 |
| Quadratic | <0.05 | 0.9939 | 0.9922 |

**Table 7.** BTE fit summary.

| Source | *p*-Value | Adjusted $R^2$ | Predicted $R^2$ |
|---|---|---|---|
| Linear | <0.05 | 0.9187 | 0.9161 |
| 2FI | <0.05 | 0.9368 | 0.9335 |
| Quadratic | <0.0001 | 0.9940 | 0.9958 |

### *5.2. Analysis of Variance and Predicting Equations*

Analysis of variance (ANOVA) is a statistical tool used for assessing the statistical significance of the model, individual terms, and interactions. It provides a detailed understanding of the regression model as the interactions between the factors and the responses can be explicitly comprehended. Tables 8 and 9 provide the ANOVA for the quadratic models of BSFC and BTE. The model F values of 383.56 and 1298.30 for BSFC and BTE imply that models are significant. The model terms have been abbreviated as A—Speed, B—HHO flow rate, and C—load. The $p$ values less than 0.0500 indicate the significance of model terms. In the case of BSFC, A, B, C, AC, BC, $A^2$, and $C^2$ are significant model terms. The $p$ values of the terms AB, BC, and $C^2$ indicate the model terms that are insignificant. The accuracy of the models under consideration has been verified using the diagnostic predicted vs actual and residual vs run plots as shown in Figure 12a–d. Figure 12a,b demonstrates that for BSFC and BTE, the RSM predicted values are in close agreement with the ANN values, indicated by the colored data point falling on the linear inclined line. Similarly, the deviation of RSM and actual (ANN) values were in the narrow residual range of [−3.7428, +3.7428], as depicted in Figure 12c,d. The even distribution atop and below the reference axis, for both the cases, signals the statistical significance of BSFC and BTE RSM models. The response surfaces of BSFC and BTE variation with engine speed, HHO percentage, and engine load are shown in Figures 13 and 14 respectively. It is visible that all the design factors had a significant effect on responses. The dark and light dots on the response surfaces shows the design points above and below predicted values, respectively.

**Table 8.** ANOVA for BSFC.

| Source | Sum of Squares | Df | Mean Square | F-Value | *p*-Value |
|---|---|---|---|---|---|
| Model | 11.51 | 9 | 1.28 | 383.56 | <0.0001 |
| A-Speed | 1.06 | 1 | 1.06 | 317.29 | <0.0001 |
| B-Flow rate | 0.0357 | 1 | 0.0357 | 10.71 | <0.0001 |
| C-Load | 7.75 | 1 | 7.75 | 2324.75 | <0.0001 |
| AB | 0.0000 | 1 | 0.0000 | 0.0032 | 0.9551 |
| AC | 0.3384 | 1 | 0.3384 | 101.49 | <0.0001 |
| BC | 0.0156 | 1 | 0.0156 | 4.68 | 0.0317 |
| $A^2$ | 0.0606 | 1 | 0.0606 | 18.18 | <0.0001 |
| $B^2$ | 0.0001 | 1 | 0.0001 | 0.0240 | 0.8771 |
| $C^2$ | 2.25 | 1 | 2.25 | 674.88 | <0.0001 |

**Table 9.** ANOVA for BTE.

| Source | Sum of Squares | Df | Mean Square | F-Value | *p*-Value |
|---|---|---|---|---|---|
| Model | 15229.95 | 9 | 1692.22 | 1298.30 | <0.0001 |
| A-Speed | 2724.04 | 1 | 2724.04 | 2089.93 | <0.0001 |
| B-Flow rate | 110.22 | 1 | 110.22 | 84.56 | <0.0001 |
| C-Load | 11414.41 | 1 | 11414.41 | 8757.34 | <0.0001 |
| AB | 19.49 | 1 | 19.49 | 14.95 | 0.0001 |
| AC | 262.83 | 1 | 262.83 | 201.65 | <0.0001 |
| BC | 9.03 | 1 | 9.03 | 6.93 | 0.0091 |
| $A^2$ | 31.18 | 1 | 31.18 | 23.92 | <0.0001 |
| $B^2$ | 0.3481 | 1 | 0.3481 | 0.2671 | 0.6059 |
| $C^2$ | 658.42 | 1 | 658.42 | 505.15 | <0.0001 |

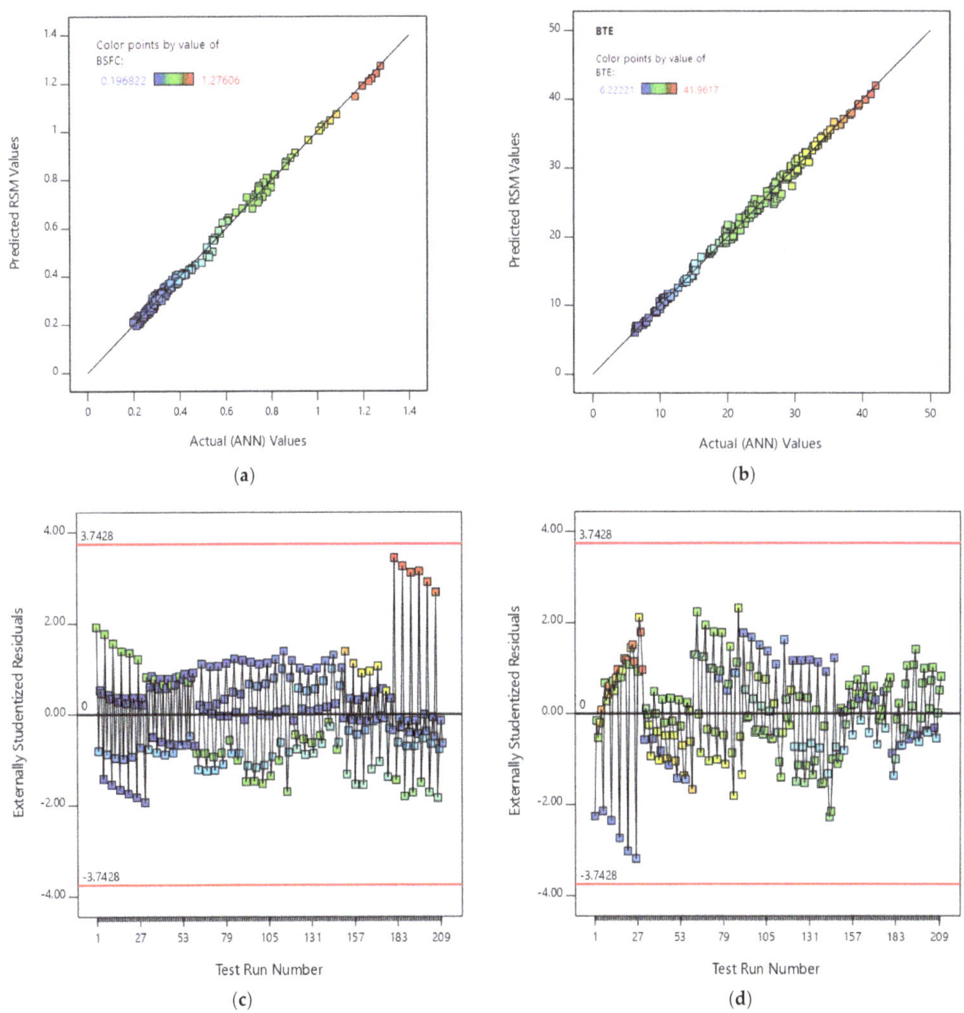


**Figure 12.** (**a**–**d**) Predicted vs actual graph for (**a**) BSFC and (**b**) BTE and residual vs. run graphs for (**a**) BSFC and BTE RSM models.

Second-order regression equations relating the input parameters and responses for estimation of performance are given by coded Equations (8) and (9). The coded alphabets A, B, and C correspond to the study design factors: speed, HHO flow rate, and load, respectively. By using the corresponding values of speed, flow rates, and engine load, in the regression equations, the values of BSFC and BTE could be accurately predicted.

$$\begin{aligned} \text{BSFC} = {} & 0.3055 + 0.0165 * \text{A} - 0.0191 * \text{B} - 0.2717 * \text{C} + 0.0005 * \text{AB} - 0.0852 * \text{AC} \\ & + 0.0178 * \text{BC} + 0.0441 * \text{A}^2 - 0.0015 * \text{B}^2 + 0.2475 * \text{C}^2 \end{aligned} \quad (8)$$

$$\begin{aligned} \text{BTE} = {} & 23.71 - 5.40 * \text{A} + 1.06 * \text{B} + 10.43 * \text{C} - 0.6690 * \text{AB} - 2.37 * \text{AC} \\ & + 0.4239 * \text{BC} + 1 * \text{A}^2 + 0.1020 * \text{B}^2 - 4.23 * \text{C}^2 \end{aligned} \quad (9)$$

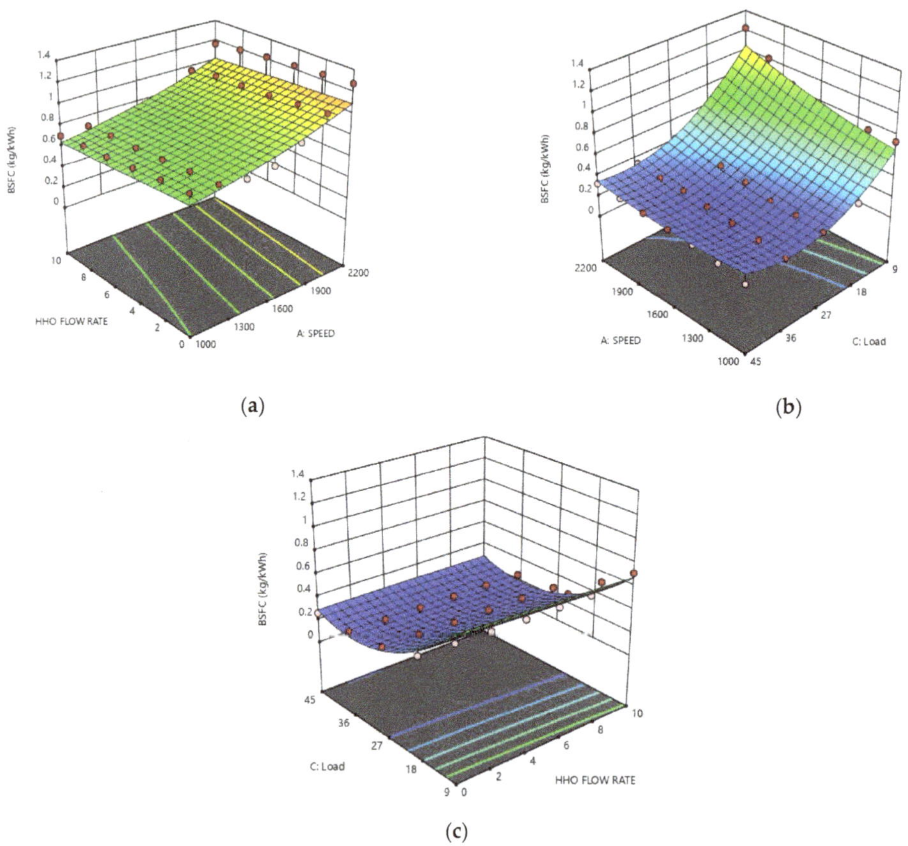


**Figure 13.** Response surfaces variation of BSFC with (**a**) flow rate and speed, (**b**) speed and load, and (**c**) load and flow rate.

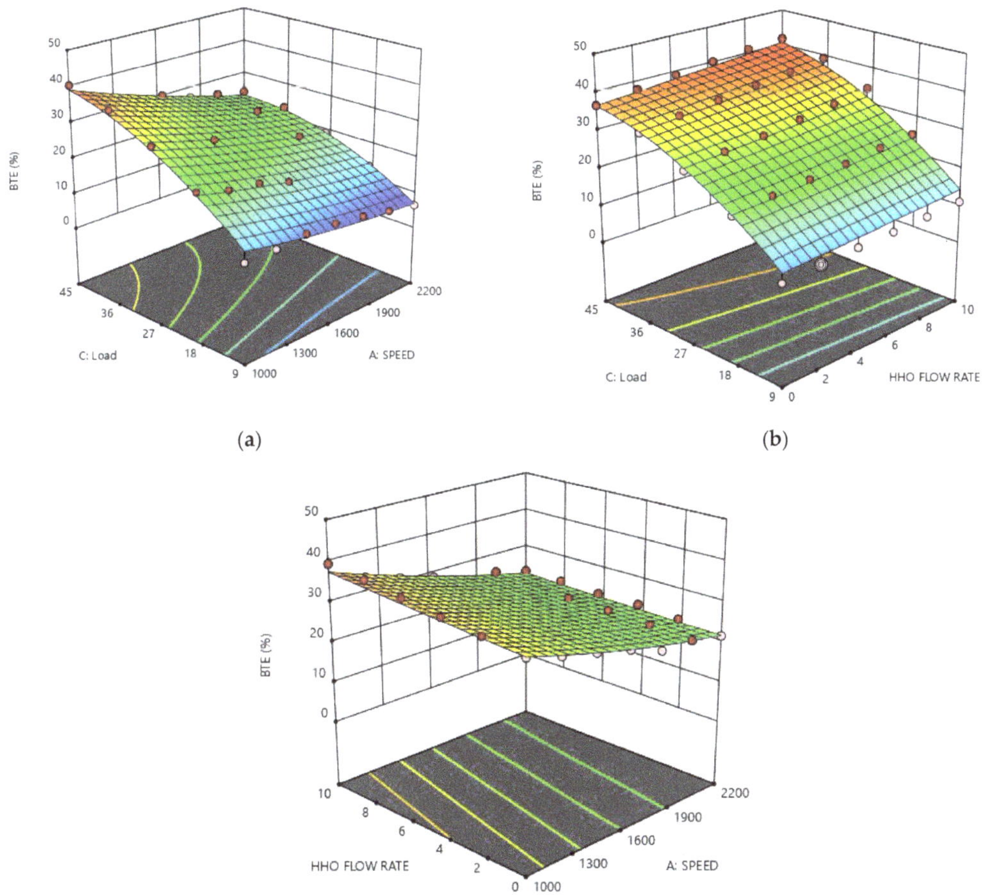


(a) (b) (c)

**Figure 14.** Response surfaces variation of BTE with (**a**) load and speed, (**b**) load and HHO, and (**c**) flow rate and speed.

### *5.3. Optimization Results and Validation*

In the current study, objective of RSM was to recognize the engine optimum working conditions. The speed, flow rate, and load were design factors, and ANN estimated values of BSFC and BTE were the outputs (responses). The design expert optimization feature demands optimal constraints to be defined for the factors and responses. Table 10 illustrates the defined constraints and setup for optimization. The goal was to optimize the engine with targets of minimizing BSFC and maximizing BTE while keeping the within range criterion for study factors.

**Table 10.** Optimization setup.

| Factors | Desired Goal | Lower Value | Upper Value | Lower Weight | Upper Weight | Importance |
|---|---|---|---|---|---|---|
| A: Speed (rpm) | Is in range | 1000 | 2200 | 1 | 1 | 3 |
| B: HHO Flow rate (LPM) | Is in range | 0 | 10 | 1 | 1 | 3 |
| C: Load (%) | Is in range | 0 | 45 | 1 | 1 | 3 |
| BSFC (kg/kWh) | Minimum | 0.196822 | 1.27606 | 1 | 1 | 3 |
| BTE (%) | Maximum | 6.22221 | 41.9617 | 1 | 1 | 3 |

The best operating parameters for engine performance came out to be 1000 rpm, 10 L per minute flow rate of HHO, and 45% engine load. The performance parameters against these optimal values of design factors are 0.301 kg/kWh BSFC and 40.939% of BTE. The composite desirability (D) is a unitless number that lies within the range of zero to one. It is a measure of favorability to which input defining factors optimize the objectives as a whole. The closer the value to the 1, the more favorable the optimization. In the current study, the composite desirability was detected to be 0.971. A value sufficiently close to 1 indicates that the employed RSM models are highly efficient and could be used to predict the optimum design factors for the efficient performance of the diesel engine.

The RSM-optimized results could be easily validated by conducting the experimental runs. Therefore, experimental observations of BSFC and BTE were recorded corresponding to optimized values of speed, flow rates, and loads, and the comparison is shown in Figure 15a,b. The experimental observation returned a value of BSFC 5.64%, less as compared with an optimized parameter. Similarly, the optimum value for brake thermal efficiency was 6.15% lower in comparison with experimental observation. With sufficient agreement between optimized and experimental observations as the basis, the RSM is viable and practically implementable.

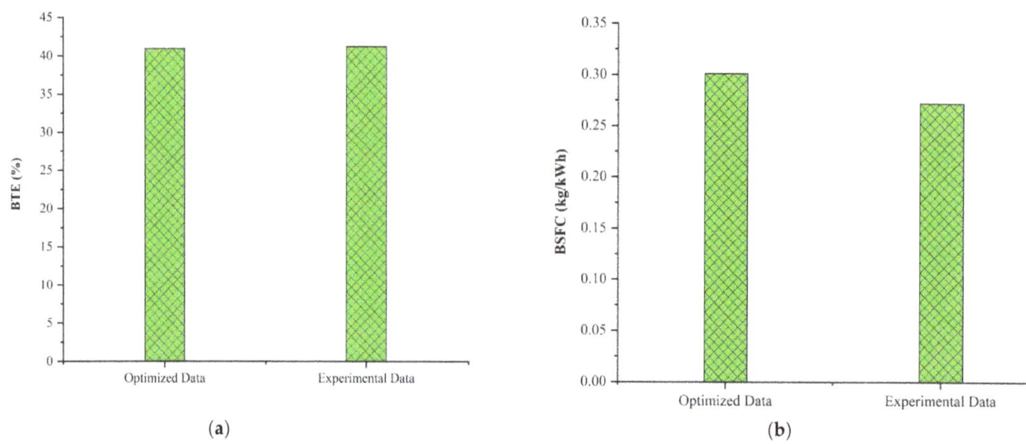


**Figure 15.** Comparison of optimized and experimental results for (**a**) BSFC and (**b**) BTE.

## 6. Comparison of ANN and RSM Models

The artificial intelligence and statistically based predicting models of BSFC and BTE seemingly have alike reliability and efficiency. However, due to the generic association of methods root task to the same domains, the comparative assessment of the two will be an ideal approach. The detailed comparison of MRE and RMSE of ANN and RSM models is shown in Table 11. The statistical comparison discloses that the ANN models of BSFC and BTE have a better ability to efficiently predict parameters of an engine due to lower MRE and RMSE. ANN and RSM returned MRE values of 1.91% and 2.26% for BTE and 2.64% and 2.94% for BSFC, respectively. Similarly, the RMSE given by ANN and RSM for BSFC were 0.012 and 0.088 kg/kWh, respectively. The comparison vouched for ANN's efficiency and reliability as the statistical parameters (MRE and RMSE) of both parameters were less than its competitor.

**Table 11.** ANN and RSM comparison.

| Models | Parameters | MRE% | RMSE |
|---|---|---|---|
| ANN | BTE (%) | 1.91 | 0.27 |
| | BSFC (kg/kWh) | 2.64 | 0.012 |
| RSM | BTE (%) | 2.26 | 0.41 |
| | BSFC (kg/kWh) | 2.94 | 0.088 |

## 7. Conclusions

This study evaluated the use of oxyhydrogen gas with diesel at different flow rates in a CI engine. ANN and RSM tools were collectively used for performance prediction and optimization. The results could be summarized as:

- 10 LPM HHO with diesel was found to be most fuel efficient among all test fuels.
- HHO addition to the diesel improved BTE for all flow rates. Pure diesel showed the least BTE among all combinations of fuels.
- The correlation coefficients of training, testing, and validation of the ANN model came out to be 0.99998, 0.99988, and 0.99978 respectively. Moreover, MRE values were in the range of 1–3%.
- RSM identified all the study factors as statistically significant owing to $p$ values less than 0.005.
- Optimum operating conditions for engine were 1000 rpm, 10 LPM HHO, and 45% loading condition.
- Composite desirability of 0.971 for multi-response optimization indicated the appropriate optimization setting.
- The experimental BSFC and BTE differed by 5.64% and 6.15% from RSM-optimized values.
- The ANN model proved better than RSM due to low RMSE and MRE values.

Thus, the addition of HHO to diesel proved highly valuable for improved performance. The statistical assessment tools (R, MRE, and RMSE) revealed that the performance could be accurately predicted by ANN and RSM models. Conclusively, the HHO enrichment to the diesel is desirable for better performance and could be optimized using Artificial Intelligence and statistical methods.

The authors aim at conducting studies to investigate the effect of HHO with a stepwise increment beyond 45% loading condition along with the collective and individual ANN modelling for outputs with different algorithms and training functions.

**Author Contributions:** Conceptualization, M.U., A.A.Z. and U.B.; methodology, M.U., H.H. and F.R.; software, R.B., M.H.S. and U.B.; validation, M.A.K., M.A.M. and A.A.Z.; formal analysis, M.A.K. and M.A.M.; investigation, H.H. and M.U.; writing—original draft preparation, M.U. and M.H.S.; writing—review and editing, F.R., R.B., M.I. and M.E.M.S.; supervision, M.U.; project administration, M.A.M.; funding acquisition, M.A.K. All authors have read and agreed to the published version of the manuscript.

**Funding:** The authors would like to thank the Faculty of Engineering at the University of Malaya, Malaysia for their support through the research grant no GPF018A-2019.

**Institutional Review Board Statement:** Not applicable.

**Informed Consent Statement:** Not applicable.

**Data Availability Statement:** Not applicable.

**Conflicts of Interest:** The authors declare no conflict of interest.

## References

1. Usman, M.; Jamil, M.K.; Riaz, F.; Hussain, H.; Hussain, G.; Shah, M.H.; Qyyum, M.A.; Salman, C.A.; Lee, M. Refining and Reuse of Waste Lube Oil in SI Engines: A Novel Approach for a Sustainable Environment. *Energies* **2021**, *14*, 2937. [CrossRef]
2. Archer, D. Fate of fossil fuel $CO_2$ in geologic time. *J. Geophys. Res. Ocean.* **2005**, *110*. [CrossRef]

3. Hussain, F.; Soudagar, M.E.M.; Afzal, A.; Mujtaba, M.; Fattah, I.; Naik, B.; Mulla, M.H.; Badruddin, I.A.; Khan, T.; Raju, V.D. Enhancement in Combustion, Performance, and Emission Characteristics of a Diesel Engine Fueled with Ce-ZnO Nanoparticle Additive Added to Soybean Biodiesel Blends. *Energies* **2020**, *13*, 4578. [CrossRef]
4. Mujtaba, M.; Kalam, M.; Masjuki, H.; Gul, M.; Soudagar, M.E.M.; Ong, H.C.; Ahmed, W.; Atabani, A.; Razzaq, L.; Yusoff, M. Comparative study of nanoparticles and alcoholic fuel additives-biodiesel-diesel blend for performance and emission improvements. *Fuel* **2020**, *279*, 118434. [CrossRef]
5. Mujtaba, M.; Masjuki, H.; Kalam, M.; Noor, F.; Farooq, M.; Ong, H.C.; Gul, M.; Soudagar, M.E.M.; Bashir, S.; Rizwanul Fattah, I. Effect of Additivized Biodiesel Blends on Diesel Engine Performance, Emission, Tribological Characteristics, and Lubricant Tribology. *Energies* **2020**, *13*, 3375. [CrossRef]
6. Soudagar, M.E.M.; Banapurmath, N.R.; Afzal, A.; Hossain, N.; Abbas, M.M.; Haniffa, M.A.C.M.; Naik, B.; Ahmed, W.; Nizamuddin, S.; Mubarak, N.M. Study of diesel engine characteristics by adding nanosized zinc oxide and diethyl ether additives in Mahua biodiesel–diesel fuel blend. *Sci. Rep.* **2020**, *10*, 15326. [CrossRef]
7. Mujtaba, M.A.; Muk Cho, H.; Masjuki, H.H.; Kalam, M.A.; Farooq, M.; Soudagar, M.E.M.; Gul, M.; Afzal, A.; Ahmed, W.; Raza, A.; et al. Effect of primary and secondary alcohols as oxygenated additives on the performance and emission characteristics of diesel engine. *Energy Rep.* **2021**, *7*, 1116–1124. [CrossRef]
8. Kumar, M.S.; Ramesh, A.; Nagalingam, B. Use of hydrogen to enhance the performance of a vegetable oil fuelled compression ignition engine. *Int. J. Hydrog. Energy* **2003**, *28*, 1143–1154.
9. Soudagar, M.E.M.; Afzal, A.; Safaei, M.R.; Manokar, A.M.; El-Seesy, A.I.; Mujtaba, M.A.; Samuel, O.D.; Badruddin, I.A.; Ahmed, W.; Shahapurkar, K.; et al. Investigation on the effect of cottonseed oil blended with different percentages of octanol and suspended MWCNT nanoparticles on diesel engine characteristics. *J. Therm. Anal. Calorim.* **2020**. [CrossRef]
10. Khan, H.; Soudagar, M.E.M.; Kumar, R.H.; Safaei, M.R.; Farooq, M.; Khidmatgar, A.; Banapurmath, N.R.; Farade, R.A.; Abbas, M.M.; Afzal, A. Effect of nano-graphene oxide and n-butanol fuel additives blended with diesel—Nigella sativa biodiesel fuel emulsion on diesel engine characteristics. *Symmetry* **2020**, *12*, 961. [CrossRef]
11. Takaishi, T.; Numata, A.; Nakano, R.; Sakaguchi, K. Approach to high efficiency diesel and gas engines. *Mitsubishi Heavy Ind. Rev.* **2008**, *45*, 21–24.
12. Lanjewar, P.B.; Rao, R.; Kale, A. Assessment of alternative fuels for transportation using a hybrid graph theory and analytic hierarchy process method. *Fuel* **2015**, *154*, 9–16. [CrossRef]
13. Mujtaba, M.; Cho, H.M.; Masjuki, H.; Kalam, M.; Ong, H.; Gul, M.; Harith, M.; Yusoff, M. Critical review on sesame seed oil and its methyl ester on cold flow and oxidation stability. *Energy Rep.* **2020**, *6*, 40–54. [CrossRef]
14. S Gavhane, R.; M Kate, A.; Pawar, A.; Safaei, M.R.; M Soudagar, M.E.; Mujtaba Abbas, M.; Muhammad Ali, H.; R Banapurmath, N.; Goodarzi, M.; Badruddin, I.A. Effect of Zinc Oxide Nano-Additives and Soybean Biodiesel at Varying Loads and Compression Ratios on VCR Diesel Engine Characteristics. *Symmetry* **2020**, *12*, 1042. [CrossRef]
15. Soudagar, M.E.M.; Mujtaba, M.A.; Safaei, M.R.; Afzal, A.; V, D.R.; Ahmed, W.; Banapurmath, N.R.; Hossain, N.; Bashir, S.; Badruddin, I.A.; et al. Effect of Sr@ZnO nanoparticles and Ricinus communis biodiesel-diesel fuel blends on modified CRDI diesel engine characteristics. *Energy* **2021**, *215*, 119094. [CrossRef]
16. Soudagar, M.E.M.; Khan, H.M.; Khan, T.M.Y.; Razzaq, L.; Asif, T.; Mujtaba, M.A.; Hussain, A.; Farooq, M.; Ahmed, W.; Shahapurkar, K.; et al. Experimental Analysis of Engine Performance and Exhaust Pollutant on a Single-Cylinder Diesel Engine Operated Using Moringa Oleifera Biodiesel. *Appl. Sci.* **2021**, *11*, 7071. [CrossRef]
17. Momirlan, M.; Veziroglu, T.N. The properties of hydrogen as fuel tomorrow in sustainable energy system for a cleaner planet. *Int. J. Hydrog. Energy* **2005**, *30*, 795–802. [CrossRef]
18. Litzinger, T.; Stoner, M.; Hess, H.; Boehman, A. Effects of oxygenated blending compounds on emissions from a turbocharged direct injection diesel engine. *Int. J. Engine Res.* **2000**, *1*, 57–70. [CrossRef]
19. Abdel-Aal, H.; Sadik, M.; Bassyouni, M.; Shalabi, M. A new approach to utilize hydrogen as a safe fuel. *Int. J. Hydrog. Energy* **2005**, *30*, 1511–1514. [CrossRef]
20. Gad, M.; El-Fakharany, M.; Elsharkawy, E. Effect of HHO gas enrichment on performance and emissions of a diesel engine fueled by biodiesel blend with kerosene additive. *Fuel* **2020**, *280*, 118632. [CrossRef]
21. Fayaz, H.; Mujtaba, M.A.; Soudagar, M.E.M.; Razzaq, L.; Nawaz, S.; Nawaz, M.A.; Farooq, M.; Afzal, A.; Ahmed, W.; Khan, T.M.Y.; et al. Collective effect of ternary nano fuel blends on the diesel engine performance and emissions characteristics. *Fuel* **2021**, *293*, 120420. [CrossRef]
22. Aydin, H.; Ilkılıc, C.J.A.T.E. Effect of ethanol blending with biodiesel on engine performance and exhaust emissions in a CI engine. *Appl. Therm. Eng.* **2010**, *30*, 1199–1204. [CrossRef]
23. Rimkus, A.; Matijošius, J.; Bogdevičius, M.; Bereczky, Á.; Török, Á. An investigation of the efficiency of using O2 and H2 (hydrooxile gas-HHO) gas additives in a ci engine operating on diesel fuel and biodiesel. *Energy* **2018**, *152*, 640–651. [CrossRef]
24. Yilmaz, A.C.; Uludamar, E.; Aydin, K. Effect of hydroxy (HHO) gas addition on performance and exhaust emissions in compression ignition engines. *Int. J. Hydrog. Energy* **2010**, *35*, 11366–11372. [CrossRef]
25. Kenanoğlu, R.; Baltacıoğlu, M.K.; Demir, M.H.; Özdemir, M.E. Performance & emission analysis of HHO enriched dual-fuelled diesel engine with artificial neural network prediction approaches. *Int. J. Hydrog. Energy* **2020**, *45*, 26357–26369.

26. Usman, M.; Farooq, M.; Naqvi, M.; Saleem, M.W.; Hussain, J.; Naqvi, S.R.; Jahangir, S.; Jazim Usama, H.M.; Idrees, S.; Anukam, A. Use of gasoline, LPG and LPG-HHO blend in SI engine: A comparative performance for emission control and sustainable environment. *Processes* **2020**, *8*, 74. [CrossRef]
27. Dongare, A.; Kharde, R.; Kachare, A.D. Introduction to artificial neural network. *Int. J. Eng. Innov. Technol.* **2012**, *2*, 189–194.
28. Gul, M.; Shah, A.N.; Aziz, U.; Husnain, N.; Mujtaba, M.; Kousar, T.; Ahmad, R.; Hanif, M.F. Grey-Taguchi and ANN based optimization of a better performing low-emission diesel engine fueled with biodiesel. *Energy Sources Part A Recovery Util. Environ. Eff.* **2019**, 1–14. [CrossRef]
29. Yildizhan, Ş.; Uludamar, E.; Çalık, A.; Dede, G.; Özcanlı, M. Fuel properties, performance and emission characterization of waste cooking oil (WCO) in a variable compression ratio (VCR) diesel engine. *Eur. Mech. Sci.* **2017**, *1*, 56–62. [CrossRef]
30. Yıldırım, S.; Tosun, E.; Çalık, A.; Uluocak, İ.; Avşar, E. Artificial intelligence techniques for the vibration, noise, and emission characteristics of a hydrogen-enriched diesel engine. *Energy Sources Part A Recovery Util. Environ. Eff.* **2019**, *41*, 2194–2206. [CrossRef]
31. Mujtaba, M.; Masjuki, H.; Kalam, M.; Ong, H.C.; Gul, M.; Farooq, M.; Soudagar, M.E.M.; Ahmed, W.; Harith, M.; Yusoff, M. Ultrasound-assisted process optimization and tribological characteristics of biodiesel from palm-sesame oil via response surface methodology and extreme learning machine-Cuckoo search. *Renew. Energy* **2020**, *158*, 202–214. [CrossRef]
32. Gul, M.; Zulkifli, N.W.M.; Kalam, M.A.; Masjuki, H.H.; Mujtaba, M.A.; Yousuf, S.; Bashir, M.N.; Ahmed, W.; Yusoff, M.N.A.M.; Noor, S.; et al. RSM and Artificial Neural Networking based production optimization of sustainable Cotton bio-lubricant and evaluation of its lubricity & tribological properties. *Energy Rep.* **2021**, *7*, 830–839.
33. Ghobadian, B.; Rahimi, H.; Nikbakht, A.; Najafi, G.; Yusaf, T. Diesel engine performance and exhaust emission analysis using waste cooking biodiesel fuel with an artificial neural network. *Renew. Energy* **2009**, *34*, 976–982. [CrossRef]
34. Uslu, S.; Celik, M.B. Performance and exhaust emission prediction of a SI engine fueled with I-amyl alcohol-gasoline blends: An ANN coupled RSM based optimization. *Fuel* **2020**, *265*, 116922. [CrossRef]
35. Dinjus, E.; Arnold, U.; Dahmen, N.; Höfer, R.; Wach, W. Green fuels–sustainable solutions for transportation. In *Sustainable Solutions for Modern Economies*; Royal Society of Chemistry: London, UK, 2009; Volume 4, pp. 125–129.
36. Uludamar, E.; Tosun, E.; Tüccar, G.; Yıldızhan, Ş.; Çalık, A.; Yıldırım, S.; Serin, H.; Özcanlı, M. Evaluation of vibration characteristics of a hydroxyl (HHO) gas generator installed diesel engine fuelled with different diesel–biodiesel blends. *Int. J. Hydrog. Energy* **2017**, *42*, 23352–23360. [CrossRef]
37. Masood, M.; Ishrat, M.; Reddy, A. Computational combustion and emission analysis of hydrogen–diesel blends with experimental verification. *Int. J. Hydrog. Energy* **2007**, *32*, 2539–2547. [CrossRef]
38. Uludamar, E. Effect of hydroxy and hydrogen gas addition on diesel engine fuelled with microalgae biodiesel. *Int. J. Hydrog. Energy* **2018**, *43*, 18028–18036. [CrossRef]
39. Dahake, M.; Patil, S.; Patil, S. Effect of hydroxy gas addition on performance and emissions of diesel engine. *Int. Res. J. Eng. Technol.* **2016**, *3*.
40. Sun, Z.-Y.; Liu, F.-S.; Liu, X.-H.; Sun, B.-G.; Sun, D.-W. Research and development of hydrogen fuelled engines in China. *Int. J. Hydrog. Energy* **2012**, *37*, 664–681. [CrossRef]
41. Premkartikkumar, S.; Annamalai, K.; Pradeepkumar, A. Using hydrogen as a fuel in automotive engines–an investigation. *Int. J. Innov. Technol. Res.* **2013**, *1*, 90–93.
42. Ahmed, E.; Usman, M.; Anwar, S.; Ahmad, H.M.; Nasir, M.W.; Malik, M.A.I. Application of ANN to predict performance and emissions of SI engine using gasoline-methanol blends. *Sci. Prog.* **2021**, *104*, 00368504211002345. [CrossRef]
43. Kesgin, U. Genetic algorithm and artificial neural network for engine optimisation of efficiency and NOx emission. *Fuel* **2004**, *83*, 885–895. [CrossRef]
44. Çay, Y.; Korkmaz, I.; Çiçek, A.; Kara, F. Prediction of engine performance and exhaust emissions for gasoline and methanol using artificial neural network. *Energy* **2013**, *50*, 177–186. [CrossRef]
45. Cay, Y. Prediction of a gasoline engine performance with artificial neural network. *Fuel* **2013**, *111*, 324–331. [CrossRef]
46. Yusaf, T.; Yousif, B.; Elawad, M. Crude palm oil fuel for diesel-engines: Experimental and ANN simulation approaches. *Energy* **2011**, *36*, 4871–4878. [CrossRef]
47. Sayin, C.; Ertunc, H.M.; Hosoz, M.; Kilicaslan, I.; Canakci, M. Performance and exhaust emissions of a gasoline engine using artificial neural network. *Appl. Therm. Eng.* **2007**, *27*, 46–54. [CrossRef]
48. Abdalla, A.N.; Tao, H.; Bagaber, S.A.; Ali, O.M.; Kamil, M.; Ma, X.; Awad, O.I. Prediction of emissions and performance of a gasoline engine running with fusel oil–gasoline blends using response surface methodology. *Fuel* **2019**, *253*, 1–14. [CrossRef]
49. Dey, S.; Reang, N.M.; Das, P.K.; Deb, M. Comparative study using RSM and ANN modelling for performance-emission prediction of CI engine fuelled with bio-diesohol blends: A fuzzy optimization approach. *Fuel* **2021**, *292*, 120356. [CrossRef]

*sustainability*

MDPI

*Article*

# Real-Time Implementation of an Optimized Model Predictive Control for a 9-Level CSC Inverter in Grid-Connected Mode

**Alamera Nouran Alquennah [1], Mohamed Trabelsi [1], Khaled Rayane [2], Hani Vahedi [3,*] and Haitham Abu-Rub [2]**

1 Electronic and Communications Engineering Department, Kuwait College of Science and Technology, Kuwait City 27235, Kuwait; a.alquennah@kcst.edu.kw (A.N.A.); m.trabelsi@kcst.edu.kw (M.T.)
2 Department of Electrical and Computer Engineering, Texas A and M University at Qatar, Qatar Foundation, Doha 23874, Qatar; khaled.rayane@qatar.tamu.edu (K.R.); haitham.abu-rub@qatar.tamu.edu (H.A.-R.)
3 Ossiaco Inc., Montreal, QC H3C 2G9, Canada
* Correspondence: hani.vahedi@ieee.org

**Abstract:** The Crossover Switches Cell (CSC) is a recent Single DC-Source Multilevel Inverter (SDCS-MLI) topology with boosting abilities. In grid-connected PV applications, the CSC should be controlled to inject a sinusoidal current to the grid with low THD% and unity power factor, while balancing the capacitor voltage around its reference. These two objectives can be met through the application of a finite control set model predictive control (FCS-MPC) method. Thus, this paper proposes a design of an optimized FCS-MPC for a 9-level grid-tied CSC inverter. The switching actions are optimized using the redundant switching states. The design is verified through simulations and real-time implementation. The presented results show that the THD% of the grid current is 1.73%, and the capacitor voltage is maintained around its reference with less than 0.5 V mean error. To test the reliability of the control design, different scenarios were applied, including variations in the control reference values as well as the AC grid voltage. The presented results prove the good performance of the designed controller in tracking the reference values and minimizing the steady-state errors.

**Keywords:** crossover switches cell; CSC; multilevel inverter; Packed-U-Cell; model predictive control; grid connection

**Citation:** Alquennah, A.N.; Trabelsi, M.; Rayane, K.; Vahedi, H.; Abu-Rub, H. Real-Time Implementation of an Optimized Model Predictive Control for a 9-Level CSC Inverter in Grid-Connected Mode. *Sustainability* **2021**, *13*, 8119. https://doi.org/10.3390/su13158119

Academic Editors: Alex Stojcevski, Ben Horan, Mehdi Seyedmahmoudian and Saad Mekhilef

Received: 27 May 2021
Accepted: 13 July 2021
Published: 21 July 2021



## 1. Introduction

The capacity of global renewable energy (RE) has witnessed an increase of 261 GW in 2020, which leads to 2799 GW of the total global RE capacity, where PV systems dominate, with around 25.3% of this total capacity [1]. In order to connect the PV strings to the utility grid, the DC output of the PV modules should be converted to AC, which is the function of the inverters stage. Several research works reviewed the inverter technologies for PV applications [2–5].

In the literature, it has been reported that multilevel inverters (MLIs) ensure a higher power quality compared to conventional 2-level inverters, which make them a good candidate for RE applications [5]. In various MLI topologies, the design mainly consists of the use of DC voltage sources, capacitors, and medium power semiconductor devices that operate at a reduced voltage rating. These topologies, which generate multiple DC voltage levels at the output terminals, have several advantages compared to the conventional two-level inverters such as lower switching losses, lower voltage stress on the power semiconductor devices, reduced electromagnetic interference, higher efficiency, and lower harmonic pollution and filter size [6–8]. On the other hand, the complexity of the inverter's design increases and the overall reliability of the system decreases with the increase in the number of levels. That is because each switch used in the design requires a related gate driver and a protection circuit. As a result, the increase in the power semiconductor switches in the design leads to increases in the system's cost and the control complexity [8,9]. Moreover, using multiple DC sources in MLIs topologies raises concerns regarding the increase in the power losses and malfunctioning in the system due to the unbalanced power sharing among the isolated DC sources [10]. In PV systems, the increase in the required DC sources implies an excessive number of DC-DC converters.

Therefore, interest is increasing in the reduced switches MLIs and in single DC sources MLIs (SDCS-MLIs) [11]. One of the recently developed SDC-MLIs is the Packed U cells (PUC) topology, which is classified as an asymmetric Flying Capacitors inverter (FCI) that could be used as a compromise between the cascaded H-bridges and the flying capacitor topologies [12,13]. The PUC MLI generates more voltage levels with high-power quality, while using a lower number of passive/active components and DC sources (single DC source) compared to other MLI topologies. These PUC-MLI features result in cost reductions and a smaller compact power conversion unit compared to even 2-level topologies. However, the maximum voltage level generated by PUC-MLI equals the DC source voltage which limits its employment to low power applications and makes it unsuitable for applications that require an output voltage greater than the input DC source. Furthermore, the topology provides the capacitor with one path for charging. Hence, problems may occur if there is a lack of energy and a long interval between the discharging and charging states [14].

These limitations are overcome by modifying the PUC MLI to have two crossover switches between the DC link and the capacitor; the new topology is called the Crossover Switches Cell (CSC) [14]. This modification provides another way of charging the capacitor and increases the number of levels from seven levels in PUC to nine levels in CSC. The maximum voltage level is the sum of the DC source and the capacitor's voltage; CSC has a boosting ability.

Voltage/current controllers are required in MLIs to deliver green energy/power from the source to the load/grid. Control schemes with modulators such as sinusoidal pulse width modulation (SPWM) and space vector (SVPWM) are commonly used with MLIs in general and in PV applications [15–17]. These methods are compatible with high switching frequencies. SPWM is easy to implement and does not require any optimization technique. SVPWM generates low current ripples and is easy to implement, while its complexity increases with the number of levels [6,18]. Space vector control is a fundamental frequency method that is effective for a high number of levels cases, but the lower-order harmonic components cannot be eliminated [6,7]. The selective harmonic elimination method generates signals with a low total harmonic distortion (THD%) and is suitable for high-power applications, but suffers with offline calculations. Additionally, adaptive controllers, sliding mode controllers, artificial intelligent controllers and Fuzzy logic controllers are designed for MLIs in the PV systems [19–22].

Model predictive control (MPC) schemes are based on the predicted states obtained by systems' model. MPC was applied effectively in systems with MLIs and in PV applications [23–28]. One of the main advantages of MPC is that the control action is applied directly to the system, without the need for a modulation stage [29]. Furthermore, MPC is a multi-objective control technique; several objectives can be designed in its cost function by specifying their priorities according to the application. This makes MPC a good candidate for MLIs in PV systems where the capacitors' voltage needs to be regulated according to their references in order to generate the required voltage levels and maintain a low THD% in the current that is fed to the grid.

In this paper, a finite control set MPC (FCS-MPC) is proposed for a nine-level, single phase, grid-connected CSC-MLI. The objective of the controller is to generate a synchronized grid current with a minimized THD%, while maintaining the capacitor voltage at around its reference value. The topology and the switching patterns of the CSC inverter are described in detail. Then, the mathematical model of the system and the control design steps are explained. The proposed controller is verified through simulation and real-time implementation and the results are shown and discussed to prove the acceptable dynamic performance of the designed controller.

## 2. CSC Topology and Mathematical Modelling

### 2.1. CSC Topology

The studied nine-level CSC-inverter is shown in Figure 1. The topology consists of a DC voltage source (representing the the output of the DC-DC converter fed by the solar panels in PV applications), a capacitor, and eight switches ($S_i$, $i = 1 \dots 8$), where two of them are bidirectional ($S_2$ and $S_5$). The switches between the DC link and the positive output terminal of the inverter, $S_1$ and $S_4$, work in a complementary manner, in which one of them is ON at a time. The same concept applies to the switches between the capacitor link and the negative output terminal of the inverter, $S_3$ and $S_6$. The four switches between the DC link and the capacitor link also work in a complementary way. All the valid switching patterns are presented in Table 1, where $s_i \in \{0, 1\}$ represents the switching state of the switch $S_i$. In order to have nine DC levels at the output terminal of the inverter, the capacitor voltage ($V_2$) is maintained at one third of the DC link voltage ($V_1$). Table 1 shows the $V_{AB}$ value when $V_1$ is set to 150 V and $V_2$ is set to 50 V. It is clear that the maximum and minimum output voltage will be ±200 V. Hence, CSC inverter has a boosting ability.

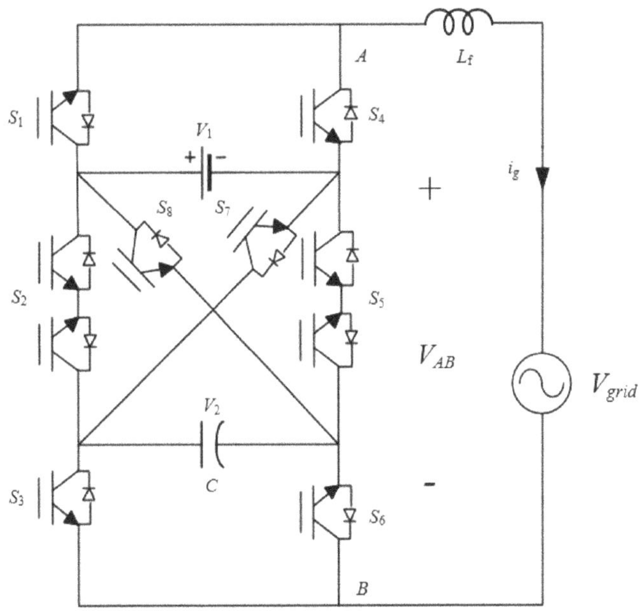


**Figure 1.** The Crossover Switches Cell (CSC) Inverter Topology.

### 2.2. Modelling

Given that C is the CSC capacitor, $L_f$ is the filtering inductor. By using Kirchhoff's voltage law (KVL) and Kirchhoff's current law (KCL) on the topology shown in Figure 1, the model state equations can be written as:

$$\dot{V}_2(t) = \frac{1}{C}(s_3 - s_2 - s_7)i_g(t) \tag{1}$$

$$\dot{i}_g(t) = \frac{1}{L_f}(V_{AB}(t) - V_g(t)) \tag{2}$$

where $\dot{V}_2(t) = \frac{dV_2}{dt}$, $\dot{i}_g(t) = \frac{di_g}{dt}$,

$$V_{AB}(t) = (s_1 - s_2 - s_8)V_1(t) + (s_2 - s_3 + s_7)V_2(t), \tag{3}$$

$V_1(t)$ is the DC voltage and $V_2(t)$ is the capacitor voltage.

**Table 1.** Switching states and the corresponding output voltage level of the CSC inverter.

| $s_1$ | $s_2$ | $s_3$ | $s_4$ | $s_5$ | $s_6$ | $s_7$ | $s_8$ | $V_{AB}$ | $V_{AB}$ (V) | Cell Capacitor |
|---|---|---|---|---|---|---|---|---|---|---|
| 1 | 0 | 0 | 0 | 0 | 1 | 1 | 0 | $V_1 + V_2$ | 200 | Charged |
| 1 | 0 | 0 | 0 | 1 | 1 | 0 | 0 | $V_1$ | 150 | Bypassed |
| 1 | 0 | 1 | 0 | 0 | 0 | 1 | 0 | $V_1$ | 150 | Bypassed |
| 1 | 0 | 1 | 0 | 1 | 0 | 0 | 0 | $V_1 - V_2$ | 100 | Discharged |
| 0 | 0 | 0 | 1 | 0 | 1 | 1 | 0 | $V_2$ | 50 | Charged |
| 1 | 1 | 0 | 0 | 0 | 1 | 0 | 0 | $V_2$ | 50 | Charged |
| 0 | 0 | 1 | 1 | 0 | 0 | 1 | 0 | 0 | 0 | Bypassed |
| 1 | 1 | 1 | 0 | 0 | 0 | 0 | 0 | 0 | 0 | Bypassed |
| 0 | 0 | 0 | 1 | 1 | 1 | 0 | 0 | 0 | 0 | Bypassed |
| 1 | 0 | 0 | 0 | 0 | 1 | 0 | 1 | 0 | 0 | Bypassed |
| 0 | 0 | 1 | 1 | 1 | 0 | 0 | 0 | $-V_2$ | −50 | Discharged |
| 1 | 0 | 1 | 0 | 0 | 0 | 0 | 1 | $-V_2$ | −50 | Discharged |
| 0 | 1 | 0 | 1 | 0 | 1 | 0 | 0 | $-V_1 + V_2$ | −100 | Charged |
| 0 | 0 | 0 | 1 | 0 | 1 | 0 | 1 | $-V_1$ | −150 | Bypassed |
| 0 | 1 | 1 | 1 | 0 | 0 | 0 | 0 | $-V_1$ | −150 | Bypassed |
| 0 | 0 | 1 | 1 | 0 | 0 | 0 | 1 | $-V_1 - V_2$ | −200 | Discharged |

## 3. Control Scheme

MPC is divided into three steps: Predicting the model, calculating the cost function and minimizing the cost function. The details of the control scheme are given in Figure 2.

Prediction of the model's state equations step depends on the discrete version of the state equations, in which the ($k$ + 1) state is predicted from the (k) state. Since the state variables' trajectory is assumed to be rectlinear over a small sampling time, the state equations can be discretized using the following relationship:

$$x(k+1) = x(k) + \frac{dx(t)}{dt}.T_s \tag{4}$$

where $x(k+1)$ is the predicted state at $(k+1)$, $x(k)$ is the measured state at $(k)$, $T_s$ is the sampling time.

At $(k)$ step, $V_g$, $i_g$, $V_1$, and $V_2$ are measured. By applying Equation (4) on Equations (1) and (2), respectively, the prediction of the capacitor voltage and the grid current values at $(k+1)$ can be found using the following equations:

$$V_2(k+1) = V_2(k) + \frac{T_s}{C}(s_3 - s_2 - s_7)i_g(k) \tag{5}$$

$$i_g(k+1) = i_g(k) + \frac{T_s}{L_f}(V_{AB}(k) - V_g(k)) \tag{6}$$

where $V_{AB}(k) = (s_1 - s_2 - s_8)V_1(k) + (s_2 - s_3 + s_7)V_2(k)$. The model is predicted for the 16 switching states given in Table 1.

For the grid-tied CSC inverter, the objective is to minimize the grid current THD% and the error between the capacitor voltage and its reference value. Hence, the cost function is designed as

$$g = \lambda_v \| V_2^*(k+1) - V_2(k+1)\|^2 + \lambda_i \| i_g^*(k+1) - i_g(k+1)\|^2, \tag{7}$$

where $\lambda_v$ and $\lambda_i$ are weighting factors, and ($i_g^*$,$V_2^*$) are the desired values for the grid current and the capacitor voltage, respectively. The cost function implies that the reference values are obtained at $(k+1)$ sample. According to [30,31], it can be assumed that $i_g^*(k+1) \approx i_g^*(k)$. Since the DC voltage $(V_1)$ is assumed to be constant, then $V_2^*(k)$ is equal to $V_2^*(k+1)$. Hence, the cost function is calculated using the following:

$$g = \lambda_v \| V_2^*(k) - V_2(k+1) \|^2 + \lambda_i \| i_g^*(k) - i_g(k+1) \|^2. \tag{8}$$

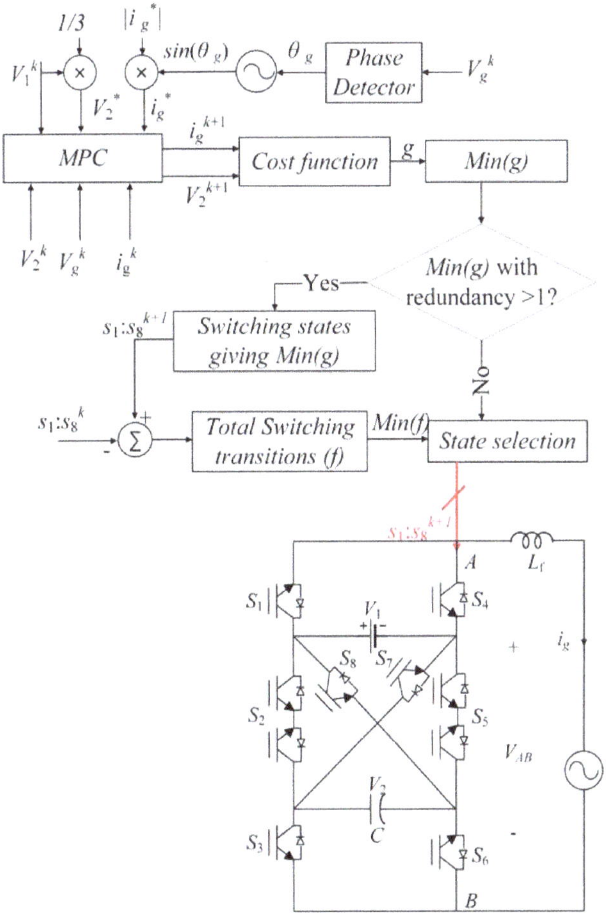


**Figure 2.** The synoptic of the proposed optimized FCS-MPC for the grid-tied 9-Level CSC inverter.

The cost function is calculated for the 16 switching states, and the switching state that will give the minimum $g$ is chosen for $(k+1)$ time step. However, in CSC inverter, as shown in Table 1, there are several redundant states that will lead to the same minimum $g$. To reduce the switching losses, the $(k)$ switching state $(s_1^k : s_8^k)$ is compared with all switching states that give $(g_{min})$, $(s_1^{k+1} : s_8^{k+1})$. The total switching transitions from $(s_1^k : s_8^k)$ to $(s_1^{k+1} : s_8^{k+1})$ are calculated, $f$. The one with the minimum switching transitions is sent to the inverter.

The weighting factors, $\lambda_v$ and $\lambda_i$ are chosen based on the objectives' priorities. In the CSC grid-connected case, a higher priority is given to the grid current THD% over the capacitor's voltage error. In our previous work [32], it was shown that to compromise between the two objectives, $\lambda_i$ is set to 10 and $\lambda_v$ is set to 5. The simulation results showed that the current THD% is 1.73% and the mean voltage error is 0.53 V.

## 4. Results and Discussion

The presented MPC for CSC inverter is tested via simulation and real-time implementation. The simulation was carried out using MATLAB/Simulink. The real-time implementation of the system was carried out using an OPAL-RT 5600 real-time simulator, enabling dynamic RT simulation responses.

The system parameters used for simulation and real-time implementation are listed in Table 2.

**Table 2.** System's parameters.

| Parameters | Value |
|---|---|
| Fundamental frequency $f_0$ | 60 Hz |
| Sampling time $T_s$ | 20 μs |
| Grid voltage peak $V_g$ | 170 V |
| Grid current peak $\|i_g^*\|$ | 5 A |
| DC source voltage $V_1$ | 150 V |
| Capacitor voltage $V_2^*$ | 50 V |
| Capacitor $C$ | 2500 μF |
| Filtering inductor $L_f$ | 6 mH |
| Current weighting factor $\lambda_i$ | 10 |
| Voltage weighting factor $\lambda_v$ | 5 |

### 4.1. Simulation Results

The effectiveness of the proposed controller in generating low-grid-current THD% is shown in Figure 3, where the THD% is 1.73%. This result is below the stated limit in IEEE 929-2000 standard for grid-connected PV systems (THD < 5%). The synchronization between the grid current and the grid voltage, the unity power factor, is shown in Figure 4. Additionally, the quality of the grid current is presented by plotting it against its reference.

Figure 5 shows the DC voltage, $V_1$, set as 150 V during the simulation. The capacitor voltage, $V_2$, is maintained at around its reference, 50 V, as shown in the same figure. The mean error of $V_2$ is 0.44 V. The third part of Figure 5 shows the generated nine levels of the inverter , $V_{AB}$, with the grid voltage, $V_g$. The results show the boosting ability of the CSC inverter, where the maximum voltage level (200 V) is the sum of the DC voltage and the capacitor voltage.

Figures 6 and 7 show the switching transitions for the eight switches ($S_1 : S_8$) of the CSC inverter. In Figure 6, the switching optimization algorithm ($f$) was used to reduce the total switching transitions, while in Figure 7, the switching optimization algorithm was skipped. Applying the switching optimization algorithm reduces the number of total switching transitions by an average of 85 transitions per cycle, and for a one-second simulation, the difference was more than 4500 transitions which is 9.3% transitions reduction. This reduction affects the conduction and switching losses in multilevel inverters, which depend on the frequency with which the switches are turned ON and OFF [33,34].

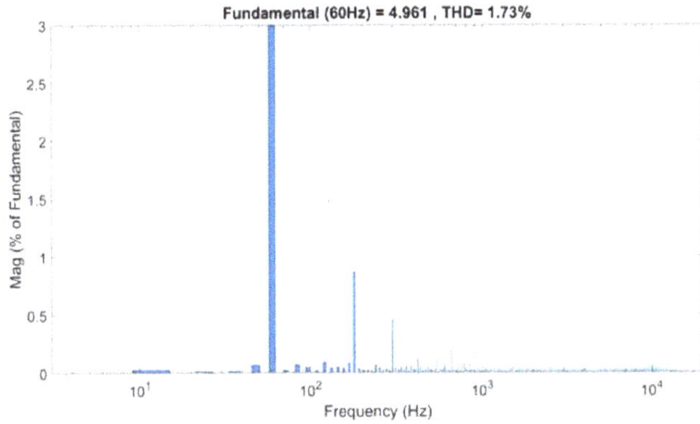


**Figure 3.** Grid current ($i_g$) THD%.

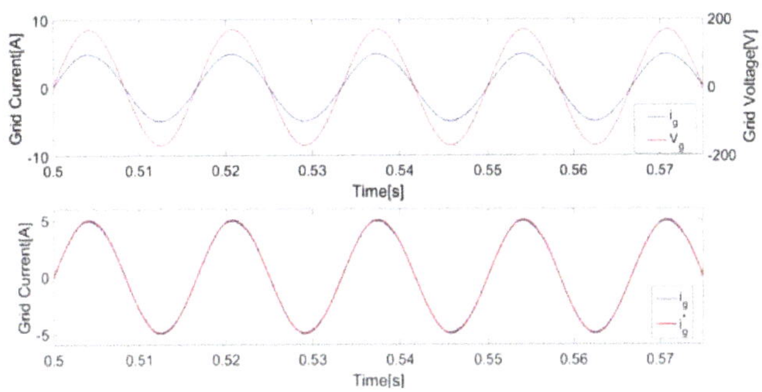


**Figure 4.** (**Upper**): The grid current ($i_g$) and the grid voltage ($V_g$), showing the synchronization between the two signals. (**Lower**): the generated grid current ($i_g$) versus its reference ($i_g^*$).

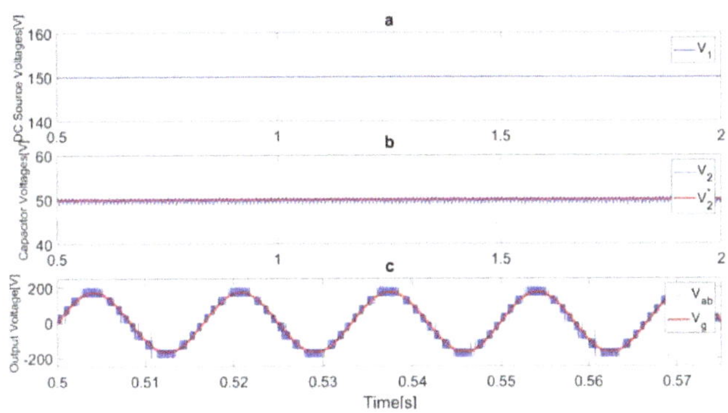


**Figure 5.** (**a**) The DC voltage ($V_1$), (**b**) The capacitor voltage ($V_2$) against its reference ($V_2^*$), and (**c**) The inverter output voltage $V_{AB}$ and the grid voltage ($V_g$).

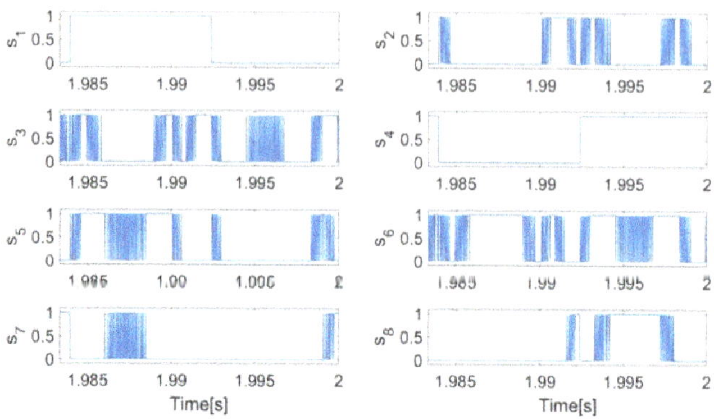

**Figure 6.** The transitions of the eight switches ($S_1 : S_8$) for one steady-state cycle when the switching optimization algorithm is used.

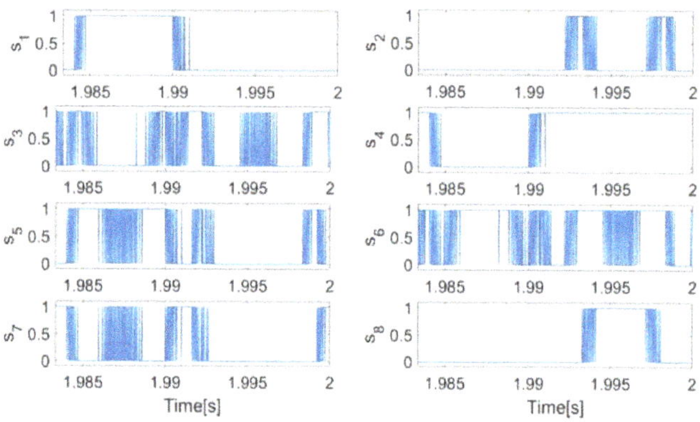

**Figure 7.** The transitions of the eight switches ($S_1 : S_8$) for one steady-state cycle without using the switching optimization algorithm.

### *4.2. Implementation Results*

The setup of the system's real-time implementation is shown in Figure 8. Figures 9–11 demonstrate the steady-state operation of the system. The nine voltage levels of the CSC inverter ($V_{AB}$) are shown in Figure 9 along with the grid voltage ($V_g$). The MPC objective is to ensure that the generated grid current ($i_g$) follows its reference ($i_g^*$). This is clearly shown in Figure 10. The phase angle that is used to generate the reference current signal ($i_g^*$) is obtained from the grid voltage signal. That is, the generated grid current ($i_g$) is synchronized with the grid voltage with unity power factor. This synchronization is demonstrated in Figure 11.

To represent the importance of choosing the proper weighting factors, a distorted case is shown through real-time implementation. In Figure 10 ($\lambda_v = 5$) and ($\lambda_i = 10$) are chosen to give a higher priority to the current error in the cost function. However, in Figure 12, high distortion is shown in the grid current when the weighting factors were chosen as ($\lambda_v = 5$) and ($\lambda_i = 1$).

Since the MPC control strategy is based on predictions of the new state value using the system's model, any mismatch between the model and the real value of the system may affect the controller's effectiveness. Therefore, the capacitor and inductor values were changed by 50% to test the robustness of the controller. Figure 13 shows the accommodation of the generated grid current by the inverter, with its reference.

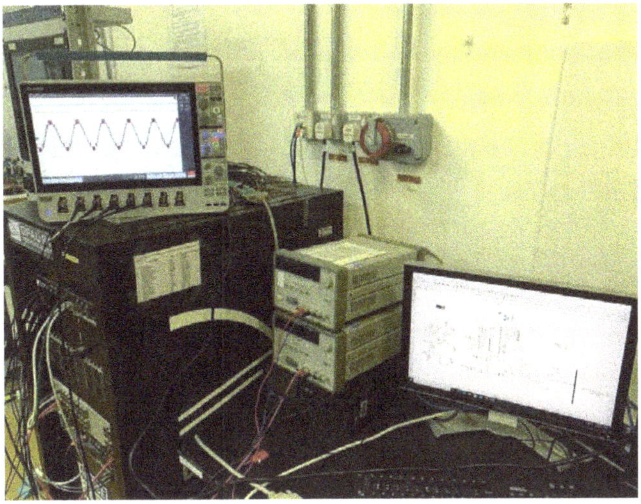

**Figure 8.** The setup of the FCS-CSC real-time implementation.

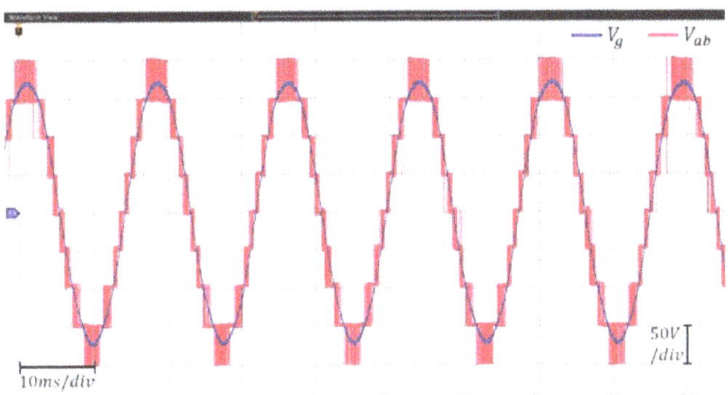


**Figure 9.** The output voltage of the CSC terminals showing the nine levels and the grid voltage.

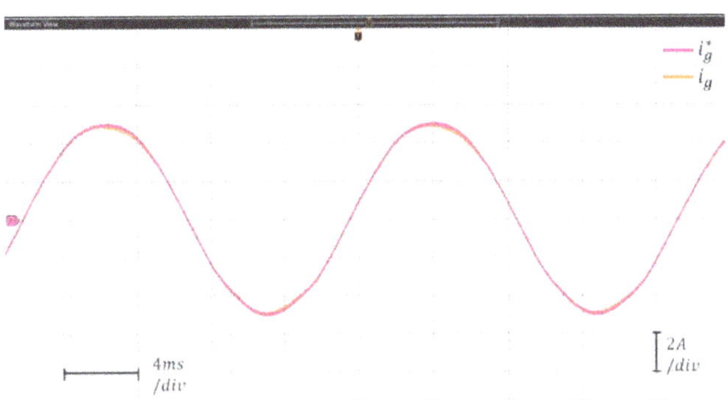


**Figure 10.** The grid currents ($i_g$) and its reference $i_g^*$ for $\lambda_v = 5$ and $\lambda_i = 10$ case.

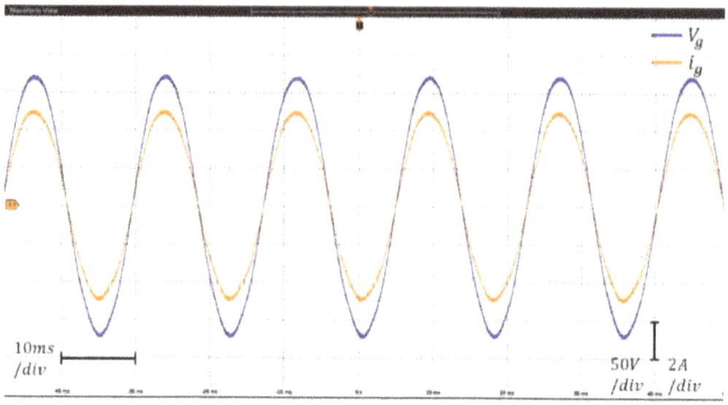


**Figure 11.** The synchronization between the grid currents ($i_g$) and the grid voltage $v_g$.

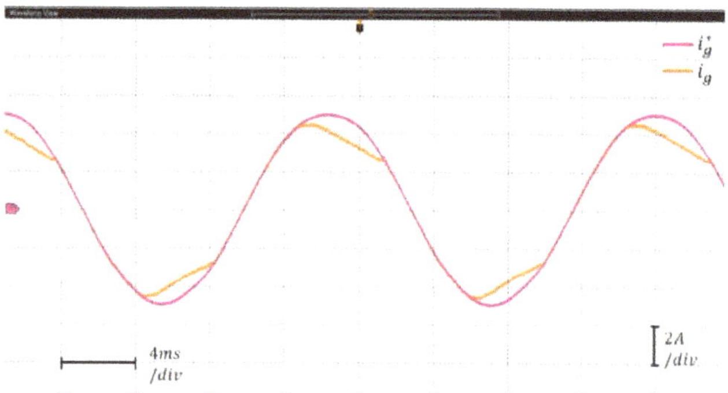


**Figure 12.** The grid current ($i_g$) and its reference $i_g^*$ for $\lambda_v = 5$ and $\lambda_i = 1$ case.

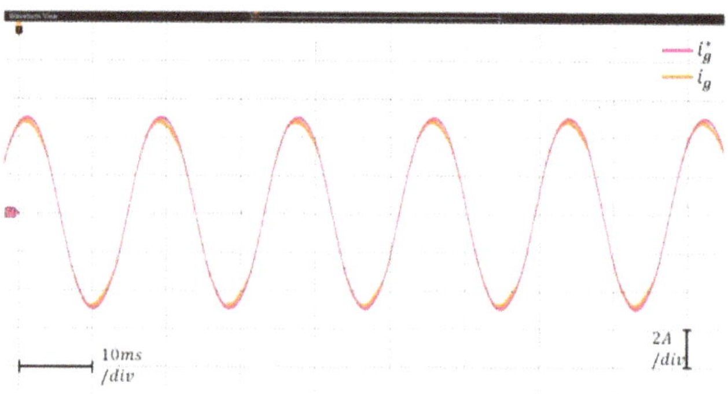


**Figure 13.** The grid current ($i_g$) and its reference ($i_g^*$) when the inductor and the capacitor values varied by ±50%.

The real-time implementation of the proposed MPC was extended to include several dynamic variation tests. Figure 14 illustrates the dynamic performance during a step-up change in the current reference peak from 5 A to 10 A. The generated grid current followed the new reference and maintained the unity power factor. In the second dynamic test, the DC voltage source was increased by (40%), i.e., $V_1$ is increased from 150 V to 210 V. As shown in Figure 15, $V_2$ accommodated the changes and increased from 50 V to 70 V ($V_2 = V_1/3$).

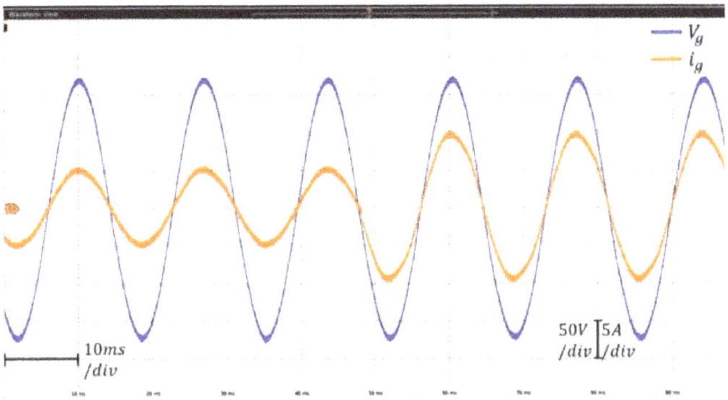


**Figure 14.** The grid voltage $V_g$ and the generated grid current $i_g$ when the grid current reference ($i_g^*$) is increased from 5 A to 10 A.

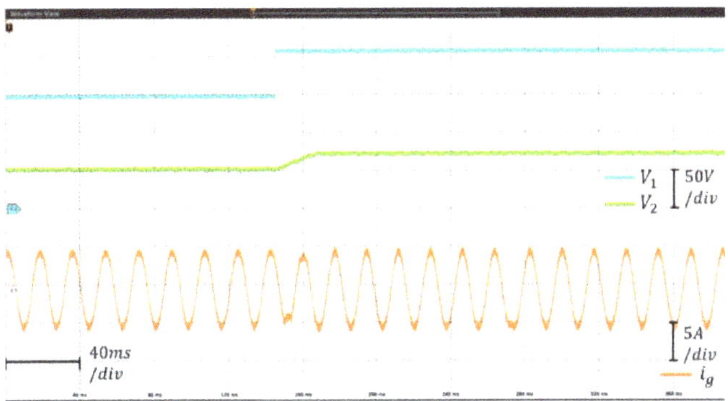


**Figure 15.** $V_1$, $V_2$, and $i_g$ when the DC Source ($V_1$) varies from 150 V to 210 V.

To ensure the reliability of the systems tied to grids and the quality of the power distribution, some power quality disturbances were introduced to systems under testing [35]. Additionally, several standards describe the criteria of the accepted electric power quality, such as EN 50160-2000 and IEC 61000-2-8-2002. Voltage sag and voltage swell are common power-quality disturbances, which were introduced to the proposed FCS-MPC for 9-level CSC inverter in this work. Voltage sag can result from short-circuit faults, a change in the load, or a sudden change in the power source [35]. To implement the system under the effect of grid voltage sag, the grid voltage $V_g$ was stepped-down by 10% from 170 V to 153 V, as shown in Figure 16. However, faults in the electrical distribution systems lead to voltage swell disturbances. Although voltage swell is less likely to occur than voltage sag, the damage it causes is greater on devices that cannot handle a voltage above their rating values [35]. To implement the voltage swell disturbance on the proposed system, the grid voltage $V_g$ was increased from 170 V to 185 V, as shown in Figure 17.

Reactive power variation was tested by introducing phase shift between the grid voltage and the grid current in two scenarios: $\pi/4$ and $\pi/6$, as shown in Figures 18 and 19, respectively.

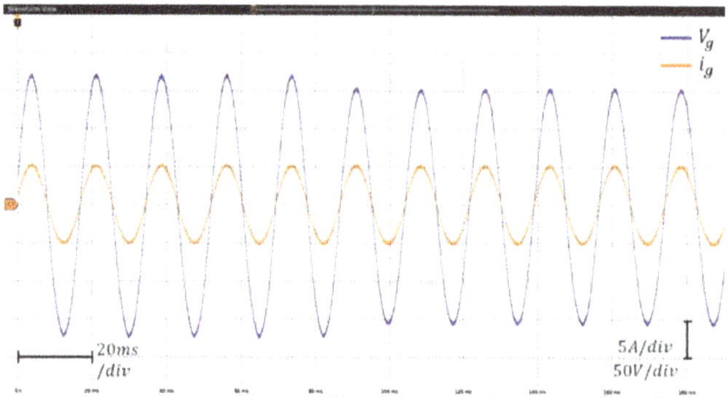


**Figure 16.** Voltage Sags—Grid Voltage ($V_g$ 170 V to 153 V), $V_g$ and $i_g$.

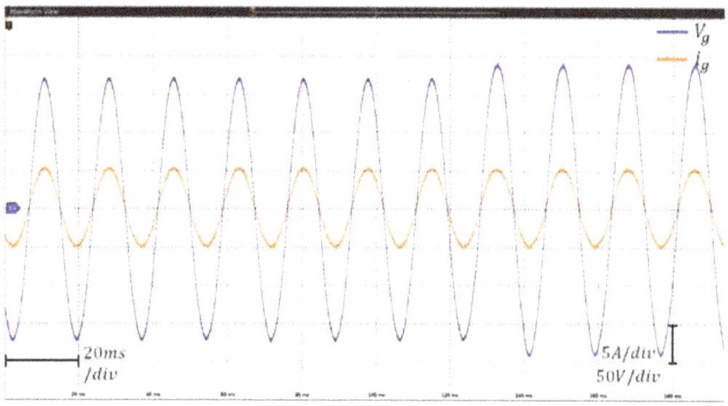


**Figure 17.** Voltage swell—Grid Voltage ($V_g$ 170 V to 185 V), $V_g$ and $i_g$.

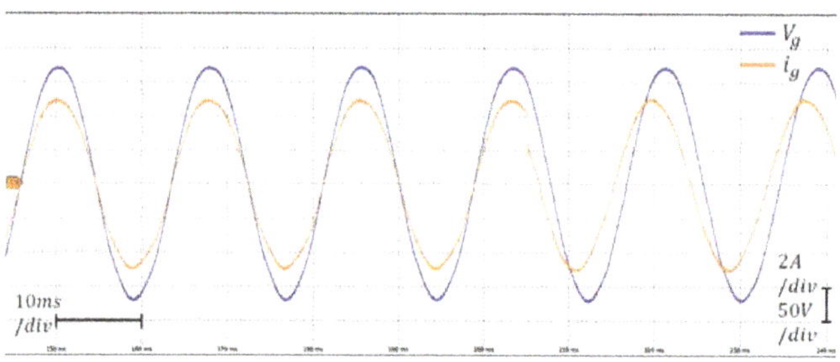


**Figure 18.** The phase shift between the grid voltage and current is $\pi/4$.

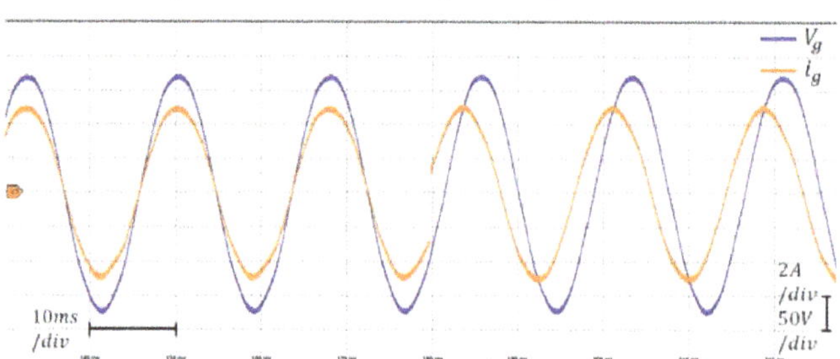


**Figure 19.** The phase shift between the grid voltage and current is $\pi/6$.

## 5. Conclusions

This paper proposes a finite control set–model predictive control (FSC-MPC) for a nine-level grid-connected single-phase crossover switches cell (CSC) multilevel inverter (MLI).

The results clearly showed the generated nine voltage levels, where the maximum voltage level was the sum of the DC source and the capacitor voltages. The cost function was designed to minimize the THD% of the generated grid current and regulate the capacitor's voltage around its reference. The simulation results showed that the current THD% is 1.73% and the mean error of the capacitor's voltage is less than 0.5 V. To connect the system to the grid, the grid current and the grid voltage should be synchronized; the power factor (PF) is one. Simulation results and real-time implementation results showed the synchronization between the two signals. The transient response of the system was tested in real-time implementation by changing the reference values of the capacitor voltage and the grid current amplitude. In both cases, the system showed a fast response to the changes and reached the new steady state. Furthermore, the results showed the ability of the system to accommodate different changes in the grid conditions, such as the grid voltage sag case, grid voltage swell case and the need for a reactive power. The control algorithm was designed to benefit from the redundant switching states in the CSC patterns by choosing, at each time step, the state that minimizes the cost function with the fewest total switching transitions. This algorithm decreased the switching transitions by 9.3% for a one-second steady-state simulation.

**Author Contributions:** A.N.A.: Conceptualization, Methodology, Software, original draft preparation. M.T.: Conceptualization, Methodology, Validation, Writing review and editing, Project administration, K.R.:Validation, Data analysis, Writing review and editing. H.V.: Formal analysis, investigation, writing review and editing. H.A.-R.: Proofreading and editing, Review and editing. All authors have read and agreed to the published version of the manuscript.

**Funding:** This research received no external funding.

**Institutional Review Board Statement:** Not applicable.

**Informed Consent Statement:** Not applicable.

**Data Availability Statement:** Not applicable.

**Conflicts of Interest:** The authors declare no conflict of interest.

## Abbreviations

The following abbreviations are used in this manuscript:

| | |
|---|---|
| MLI | Multilevel Inverter |
| THD | Total harmonic distortion |
| CSC | Crossover switches Cell |
| FCS-MPC | Finite control set-model predictive control |
| $s_i$ ($i$ 1 to 8) | CSC switches |
| $V_1$ | DC link voltage |
| $V_2$ | Capacitor voltage |
| $V_{AB}$ | CSCoutput Voltage |
| $V_g$ | Grid voltage |
| $i_g$ | Grid current |
| $T_s$ | Sampling time |

## References

1. IRENA. *Renewable Capacity Statistics 2021 International Renewable Energy Agency (IRENA)*; IRENA: Abu Dhabi, United Arab Emirates, 2021.
2. Alluhaybi, K.; Batarseh, I.; Hu, H. Comprehensive Review and Comparison of Single-Phase Grid-Tied Photovoltaic Microinverters. *IEEE J. Emerg. Select. Top. Power Electron.* **2020**, *8*, 1310–1329. [CrossRef]
3. Edwin, F.; Xiao, W.; Khadkikar, V. Topology review of single phase grid-connected module integrated converters for PV applications. In Proceedings of the IECON 2012—38th Annual Conference on IEEE Industrial Electronics Society, Montreal, QC, Canada, 25–28 October 2012; pp. 821–827.
4. Chaudhary, S.; Ahmad, Z.; Singh, S.N. Single Phase Grid Interactive Solar Photovoltaic Inverters: A Review. In Proceedings of the 2018 National Power Engineering Conference (NPEC), Madurai, India, 9–10 March 2018; pp. 1–6.

5. Kjaer, S.B.; Pedersen, J.K.; Blaabjerg, F. A review of single-phase grid-connected inverters for photovoltaic modules. *IEEE Trans. Ind. Appl.* **2005**, *41*, 1292–1306. [CrossRef]
6. Bana, P.R.; Panda, K.P.; Naayagi, R.T.; Siano, P.; Panda, G. Recently Developed Reduced Switch Multilevel Inverter for Renewable Energy Integration and Drives Application: Topologies, Comprehensive Analysis and Comparative Evaluation. *IEEE Access* **2019**, *7*, 54888–54909. [CrossRef]
7. Abu-Rub, H.; Holtz, J.; Rodriguez, J.; Baoming, G. Medium-Voltage Multilevel Converters—State of the Art, Challenges, and Requirements in Industrial Applications. *IEEE Trans. Ind. Electron.* **2010**, *57*, 2581–2596. [CrossRef]
8. Ebrahimi, J.; Babaei, E.; Gharehpetian, G.B. A New Topology of Cascaded Multilevel Converters With Reduced Number of Components for High-Voltage Applications. *IEEE Trans. Power Electron.* **2011**, *26*, 3109–3118. [CrossRef]
9. Iannati, O.; Mohammad, R.; Ebrahim, S.; Sajad, N. R. Creative design of symmetric multilevel converter to enhance the circuit's performance. *IET Power Electron.* **2015**, *8*, 96–102. [CrossRef]
10. Malinowski, M.; Gopakumar, K.; Rodriguez, J.; Pérez, M.A. A Survey on Cascaded Multilevel Inverters. *IEEE Trans. Ind. Electron.* **2010**, *57*, 2197–2206. [CrossRef]
11. Trabelsi, M.; Vahedi, H.; Abu-Rub, H. Review on Single-DC-Source Multilevel Inverters: Topologies, Challenges, Industrial Applications, and Recommendations. *IEEE Open J. Ind. Electron. Soc.* **2021**, *2*, 112–127. [CrossRef]
12. Ounejjar, Y.; Al-Haddad, K.; Gregoire, L. Packed U Cells Multilevel Converter Topology: Theoretical Study and Experimental Validation. *IEEE Trans. Ind. Electron.* **2011**, *58*, 1294–1306. [CrossRef]
13. Vahedi, H.; Trabelsi, M. *Single-DC-Source Multilevel Inverters*; Springer Nature: Cham, Switzerland, 2019; ISBN 978-3-030-15252-9. [CrossRef]
14. Vahedi, H.; Al-Haddad, K.; Ounejjar, Y. Crossover Switches Cell (CSC): A new multilevel inverter topology with maximum voltage levels and minimum DC sources. In Proceedings of the IECON 2013-39th Annual Conference of the IEEE Industrial Electronics Society, Vienna, Austria, 10–13 November 2013; pp. 54–59.
15. Nho, N.V.; Youn, M.J. Comprehensive study on space-vector-PWM and carrier-based-PWM correlation in multilevel invertors. *IEE Proc. Electr. Power Appl.* **2006**, *153*, 149–158. [CrossRef]
16. Valderrama, G.E.; Guzman, G.V.; Pool-Mazún, E.I.; Martinez-Rodriguez, P.R.; Lopez-Sanchez, M.J.; Zuñiga, J.M.S. A Single-Phase Asymmetrical T-Type Five-Level Transformerless PV Inverter. *IEEE J. Emerg. Select. Top. Power Electron.* **2018**, *6*, 140–150. [CrossRef]
17. Selvaraj, J.; Rahim, N.A. Multilevel Inverter For Grid-Connected PV System Employing Digital PI Controller. *IEEE Trans. Ind. Electron.* **2009**, *56*, 149–158. [CrossRef]
18. Rodriguez, J.; Lai, J.-S.; Peng, F.Z. Multilevel inverters: A survey of topologies, controls, and applications. *IEEE Trans. Ind. Electron.* **2002**, *49*, 724–738. [CrossRef]
19. Fri, A.; Bachtiri, R.E.; Elghzizal, A. Adaptive Control of a Multilevel Inverter (5L) Connected to the Single Phase Electrical Grid. In Proceedings of the 2018 6th International Renewable and Sustainable Energy Conference (IRSEC), Rabat, Morocco, 5–8 December 2018; pp. 1–6.
20. Foito, D.; Pires, V.F.; Cordeiro, A.; Silva, J.F. Sliding Mode Vector Control of Grid-Connected PV Multilevel Systems Based on Triple Three-Phase Two-Level Inverters. In Proceedings of the 2020 9th International Conference on Renewable Energy Research and Application (ICRERA), Glasgow, UK, 27–30 September 2020; pp. 399–404.
21. Revana, G.; Kota, V.R. Closed loop artificial neural network controlled PV based cascaded boost five-level inverter system. In Proceedings of the 2017 International Conference on Green Energy and Applications (ICGEA), Singapore, 25–27 March 2017; pp. 11–17.
22. Cecati, C.; Ciancetta, F.; Siano, P. A Multilevel Inverter for Photovoltaic Systems With Fuzzy Logic Control. *IEEE Trans. Ind. Electron.* **2010**, *57*, 4115–4125. [CrossRef]
23. Kouro, S.; Cortes, P.; Vargas, R.; Ammann, U.; Rodriguez, J. Model Predictive Control—A Simple and Powerful Method to Control Power Converters. *IEEE Trans. Ind. Electron.* **2009**, *56*, 1826–1838. [CrossRef]
24. Richter, S.; Mariéthoz, S.; Morari, M. High-speed online MPC based on a fast gradient method applied to power converter control. In Proceedings of the 2010 American Control Conference, Baltimore, MD, USA, 30 June–2 July 2010; pp. 4737–4743.
25. Trabelsi, M.; Retif, J.M.; Lin, S.X.; Brun, X.; Morel, F.; Bevilacqua, P. Hybrid control of a three-cell converter associated to an inductive load. In Proceedings of the 2008 IEEE Power Electronics Specialists Conference, Rhodes, Greece, 15–19 June 2008; pp. 3519–3525.
26. Trabelsi, M.; Bayhan, S.; Ghazi, K.A.; Abu-Rub, H.; Ben-Brahim, L. Finite-Control-Set Model Predictive Control for Grid-Connected Packed-U-Cells Multilevel Inverter. *IEEE Trans. Ind. Electron.* **2016**, *63*, 7286–7295. [CrossRef]
27. Trabelsi, M.; Bayhan, S.; Refaat, S.S.; Abu-Rub, H.; Ben-Brahim, L. Multi-objective model predictive control for grid-tied 15-level packed U cells inverter. In Proceedings of the 2016 18th European Conference on Power Electronics and Applications (EPE'16 ECCE Europe), Karlsruhe, Germany, 5–9 September 2016; pp. 1–7.
28. Karamanakos, P.; Geyer, T. Guidelines for the design of finite control set model predictive controllers. *IEEE Trans. Power Electron.* **2019**, *35*, 7434–7450. [CrossRef]
29. Rodriguez, J. State of the Art of Finite Control Set Model Predictive Control in Power Electronics. *IEEE Trans. Ind. Inform.* **2013**, *9*, 1003–1016. [CrossRef]

30. Rodriguez, J.; Pontt, J.; Silva, C.A.; Correa, P.; Lezana, P.; Cortés, P.; Ammann, U. Predictive current control of a voltage source inverter. *IEEE Trans. Ind. Electron.* **2007**, *54*, 495–503. [CrossRef]
31. Kukrer, O. Discrete-time current control of voltage-fed three-phase PWM inverters. *IEEE Trans. Power Electron.* **1996**, *11*, 260–269. [CrossRef]
32. Alquennah, A.N.; Trabelsi, M.; Vahedi, H. FCS-MPC of Grid-Connected 9-Level Crossover Switches Cell Inverter. In Proceedings of the IECON 2020 the 46th Annual Conference of the IEEE Industrial Electronics Society, Singapore, 18–21 October 2020; pp. 2430–2435.
33. Zhou, Z.; Khanniche, M.S.; Igic, P.; Kong, S.T.; Towers, M.; Mawby, P.A. A fast power loss calculation method for long real-time thermal simulation of IGBT modules for a three-phase inverter system. In Proceedings of the 2005 European Conference on Power Electronics and Applications, Dresden, Germany, 11–14 September 2005; pp. 9–10.
34. Kim, T.; Kang, D.; Lee, Y.; Hyun, D. The analysis of conduction and switching losses in multi-level inverter system. In Proceedings of the 2001 IEEE 32nd Annual Power Electronics Specialists Conference (IEEE Cat. No. 01CH37230), Vancouver, BC, Canada, 17–21 June 2001; pp. 1363–1368.
35. Krarti, M. *Chapter 4-Utility Rate Structures and Grid Integration, Optimal Design and Retrofit of Energy Efficient Buildings*; Butterworth-Heinemann: Oxford, UK, 2018; pp. 189–245.

sustainability MDPI

Article

# The Way towards an Energy Efficient Transportation by Implementation of Fuel Economy Standards: Fuel Savings and Emissions Mitigation

**Ahmad Zuhairi Muzakir [1,*], Eng Hwa Yap [2] and Teuku Meurah Indra Mahlia [3,*]**

1 Transdisciplinary School (TD School), University of Technology Sydney, 15 Broadway, Ultimo, Sydney, NSW 2007, Australia
2 School of Intelligent Manufacturing Ecosystem, XJTLU Entrepreneur College (Taicang), Xi'an Jiaotong-Liverpool University, 111 Ren'ai Road, Suzhou Industrial Park, Suzhou 215123, China; EngHwa.Yap@xjtlu.edu.cn
3 Centre for Green Technology, Faculty of Engineering & Information Technology, University of Technology Sydney, 15 Broadway, Ultimo, Sydney, NSW 2007, Australia
* Correspondence: ahmadzuhairi.binmuzakir@student.uts.edu.au (A.Z.M.); TMIndra.Mahlia@uts.edu.au (T.M.I.M.)

**Citation:** Muzakir, A.Z.; Yap, E.H.; Mahlia, T.M.I. The Way towards an Energy Efficient Transportation by Implementation of Fuel Economy Standards: Fuel Savings and Emissions Mitigation. *Sustainability* **2021**, *13*, 7348. https://doi.org/10.3390/su13137348

Academic Editors: Mehdi Seyedmahmoudian, Alex Stojcevski, Ben Horan and Saad Mekhilef

Received: 14 May 2021
Accepted: 22 June 2021
Published: 30 June 2021



**Abstract:** Final energy use in Malaysia by the transport sector accounts for a consistent share of around 40% and even more in some years within the past two decades. Amongst all modes of transport, land transport dominates and within land transport, private travels are thought to be the biggest contributor. Personal mobility is dominated by the use of conventional internal-combustion-engine-powered vehicles (ICE), with the ownership trend of private cars has not shown any signs of tapering-off. Fuel consumption by private cars is currently not governed by a national policy on fuel economy standards. This is in contrast against not only the many developed economies, but even amongst some of the ASEAN neighbouring countries. The lack of fuel economy standards has resulted in the loss of potentially tremendous savings in fuel consumption and emission mitigation. This study analysed the increase in private vehicle stock to date, the natural fuel economy improvements brought by technology in a business as usual (BAU) situation, and the additional potential energy savings as well as emissions reduction in the ideal case of mandatory fuel economy standards for motor vehicles, specifically cars in Malaysia. The model uses the latest available data, relevant and most current parameters for the simulation and projection of the future scenario. It is found that the application of the fuel economy standards policy for cars in Malaysia is long overdue and that the country could benefit from the immediate implementation of fuel economy standards.

**Keywords:** fuel economy; fuel consumption; energy savings; emissions mitigation; $CO_2$ emissions; Malaysia

## 1. Introduction

The contribution of the transport sector to the final energy consumption of Malaysia is among the highest across all sectors of energy use. Final energy use in the transport sector has shown to be the most urgent issue to be addressed by the Malaysian government. Since the late 1970s, along with industrial sector use, it has almost the same share until 2008 when a divergent trend began to appear, and the transport sector's consumption continued to rise exponentially while industrial sector energy demand mellowed (Figure 1). In 2014, the share of final energy use by the transport sector breached 46%, the highest in history and was still hovering above 40% in the year 2017 (Figure 2).

While the transport sector comprises the land, marine and air sector, this analysis focuses on land transport, primarily the use of petrol fuel in the internal combustion engine (ICE) motor vehicles, specifically cars. The increase in the rate of motorisation, including light-duty vehicles (LDV) or cars, has been steady since early the 1990s [3,4]. This focus is due to the enormous growth in car numbers, from around 4 million in 2000 to almost 13 million units in 2016 [5]. In addition, this segment of land transport is the biggest user of energy in the sector. Therefore, addressing energy use by the ever-increasing fleet of cars is imperative to reduce fuel consumption and mitigate its ensuing emissions. In this study, this is achieved by improving the fuel economy of cars.

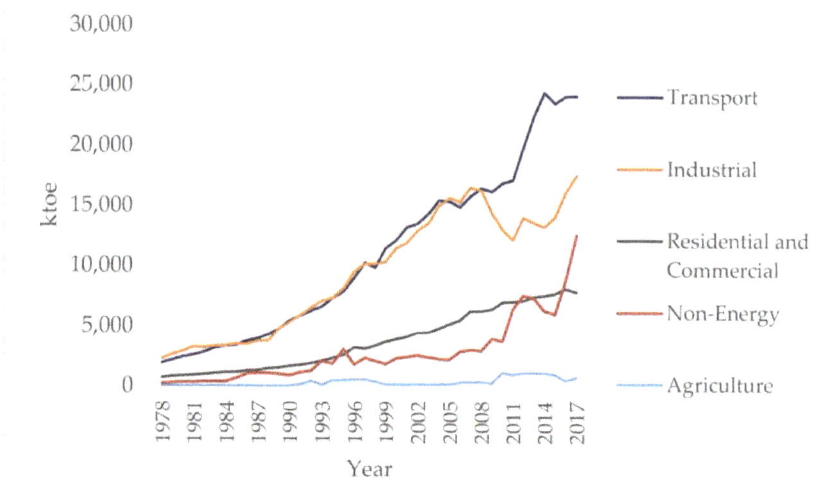


**Figure 1.** Final energy demand by sectors (ktoe), 1978–2017 [1].

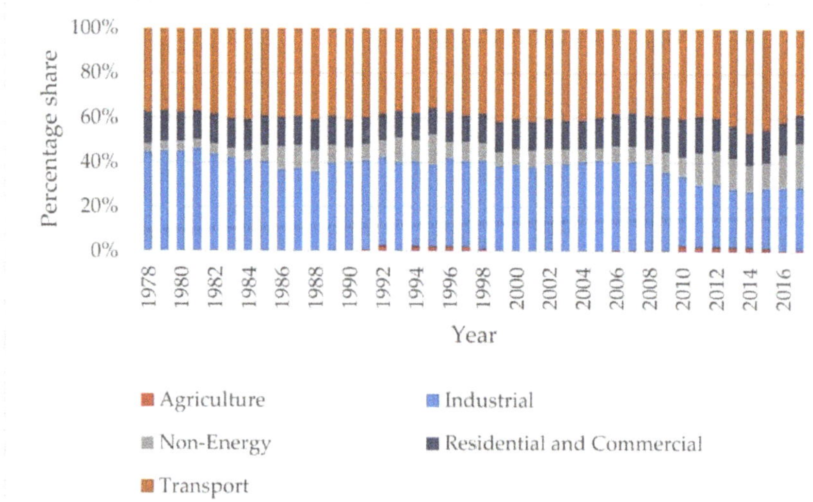


**Figure 2.** The percentage share of final energy demand by sectors, 1978–2017 [2].

For this study, we define FE as a measure of how energy efficient a motor vehicle is, commonly understood as the rate of its fuel consumption measured by calculating the amount of fuel used for every unit distance travelled [6]. FE is also driven by essential factors, including powertrain efficiency to convert fuel energy to functional work at the wheels, vehicle weight, speed, aerodynamics, tyres rolling resistance and many more [6]. However, the simple idea of energy use per unit distance moved is the working definition adopted by governments and international organisations worldwide in their reports [7–9].

There are many ways to improve the FE situation, and these include FE standards, which is a regulatory measure; fuel labelling, which is an information and awareness measure; innovation in vehicle technology; and fiscal measures [7,10,11]. Some of these have been implemented in some developed economies such as Australia, Canada, the EU and the US, with some early adopters in Asia, including China, India, Japan and South Korea [7,8,12,13]. In the Southeast Asian (SEA) region, Singapore, Vietnam and Thailand had introduced a vehicle fuel economy labelling scheme in 2012, 2014 and 2015, respectively [14], whereas fuel economy labels are voluntary in Indonesia. While no ASEAN member states have mandatory FE standards, fuel consumption or $CO_2$ emissions policies, Singapore and Thailand have fiscal policies on vehicles based on their emissions [9].

The focus of this study is the benefits of having a Fuel Economy (FE) standard, which improves the fuel economy of these vehicles by a mandatory measure [10,15]. FE standard is a type of regulation that sets a limit to vehicle fuel consumption for new vehicles entering the market when the standard is in place [7,9]. This is done by the introduction of specific regulations by the government, for example, the Corporate Average Fuel Economy Standards (CAFE) in the US [16,17] and the 'Top Runner' energy efficiency program in Japan [8,11]. These regulations compel the vehicle manufacturers to meet the FE target set by the regulator by making their vehicles more fuel-efficient, not at the individual vehicle level, due to factors that drive FE described above. However, it is designed as a fleet-wide average to allow for a flexible mix of various models introduced into the market, like the US CAFE [8]. It is a fact that Malaysia has yet to have implemented FE standard measure for its car market. Implementing a FE standard policy for cars in Malaysia is needed to reduce its ever-increasing fuel use and emissions in the transport sector, which depends on the dedication and will of the government to implement this measure. This study analyses and discuss just how much energy can be saved and emissions can be curbed by this measure. Without FE standard policy, there is no push for the automotive industry to introduce new car models into the market with the best fuel-efficient technology. If this is coupled with the fuel price situation, which is subsidised in the form of sales tax exemption, unnecessary fuel use will continue to prevail [18], at the expense of the national fiscal situation, health of the people and the environment. By introducing this policy, Malaysia has the opportunity to address these pressing issues.

## 2. Methods

For this study, we have adapted the method developed by [19] to investigate the impact of adopting a fuel economy standards policy on passenger vehicles. We employed many of the equations and explain the principles of calculations in the subsequent sections. We have listed the symbols employed in the Nomenclature list. In short, we will first forecast the number of cars and fuel consumption amount using a polynomial curve-fitting method of the latest published data. These are used to determine the average fuel use per unit distance travel (the FE of the car) for each year in the available and forecasted data. There will be a natural improvement of FE, even without the imposition of FE standard due to normal automotive technology advancement. We forecast the natural improvement of FE and the corresponding fuel use as a business-as-usual (BAU) scenario. We then forecast the number of cars affected by the mandatory FE policy (STD). The affected cars will be imposed a mandatory FE number, based on percentage reduction of BAU FE during the first year of implementation. We then calculate the difference of fuel use under BAU and STD scenarios as fuel savings and its avoided emissions. This method is suitable for fuel use analysis at the macro level, where we do not have granular insights into the respective car segment. The flexibility of this method was utilised by [20] in their study to calculate fuel savings. This study includes the added analysis of greenhouse gas emissions mitigation, not previously calculated by [20].

We sourced input data for the model from various government reports, statistics and previous literature. The numbers of privately owned vehicles were sourced from [5,20]. Energy consumption in the form of petrol fuel data was sourced from [1]. We only include vehicles that run on petrol (gasoline) for this study. The focus on petrol was based on the substantial number of petrol-powered ICE cars (taken to be 89% overall) compared with non-petrol-powered vehicles [21]. The annual petrol fuel consumption (1990–2018) and the corresponding total number of cars (1990 to 2016) are taken from various sources and demonstrated in Table 1.

**Table 1.** The annual petrol fuel consumption and number of cars [1,5,20].

| Year [1,2,3] | Petrol Fuel Consumption (ktoe) | Cars (Units) |
|---|---|---|
| 1990 | 2901 | 1,678,980 |
| 1991 | 3135 | 1,824,679 |
| 1992 | 3326 | 1,942,016 |
| 1993 | 3666 | 2,088,300 |
| 1994 | 4139 | 2,302,547 |
| 1995 | 4548 | 2,553,574 |
| 1996 | 5205 | 2,886,536 |
| 1997 | 5586 | 3,271,304 |
| 1998 | 5854 | 3,452,854 |
| 1999 | 6793 | 3,787,047 |
| 2000 | 6387 | 4,145,982 |
| 2001 | 6827 | 4,557,992 |
| 2002 | 6948 | 5,001,273 |
| 2003 | 7360 | 5,426,026 |
| 2004 | 7839 | 5,898,142 |
| 2005 | 8211 | 6,473,261 |
| 2006 | 7517 | 6,941,996 |
| 2007 | 8600 | 7,419,643 |
| 2008 | 8842 | 7,966,525 |
| 2009 | 8766 | 8,506,080 |
| 2010 | 9560 | 9,114,920 |
| 2011 | 8155 | 9,721,447 |
| 2012 | 10,843 | 10,354,678 |
| 2013 | 12,656 | 10,535,575 |
| 2014 | 12,705 | 11,028,296 |
| 2015 | 12,804 | 11,871,696 |
| 2016 | 13,411 | 12,997,839 |
| 2017 | 13,437 | - |
| 2018 | 13,041 | - |

[1] Vehicle numbers 1990–2008 from [20], [2] Vehicle numbers 2009–2016 from [5], [3] Fuel consumption 1990–2018 from [1].

## 2.1. Projection of Petrol Fuel Consumption and Motor Vehicle Numbers

The basis of reduction in petrol fuel consumption and its corresponding emissions realised by the FE standards implementation hinges upon two important factors, namely the annual fuel consumption and motor vehicle numbers. The polynomial regression is instrumental and reliable in projecting future values beyond the presently available data. We define variable x as the number of the year, whereas variable y is the number of cars and petrol fuel consumption as a function of available data x. Polynomial regression enables the best fit line to fit available data points to make future predictions. The following equation represents a polynomial function of order k in x used in this study:

$$Y = C_0 + C_1 x + C_2 x^2 + \ldots + C_k x^k \quad (1)$$

## 2.2. Potential Fuel Savings Calculations

### 2.2.1. Base Year Baseline Fuel Consumption, $BFC_{Y_B}$

The baseline fuel consumption is the current state of affairs, also called the BAU situation. The base year $Y_B$ is taken as the year 2018 as the latest of the real data available. It is easy to determine the baseline fuel consumption for products with standards already implemented, taken as the standard or the rating level. Since Malaysia has no fuel FE standard for cars, we assumed that the baseline fuel consumption for cars is equal to the annual average of fuel consumption of cars. The total fuel consumption (petrol) in litres divided by the numbers of petrol-powered ICE cars in Malaysia, as per the following equation:

$$BFC_{Y_B} = \frac{FC_i}{NV_i}\ (L) \tag{2}$$

2.2.2. Average Annual Fuel Economy Rating, $FER_i$

We calculate the fuel economy of a motor vehicle by averaging the distance travelled by the unit of fuel consumed, typically measured in either miles per gallon (mpg) or kilometres per litres (km/L). The average annual kilometres travelled by car is multiplied by the total number of cars divided by the total fuel consumption in litres. The average fuel economy rating is then:

$$FER_i = AM \times \frac{NV_i}{FC_i}\ (km/L) \tag{3}$$

2.2.3. Annual Fuel Economy Improvement, $AFI_i$

This parameter is the overall percentage improvement of all cars' fuel consumption on a year-on-year basis. This results from natural technological advancement in automotive technology that enables the cars, overall to travel the same average distance with less fuel. This parameter is represented by the following equation:

$$AFI_i = \left[\frac{FER_i - FER_{i-1}}{FER_{i-1}}\right] \times 100\ (\%) \tag{4}$$

2.2.4. Future Baseline Fuel Consumption, $BFC_{Y_s}$

We define this parameter as the baseline for petrol fuel use by the whole car population in the policy implementation year ($Y_s$) in a BAU scenario. This parameter is predicted from the projection of the fuel consumption that experiences natural fuel economy improvement over the years. The $BFC_{Y_B}$ is applied a compounding interest function whereby the interest rate is taken as the average of the annual fuel economy improvement ($AFI_{avg}$) (throughout the years of available data), over the number of years from the $Y_B$ and $Y_S$. $BFC_{Y_s}$ is represented by:

$$BFC_{Y_s} = BFC_{Y_B} \times \left(1 + AFI_{avg}\right)^{(Y_s - Y_B)}\ (L) \tag{5}$$

2.2.5. Fuel Consumption under FE Standard Implementation, $SFC_{Y_S}$

The fuel consumption under FE standard implementation is the discounted value of the $BFC_{Y_s}$ of the percentage reduction of fuel use applied under the FE standard. It is the FE improvement from the future baseline fuel consumption, demonstrated as follows:

$$SFC_{Y_S} = BFC_{Y_s} \times (1 - \eta_s)\ (L) \tag{6}$$

2.2.6. Initial Unit Fuel Savings, $UFS_{Y_S}$

Initial unit fuel savings is the difference between the baseline fuel consumption in the first year FE standard is rolled out (BAU, in the absence of FE standard) and the reduced petrol use of the cars under the implementation of the FE standard (applicable to the affected vehicles under the standard). The expression for the initial unit fuel savings is as follows:

$$UFS_{Y_S} = BFC_{Y_s} - SFC_{Y_s}\ (L) \tag{7}$$

2.2.7. Shipment, $Sh_i$

We adapted the concept of 'shipment' from [19]. This parameter is a description of the included stock of cars under the FE standard implementation, as not all cars in the first year FE standard is rolled out is included by the policy, namely the previous year's model of the cars. The number of cars affected by the FE standard is the sum of the difference between the number of cars in the current and the past year (the newly registered cars in the current year), and the replacement stock of the scrapped cars the same year (due to reaching its end-of-life). For example, if the general lifespan L of the vehicles is ten years, then these cars will be scrapped in 10 years time, and the total replacement for these cars will be back in the system in the 11th year. The following expression demonstrates the concept of shipment of the cars:

$$Sh_i = (NV_i - NV_{i-1}) + NV_{i\ L}\ (\text{units}) \quad (8)$$

#### 2.2.8. Overall Fuel Economy Improvement, $TI_{Y_s}$

We define the overall fuel economy improvement as a measure of the initial unit fuel savings from the future baseline fuel (in $Y_s$). The parameter is expressed as:

$$TI_{Y_s} = \frac{UFS_{Y_S}}{BFC_{Y_s}} \times 100\ (\%) \quad (9)$$

#### 2.2.9. Scaling Factor, $SF_i$

The scaling factor is a concept of the natural decrease of fuel consumption of the overall available cars in the country. This parameter is enabled by natural technological advances in the automotive industry, making the cars more fuel-efficient over time, even without the enforcement of an FE standard. Scaling factor reduces the initial unit fuel savings of the cars over the effective span of the policy implementation in a linear manner. In each year after the implementation of the FE standard, this parameter affects the unit fuel savings in that particular year. The scaling factor is expressed as:

$$SF_i = 1 - (Y_{Sh_i} - Y_s)\ \frac{AFI_{avg}}{TI_{Y_s}}\ (\text{dimensionless}) \quad (10)$$

#### 2.2.10. Unit Fuel Savings, $UFS_i$

This parameter is the value of the unit fuel savings for each year after the implementation of FE standard. Due to the natural technological advancement in the automotive industry as described above, this value is adjusted with the scaling factor $SF_i$ annually, and expressed as:

$$UFS_i = SF_i \times UFS_{Y_S}\ (L) \quad (11)$$

#### 2.2.11. Shipment Survival Factor, $SSF_i$

The $SSF_i$ is a concept of the common survival rate of a product in light of its average lifespan L. The concept is introduced in [19,22]. The 'shipment' of the cars will survive 100% up to 2/3 of its lifespan L. If the age of the car's stock is more than 2/3 of average lifespan L but less than 1 1/3 of the average lifespan L, the survival rate is expressed as [2 − Age × 1.5/(Average Life)]. For the age of over 4/3 of its average lifespan L, 0% of the stock survives. This factor can be graphically demonstrated as per Figure 3.

**Figure 3.** The relationship between the Age/Average lifespan of a product with Product Survival.

#### 2.2.12. Affected Stock, $AS_i$

We define the affected stock of cars for the adherence to the FE standards as the shipment of cars in the specific year multiplied by the shipment survival factor, plus the number of cars under the standards in the previous year. Therefore, the expression for the parameter is as follows:

$$AS_i = (Sh_i \times SSF_i) + AS_{i-1} \text{ (unit)} \quad (12)$$

#### 2.2.13. Fuel Savings, $FS_i$

The fuel savings are the actual savings of fuel consumed under the FE standard implementation. It is determined by the unit fuel savings and the applicable stock and expressed as:

$$FS_i = \sum_{i=Y_s}^{T} (AS_i \times UFS_i) \text{ (L)} \quad (13)$$

### 2.3. *Potential Emissions Reduction, $ER_i$*

Emissions can potentially be reduced when there is substantial fuel saving resulting from the FE standard implementation. The most common tailpipe emissions of cars include methane ($CH_4$), carbon monoxide (CO), carbon dioxide ($CO_2$), nitrous oxide ($N_2O$), nitrogen oxides ($NO_X$) and sulphur dioxide ($SO_2$). The tailpipe emissions avoided are calculated from the total fuel savings and the emission factors of the respective gases per unit litre of petrol. The emissions reduction is therefore expressed by:

$$ER_i = FS_i \times \left(Em_{CH_4} + Em_{CO} + Em_{CO_2} + Em_{N_2O} + Em_{NO_x} + Em_{SO_2}\right) \text{ (kg)} \quad (14)$$

## 3. Results and Discussion

Based on the method described, we demonstrate sample calculations and the results obtained in this section.

### 3.1. *Data Analysis*

The forecasted fuel consumption for private vehicles was calculated with Equation (1). The polynomial regression method was used on the dataset in Table 1. The mathematical equation for the curve fitted plot is shown below, and the plot is shown in Figure 4.

$$y = 4.168x^2 - 16{,}325x + 15{,}986{,}198 \quad R^2 = 0.9579 \quad (15)$$

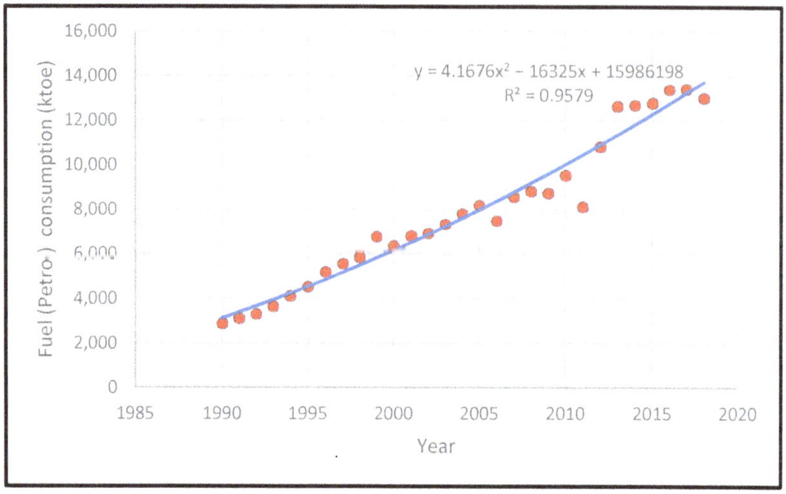


**Figure 4.** The prediction of petrol fuel consumption for cars with polynomial regression.

The forecasted number of cars can be predicted using the same polynomial regression method and Equation (1) on the dataset in Table 1. The polynomial expression for the curve fitted plot of vehicle numbers is shown below, and the plot is shown in Figure 5.

$$y = 10{,}004.124x^2 - 39{,}644{,}786x + 39{,}277{,}331{,}102 \quad R^2 = 0.9983 \tag{16}$$

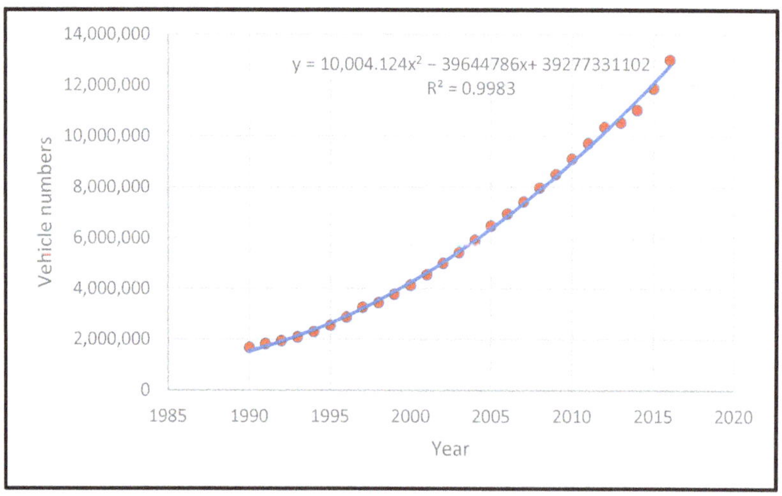


**Figure 5.** The prediction of the number of cars using polynomial regression.

We tabulated the forecasted petrol fuel consumption of cars and the number of cars in Malaysia from 2010 until 2020 by using the polynomial regression equation in Table 2. Since the subsequent fuel economy calculations will be in litres, this study converted the data on energy use published by the Energy Commission in toe (or ton oil equivalent, which is the measure of the energy contained in a metric ton of crude oil) into the appropriate unit of measurement. Therefore, the study adopted the conversion factor whereby 1 ktoe equals the net calorific value of 43.9614 TJ for petrol [1].

**Table 2.** The forecasted number of cars running on petrol and its petrol fuel consumption.

| Year | Car Petrol Fuel Consumption (ktoe) | Car Petrol Fuel Consumption (Litres) | Number of Cars |
|---|---|---|---|
| 2019 | 14,222 | 18,649,416,515 | 13,285,168 |
| 2020 | 14,730 | 19,315,158,648 | 13,963,231 |
| 2021 | 15,246 | 19,991,830,550 | 14,659,102 |
| 2022 | 15,770 | 20,679,432,222 | 15,372,780 |
| 2023 | 16,303 | 21,377,963,664 | 16,104,266 |
| 2024 | 16,844 | 22,087,424,876 | 16,853,558 |
| 2025 | 17,394 | 22,807,815,857 | 17,620,658 |
| 2026 | 17,951 | 23,539,136,608 | 18,405,566 |
| 2027 | 18,517 | 24,281,387,128 | 19,208,281 |
| 2028 | 19,092 | 25,034,567,418 | 20,028,803 |
| 2029 | 19,674 | 25,798,677,478 | 20,867,132 |
| 2030 | 20,265 | 26,573,717,307 | 21,723,269 |
| 2031 | 20,865 | 27,359,686,906 | 22,597,213 |
| 2032 | 21,473 | 28,156,586,275 | 23,488,965 |
| 2033 | 22,089 | 28,964,415,413 | 24,398,523 |
| 2034 | 22,713 | 29,783,174,321 | 25,325,890 |
| 2035 | 23,346 | 30,612,862,999 | 26,271,063 |

#### 3.1.1. Potential Fuel Savings Calculation

The year 2018 was taken as the base year for the baseline fuel consumption calculation. The calculation used Equation (2) and shown below:

$$BFC_{2018} = \frac{17,100,410,816}{12,624,912} = 1354 \text{ L}$$

A total of 17,100,410,816 litres of petrol were consumed in the year 2018. We derived this number from published petroleum products final energy use data for 2018, reported in kilotonnes of oil equivalent (ktoe) unit. We then converted the value to the unit litres by adopting the conversion factor for toe to GJ and GJ to litres of petrol [1,23], whereby one ktoe of energy is equal to 1,311,280.64 L of gasoline (petrol) [23].

There were 12,624,912 cars using petrol fuel in the year 2018. This number represented 89% of the overall motor vehicle numbers for the year. The overall motor vehicle numbers were derived from the polynomial expression in Equation (1). The share of 89% for gasoline (petrol) powered internal combustion engine (ICE) cars (out of the overall total) were adopted from the work of [21]. Therefore, we assumed that petrol ICE cars are 89% of the total number of cars throughout the simulation years for this study.

We used Equations (3) and (4), respectively to calculate the overall fuel economy ratio—FER in km/L—for each year between 1990 and 2018, and the annual fuel economy improvement (AFI), by using the petrol consumption (in litres) and the number of petrol cars, as demonstrated in Table 3. Another critical assumption for this calculation was the average annual distance travelled per car of 20,000 km. We then calculated the average of the AFI ($AFI_{avg}$), which was 2.64% based on each known AFI from the year 1991 to 2018. Consequently, we used the $AFI_{avg}$ value in Equation (5) to forecast the baseline fuel consumption during the first year of the FE standards roll-out ($BFC_{Y_s}$—in the year 2025). For this case, based on known BFC in the year 2018, the baseline fuel consumption in the implementation year of the standard (2025) is shown below:

$$BFC_{Y_s} = BFC_{2025} = 1354 \times (1 + 2.64\%)^{(2025-2018)} = 1625.12 \text{ L/year}$$

Table 3. Fuel Economy Ratio, Annual Fuel Economy Improvement (AFI) and Average AFI.

| Year | FER (km/L) | AFI (%) |
|---|---|---|
| 1990 | 7.86 | |
| 1991 | 7.90 | 0.57 |
| 1992 | 7.93 | 0.32 |
| 1993 | 7.73 | −2.44 |
| 1994 | 7.55 | −2.34 |
| 1995 | 7.62 | 0.93 |
| 1996 | 7.53 | −1.23 |
| 1997 | 7.95 | 5.60 |
| 1998 | 8.01 | 0.72 |
| 1999 | 7.57 | −5.48 |
| 2000 | 8.81 | 16.44 |
| 2001 | 9.06 | 2.85 |
| 2002 | 9.77 | 7.81 |
| 2003 | 10.01 | 2.42 |
| 2004 | 10.24 | 2.29 |
| 2005 | 10.70 | 4.54 |
| 2006 | 12.54 | 17.14 |
| 2007 | 11.71 | −6.58 |
| 2008 | 12.23 | 4.43 |
| 2009 | 13.17 | 7.70 |
| 2010 | 12.94 | −1.74 |
| 2011 | 16.18 | 25.03 |
| 2012 | 12.96 | −19.89 |
| 2013 | 11.30 | −12.83 |
| 2014 | 11.78 | 4.27 |
| 2015 | 12.59 | 6.82 |
| 2016 | 13.16 | 4.53 |
| 2017 | 13.60 | 3.38 |
| 2018 | 14.77 | 8.56 |
| Average | | 2.64% |

The remaining analysis required some other data and statistics for the basis of assumptions used. There are many improvements needed in the data recording, maintenance and reporting for the transport sector in Malaysia. In lieu of the lack of data, these data estimates were nevertheless adapted from [24,25] and summarised in Table 4.

Table 4. Input data for calculation of potential fuel savings.

| Description | Values |
|---|---|
| Implementation Year | 2025 |
| Average Lifespan | 10 years |
| $BFC_{Y_s}$ ($Y_s = 2025$) | 1625.12 L/year |
| Target FE efficiency improvement | 10% |
| Standards fuel consumption | 1333 L/year |
| Annual mileage | 20,000 km/year |
| Average Annual Fuel Economy Improvement (AFI) | 2.64% |

The potential fuel savings calculation results realised by enforcement of FE standard on cars in Malaysia (beginning year 2025) is outlined in Table 5. As can be seen, the efficacy of the policy lasts for only a few years before the natural improvement of the AFI, due to the advancement of automotive technology, catches up with the target fuel savings of the standards. Based on the previous data, it was assumed that the annual AFI will improve at 2.64% on average, without the FE standard policy in place. Therefore, if the fixed FE standard is not revised to the latest relevant base year, the FE standard's savings will cease to be relevant a few years after its implementation. As a demonstration of this point, based on Table 5 and Figure 6, the FE standard of 15% reduction based on the year 2025 will be effective for six years, up to the year 2030.

**Table 5.** The potential fuel savings calculation results.

| Year | Shipment ('000) | Applicable Stock ('000) | Scaling Factor | Unit Fuel Savings (L) | Potential Fuel Savings (L) |
|---|---|---|---|---|---|
| 2025 | 10,565,809 | 10,565,809 | 1.00 | 162.51 | 21,090,749,294 |
| 2026 | 11,568,077 | 22,133,886 | 0.82 | 133.95 | 20,574,304,446 |
| 2027 | 11,982,463 | 34,116,349 | 0.65 | 105.39 | 20,685,925,377 |
| 2028 | 12,624,912 | 46,741,261 | 0.47 | 76.83 | 21,443,597,561 |
| 2029 | 13,285,168 | 60,026,429 | 0.30 | 48.26 | 22,901,509,777 |
| 2030 | 13,963,231 | 73,989,660 | 0.12 | 19.70 | 25,115,885,234 |

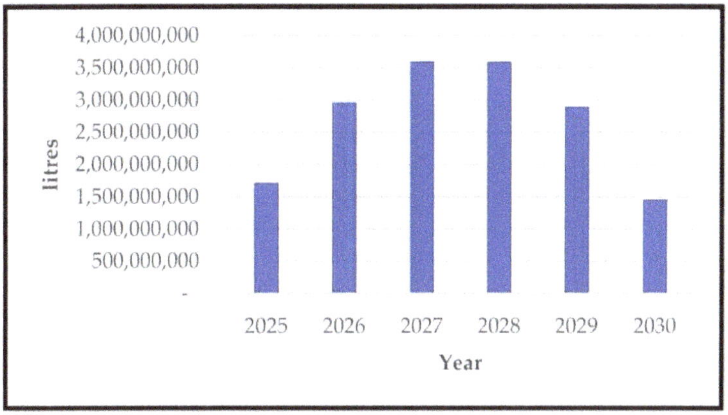


**Figure 6.** The prediction of annual fuel savings for cars.

It can be seen in Figure 6 that substantial savings will begin in the first year of the FE standard implementation and continues to increase as more applicable stock gets into the system after the year 2025. After that, however, this effect starts to taper off four years into the FE standard implementation until it ceases to be relevant after the year 2030. This situation happens as the effect of reducing scaling factor kicks in due to the natural increase of the technological advancement in automotive technologies that increases the fuel efficiencies of cars against the FE standard.

The comparison between annual fuel consumption in a BAU situation and fuel consumption under FE standard implementation is shown in Figure 7, whereby STD is the potential fuel consumption at the much-reduced level under the FE standard. The total cumulative savings during the years the FE standard policy is effective is more than 16.2 billion litres of petrol or more than 12,300 ktoe. It is nice to be aware that these savings are based on a minimum of 15% efficiency improvement. With continuous technological improvements, the fuel savings for the future period can be better.

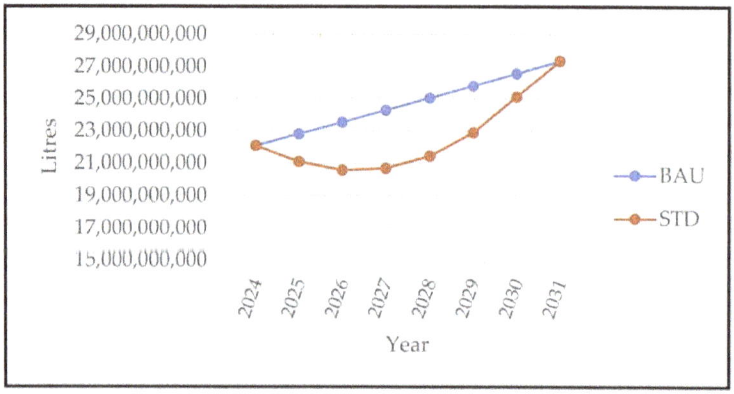


**Figure 7.** The prediction of annual fuel savings for cars.

### 3.1.2. Potential Emissions Mitigation

The fuel savings to be achieved may result in tailpipe emissions reduction, which is beneficial to the global environment. Tailpipe emissions from gasoline (petrol) comprise $CH_4$, CO, $CO_2$, $N_2O$, $NO_X$ and $SO_2$. The amount of emissions avoided is a function of the emission factors and the amount of petrol saved. We adapted the emission factors from [20,26] in this study. We did some necessary unit conversions as some factors were originally in the units of gallons, and the emission factors are eventually in the form of kg/L or g/L. Table 6 outlines the corresponding emission factors used in this study.

It is essential to understand these from the lens of its respective Global Warming Potential (GWP), in the normalised units of a reference gas, in this case, the $CO_2$ in the form of carbon dioxide equivalent ($CO_2$ eq). Each gas has its GWP factor that measures its propensity to the global warming effects, which depends on the time horizon of 100 years. It is interesting to note that depending on the different time horizons adopted, the GWP factor varies. However, the parties to the Conference of the Parties (COP) to the United Nations Framework Convention on Climate Change (UNFCCC) has adopted the 100-year time horizon since the Kyoto Protocol and reaffirmed in the IPCC Second Assessment Report [27] and IPCC Fifth Assessment Report [28]. We outlined the GWP of the respective gases in Table 6. It is worth noting that CO, $SO_2$ and $NO_x$ are considered indirect greenhouse gases, as compared to $CO_2$, $CH_4$ and $N_2O$, which has direct global warming potential. Therefore, we excluded the effects of CO, $SO_2$ and $NO_x$ on global warming from this study as indirect greenhouse gases can be highly uncertain, compared with direct GWPs, believed to be highly accurate [29].

**Table 6.** The emission factor for motor gasoline (petrol) and Global Warming Potential (GWP) of gases.

| Type of Emission | Emission Factor [1,2] | Emission Factor | GWP [3] |
|---|---|---|---|
| $CO_2$ | 8.78 kg/gal | 2.319 kg/L | 1 |
| $CH_4$ | 0.38 g/gal | 0.100 g/L | 21 |
| $N_2O$ | 0.08 g/gal | 0.021 g/L | 310 |
| CO | 3.49086 kg/GJ | 116.400 g/L | indirect |
| $SO_2$ | 0.00228 kg/GJ | 0.076 g/L | indirect |
| $NO_X$ | 1.36876 kg/GJ | 45.630 g/L | indirect |

[1] Emission factor for $CO_2$, $CH_4$, and $N_2O$ from [26]; [2] Emission factor for CO, $SO_2$, and $NO_X$ from [20]; [3] GWP from [30].

Table 7 shows the result of the emissions avoided throughout FE standard implementation. Consequently, we applied the GWP factor to $CO_2$, $CH_4$ and $N_2O$, and greenhouse gas emission avoidance over the FE standard period, as demonstrated in Table 8.

**Table 7.** The emissions avoidance calculation results.

| Year | $CO_2$ (Ton) | $CH_4$ (kg) | $N_2O$ (kg) | CO (kg) | $SO_2$ (kg) | $NO_X$ (kg) |
|---|---|---|---|---|---|---|
| 2025 | 3,982,619 | 172,368 | 36,288 | 199,866,548 | 130,497 | 78,349,747 |
| 2026 | 6,876,726 | 297,626 | 62,658 | 345,106,464 | 225,327 | 135,285,292 |
| 2027 | 8,339,428 | 360,932 | 75,986 | 418,511,748 | 273,255 | 164,060,920 |
| 2028 | 8,329,009 | 360,481 | 75,891 | 417,988,891 | 272,914 | 163,855,955 |
| 2029 | 6,719,783 | 290,833 | 61,228 | 337,230,320 | 220,185 | 132,197,762 |
| 2030 | 3,381,342 | 146,345 | 30,809 | 169,691,653 | 110,795 | 66,520,877 |

[1] Emission factor for $CO_2$, $CH_4$, and $N_2O$ from [26]; [2] Emission factor for CO, $SO_2$, and $NO_X$ from [20]; [3] GWP from [30].

**Table 8.** The greenhouse gas emissions avoidance.

| Year | $CO_2$ (Ton) | $CH_4$ (kg $CO_{2\,eq}$) | $N_2O$ (kg $CO_{2\,eq}$) |
|---|---|---|---|
| 2025 | 3,982,619 | 3,619,738 | 762,050 |
| 2026 | 6,876,726 | 6,250,145 | 1,315,820 |
| 2027 | 8,339,428 | 7,579,571 | 1,595,699 |
| 2028 | 8,329,009 | 7,570,102 | 1,593,706 |
| 2029 | 6,719,783 | 6,107,502 | 1,285,790 |
| 2030 | 3,381,342 | 3,073,247 | 646,999 |

GHG emissions avoidance can be substantial, especially for $CO_2$, while $CH_4$ and $N_2O$ can be negligible relative to the $CO_2$ scale, as demonstrated by Figure 8. Total $CO_2$ emissions reduction is 37.6 million tons, while $CH_4$ and $N_2O$ account for 41,400 tons of CO2 equivalent. Nevertheless, these should count towards the GHG reduction potential as each contribution counts for Malaysia's commitments to reducing GHG emissions.

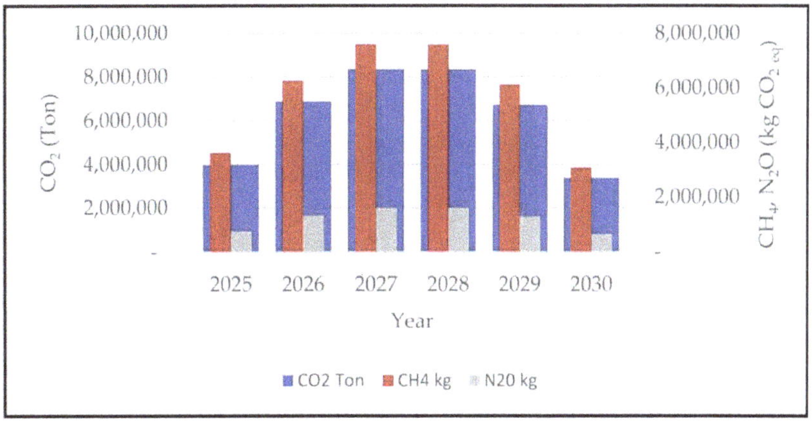


**Figure 8.** The greenhouse gas emissions avoidance under the FE standard implementation.

## 4. Conclusions

The analysis in this study for the implementation of the FE standard in the year 2025 is fortunately timed with the commitments of the Malaysian government in reducing its GHG emissions by the year 2030. This study forecasted the stock of cars in the study period and its corresponding fuel savings and emissions mitigation under the FE standard implementation. The key findings that we have found are that, in the period of implementation, fuel savings of 16.2 billion litres of petrol or more than 12,300 ktoe can be achieved, along with the reduction in at least 37.6 million tons $CO_2$ equivalent GHG emissions. In Malaysia's official projection to the UNFCCC, under the BAU scenario, the GHG emissions up to the year 2030 (from 2005) is 549,535 Gg $CO_{2\,eq}$ (549.535 million Ton $CO_{2\,eq}$), while the mitigation plan scenario is expected to lower this value to 510,205 Gg $CO_{2\,eq}$ (510.205 million Ton

$CO_{2\,eq}$). The reduction of the overall 39.3 million Ton $CO_{2\,eq}$ pledged by Malaysia in its Third National Communication and Second Biennial Update Report to the UNFCCC seems within reach with just this FE standard implementation. These certainly will do well for Malaysia in meeting its commitments to the international community.

The implementation of a FE standard policy for cars in Malaysia is a promising policy to help Malaysia reduce its energy use from the transport sector. This step could be one of the most effective measures, among other FE initiatives [12], nudged positively by the discussion and public discourse of the policy that has happened at various levels within Malaysia and regionally [8,9,31]. However, Malaysia still has a lot to do before the implementation of the FE standard can be realised.

Malaysia has policy documents that outline the intention to have the FE standard implementation timed nicely within the timeframe of this analysis [32–34]. Specifically, the Ministry of Transport (MOT) (the ministry in charge of transport policies and regulations) plan to formulate and implement a fuel economy policy between the year 2019 and 2030 [34]. In addition, a further commitment was made by the Ministry of International Trade and Industry (MITI) (the ministry in charge of the development of automotive industry), "pledged to reduce carbon emission by improving fuel economy level in Malaysia by 2025 in line with the ASEAN Fuel Economy Roadmap of 5.3 Lge/100 km" [33]. Both the government automotive and transport policy statements [33,34] for the FE as outlined above indicate that Malaysia is on the right track towards the realisation of the policy.

Despite all these, Malaysia needs to designate a body focusing on the technical aspects and regulatory matters to realise this policy [35]. While various government agencies are related to road transport, prior existing jurisdictions rendered the policy fall in between the cracks, as no specific government agency in Malaysia is responsible for both energy use and transport under its roof. For the technical aspect, one of the actions required involves the driving test cycle suitable for the local situation for measuring the right FE situation. The IEA has outlined the policy pathway and critical actions to implement FE policies, including deciding on the form of standard, target values, introducing a mechanism for increased vehicle weights as part of the policy design process, before implementing and monitoring the progress of the policy implementation [11]. The implementation of FE standard itself should regularly be updated as natural improvements happen over time, rendering the standard obsolete. In addition, conflicting priorities like the encouragement of car ownership as a support to the national automotive industry [3] and curbing energy usage from car use through the implementation of FE standard may impact the competitiveness of the national car industry. This is where Malaysia should resolve its will so that the implementation of the FE standard becomes a reality.

**Author Contributions:** Conceptualization, A.Z.M. and T.M.I.M.; Data curation, A.Z.M.; Formal analysis, A.Z.M. and T.M.I.M.; Investigation, A.Z.M. and E.H.Y.; Methodology, T.M.I.M.; Supervision, E.H.Y. and T.M.I.M.; Validation, T.M.I.M.; Visualization, A.Z.M.; Writing—original draft, A.Z.M.; Writing—review and editing, A.Z.M. and T.M.I.M. All authors have read and agreed to the published version of the manuscript.

**Funding:** This research received no external funding.

**Institutional Review Board Statement:** Not applicable.

**Informed Consent Statement:** Not applicable.

**Data Availability Statement:** Data available upon request.

**Conflicts of Interest:** The authors declare no conflict of interest.

## Nomenclature

| List of symbols | |
|---|---|
| AM | Annual mileage (km) |
| $AS_i$ | Affected stock of cars in the year i (unit) |
| $AS_{i-1}$ | Affected stock of cars in the year i−1 (unit) |
| $AFI_i$ | Annual fuel economy improvement in the year i (%) |
| $AFI_{avg}$ | Average annual fuel economy improvement (%) |
| $BFC_{Y_B}$ ($BFC_{2018}$) | Base year baseline fuel consumption (2018 baseline fuel consumption) (L) |
| $BFC_{Y_s}$ ($BFC_{2025}$) | Future baseline fuel consumption in the year policy is implemented (2025) (L) |
| $Em_{CH_4}$ | Emission factor for $CH_4$ (g/L) |
| $Em_{CO}$ | Emission factor for CO (g/L) |
| $Em_{CO_2}$ | Emission factor for $CO_2$ (kg/L) |
| $Em_{N_2O}$ | Emission factor for $N_2O$ (g/L) |
| $Em_{NO_x}$ | Emission factor for $NO_x$ (g/L) |
| $Em_{SO_2}$ | Emission factor for $SO_2$ (g/L) |
| $ER_i$ | Potential emissions reduction in the year i (kg) |
| $FC_i$ | Fuel consumption in the year i (L) |
| $FER_i$ | Average annual fuel economy rating in the year i (km/L) |
| $FER_{i-1}$ | Average annual fuel economy rating in the year i−1 (km/L) |
| $FS_i$ | Fuel savings in the year i (L) |
| L | Lifespan of the vehicles (year) |
| $NV_i$ | Number of vehicles in the year i (unit) |
| $NV_{i-1}$ | Number of vehicles in the year i−1 (unit) |
| $NV_{i\text{-}L}$ | Number of vehicles in the year i-L (unit) |
| $\eta_s$ | Percentage reduction of fuel use as the result of FE standard (%) |
| $SFC_{Y_S}$ | Fuel consumption under FE standard implementation (L) |
| $Sh_i$ | Shipment (included stock of cars under FE standard implementation) |
| $SF_i$ | Scaling factor in the year i |
| $SSF_i$ | Shipment survival factor in the year i |
| $TI_{Y_s}$ | Overall fuel economy improvement (%) |
| $UFS_{Y_S}$ | Initial unit fuel savings in the first-year roll-out of the standard (L) |
| $UFS_i$ | Unit fuel savings in the year i (L) |
| x | Variable x in polynomial expression, year |
| Y | Variable Y in polynomial expression, (number of cars or petrol fuel consumption) |
| $Y_B$ | Base year |
| $Y_S$ | Year when FE standard is implemented |
| $Y_{Shi}$ | Year of the Shipment in year i |
| ICE | Internal combustion engine |
| ASEAN | Association of Southeast Asian Nations |
| BAU | Business-as-usual |
| $CH_4$ | Methane |
| CO | Carbon monoxide |
| $CO_2$ | Carbon dioxide |
| $CO_{2\,eq}$ | Carbon dioxide equivalent |
| COP | Conference of the Parties |
| $N_2O$ | Nitrous oxide |
| $NO_x$ | Nitrogen oxides |
| $SO_2$ | Sulphur dioxide |
| DSM | Demand Side Management |
| EPU | Economic Planning Unit |
| EU | European Union |
| FE | Fuel economy |
| GHG | Greenhouse gas |
| GJ | Giga Joule |
| GWP | Global warming potential |
| IPCC | Intergovernmental Panel on Climate Change |
| ktoe | Kilo tonnes of oil equivalent |

| | |
|---|---|
| LDV | Light-duty vehicles |
| MITI | Ministry of International Trade and Industry |
| MOT | Ministry of Transport |
| SEA | Southeast Asia |
| toe | Ton oil equivalent |
| UNFCCC | United Nations Framework Convention on Climate Change |
| US | United States of America |
| toe | Ton oil equivalent |

## References


1. The Energy Commission. *National Energy Balance 2018*; The Energy Commission: Putrajaya, Malaysia, 2020.
2. The Energy Commission. *Final Energy Demand by Sectors*; Malaysia Energy Information Hub: Putrajaya, Malaysia, 2019. Available online: https://meih.st.gov.my/statistics?p_auth=SDzwojf5&p_p_id=Eng_Statistic_WAR_STOASPublicPortlet&p_p_lifecycle=1&p_p_state=maximized&p_p_mode=view&p_p_col_id=column-1&p_p_col_pos=1&p_p_col_count=2&_Eng_Statistic_WAR_STOASPublicPortlet_execution=e1s1&_Eng_Sta (accessed on 4 December 2019).
3. Kasipillai, J.; Chan, P. Travel Demand Management: Lessons for Malaysia. *J. Public Transp.* **2015**, *11*, 41–55. [CrossRef]
4. Ng, W.-S.S. Urban Transportation Mode Choice and Carbon Emissions in Southeast Asia. *Transp. Res. Rec.* **2018**, *2672*, 54–67. [CrossRef]
5. MAI. Vehicle Registration Statistics for Private Vehicle, Goods Vehicle and Others by Year. 2021. Available online: https://www.data.gov.my/data/ms_MY/dataset/vehicle-registration-statistics-for-private-vehicle-goods-vehicle-and-others (accessed on 15 May 2021).
6. Cleveland, C.J.; Morris, C. *Dictionary of Energy*, 2nd ed.; Morris, C., Ed.; Elsevier: Boston, MA, USA, 2015; pp. 217–246.
7. IEA. *Fuel Economy of Road Vehicles*; IEA: Paris, France, 2011.
8. GFEI. *Improving Vehicle Fuel Economy in the ASEAN Region*; GFEI: London, UK, 2010.
9. ASEAN Secretariat. *ASEAN Fuel Economy Roadmap for the Transport Sector 2018–2025: With Focus on Light-Duty Vehicles*; ASEAN Secretariat: Jakarta, Indonesia, 2019.
10. Atabani, A.; Badruddin, I.A.; Mekhilef, S.; Silitonga, A. A review on global fuel economy standards, labels and technologies in the transportation sector. *Renew. Sustain. Energy Rev.* **2011**, *15*, 4586–4610. [CrossRef]
11. IEA. *Improving the Fuel Economy of Road Vehicles, a Policy Packace*; IEA: Paris, France, 2012.
12. Lipu, M.S.H.; Hannan, M.A.; Karim, T.F.; Hussain, A.; Saad, M.H.M.; Ayob, A.; Miah, M.S.; Mahlia, T.M.I. Intelligent algorithms and control strategies for battery management system in electric vehicles: Progress, challenges and future outlook. *J. Clean. Prod.* **2021**, *292*, 126044. [CrossRef]
13. Clerides, S.; Zachariadis, T. The effect of standards and fuel prices on automobile fuel economy: An international analysis. *Energy Econ.* **2008**, *30*, 2657–2672. [CrossRef]
14. Yang, Z. Overview of global fuel economy policies. In Proceedings of the 2018 AP CAP Joint Forum and Clean Air Week Theme: Solutions Landscape for Clean Air, Bangkok, Thailand, 19–23 March 2018.
15. Mallick, P. *Materials, Design and Manufacturing for Lightweight Vehicles*, 2nd ed.; Mallick, P., Ed.; Woodhead Publishing: Cambridge, UK, 2020.
16. Crandall, R.W. Policy Watch: Corporate Average Fuel Economy Standards. *J. Econ. Perspect.* **1992**, *6*, 171–180. [CrossRef]
17. Portney, P.R.; Parry, I.W.; Gruenspecht, H.K.; Harrington, W. Policy Watch: The Economics of Fuel Economy Standards. *J. Econ. Perspect.* **2003**, *17*, 203–217. [CrossRef]
18. Ying, L.S.; Harun, M. Fuel subsidy abolition and performance of the sectors in Malaysia: A computable general equilibrium approach. *Malays. J. Econ. Stud.* **2019**, *56*, 303–326.
19. Mahlia, T.; Masjuki, H.; Choudhury, I. Theory of energy efficiency standards and labels. *Energy Convers. Manag.* **2002**, *43*, 743–761. [CrossRef]
20. Aizura, A.B.; Mahlia, T.M.I.; Masjuki, H.H. Potential fuel savings and emissions reduction from fuel economy standards implementation for motor-vehicles. *Clean Technol. Environ. Policy* **2009**, *12*, 255–263. [CrossRef]
21. Ishak, A. Vehicle Inspection in Malaysia. In Proceedings of the Regional Workshop on Vehicle Inspection and Maintenance Policy in Asia, Bangkok, Thailand, 10–12 December 2001.
22. Sanchez, M.; Koomey, J.; Moezzi, M.; Meier, A.; Huber, W. *Miscellaneous Electricity Use in the US Residential Sector*; Lawrence Berkeley National Laboratory: Berkeley, CA, USA, 1998.
23. US Energy Infomation Administration. Units and Calculators Explained—Energy Conversion Calculators. 2020. Available online: https://www.eia.gov/energyexplained/units-and-calculators/energy-conversion-calculators.php (accessed on 4 February 2021).
24. Azman, Z.; Radin, D.; Ahmad, R.; Ramdzani, A.; Chung, I. The estimation of carbon dioxide ($CO_2$) emissions from the transport sector in Malaysia (2000–2020). In Proceedings of the 9th AIM International Workshop, National Institute for Environmental Studies, Tsukuba, Japan, 12–13 March 2004.
25. Masjuki, H.H.; Karim, M.; Mahlia, T. *Energy Use in the Transportation Sector of Malaysia*; University of Malaya: Kuala Lumpur, Malaysia, 2004.

26. US Environmental Protection Agency. Emission Factors for Greenhouse Gas Inventories. 2014. Available online: https://www.epa.gov/sites/production/files/2015-07/documents/emission-factors_2014.pdf (accessed on 6 May 2021).
27. IPCC. *Global Warming of 1.5 °C. An IPCC Special Report*; IPCC: Geneva, Switzerland, 2018.
28. Myhre, G.; Stocker, T.F.; Qin, D.; Plattner, G.-K.; Tignor, M.M.B.; Allen, S.K.; Boschung, J.; Nauels, A.; Xia, Y.; Bex, V.; et al. Anthropogenic and Natural Radiative Forcing. In *Climate Change 2013: The Physical Science Basis. Contribution of Working Group I to the Fifth Assessment Report of the Intergovernmental Panel on Climate Change*; IPCC: Geneva, Switzerland, 2013.
29. IPCC. *Climate Change 2001: The Scientific Basis. Contribution of Working Group I to the Third Assessment Report of the Intergovernmental Panel on Climate Change*; IPCC: Geneva, Switzerland, 2001.
30. UNFCC. Global Warming Potentials (IPCC Second Assessment Report). 2021. Available online: https://unfccc.int/process/transparency-and-reporting/greenhouse-gas-data/greenhouse-gas-data-unfccc/global-warming-potentials (accessed on 6 May 2021).
31. Briggs, H.G.; Leong, H.K. *ASEAN-German Technical Cooperation | Energy Efficiency and Climate Change Mitigation in the Land Transport Sector Malaysia Stocktaking Report on Sustainable Transport and Climate Change Data, Policy, and Monitoring*; GIZ: Kuala Lumpur, Malaysia, 2016.
32. EPU. *Final Report on Demand Side Management (DSM) Preliminary Study*; EPU: Putrajaya, Malaysia, 2017.
33. Ministry of International Trade and Industry (MITI). *National Automotive Policy 2020*; Ministry of International Trade and Industry: Kuala Lumpur, Malaysia, 2020.
34. Ministry of Transport Malaysia. *National Transport Policy 2019–2030*; Ministry of Transport Malaysia: Putrajaya, Malaysia, 2019.
35. Indati Mustapa, S.; Ali Bekhet, H. International Journal of Energy Economics and Policy Investigating Factors Affecting $CO_2$ Emissions in Malaysian Road Transport Sector. *Int. J. Energy Econ. Policy* **2015**, *5*, 1073–1083.

sustainability

MDPI

Article

# Thermodynamic and Energy Efficiency Analysis of a Domestic Refrigerator Using $Al_2O_3$ Nano-Refrigerant

Farhood Sarrafzadeh Javadi [1,2] and Rahman Saidur [3,4,*]

1 Department of Mechanical Engineering, Faculty of Engineering, University of Malaya, Kuala Lumpur 50603, Malaysia; fjavadi2001@gmail.com or farhood.javadi@adriansvc.com
2 Center of Technology and Innovation, Adrian Integrated Services Sdn Bhd, Kuala Lumpur 53300, Malaysia
3 Research Center for Nano-Materials and Energy Technology (RCNMET), School of Engineering and Technology, Sunway University, Petaling Jaya 47500, Malaysia
4 Department of Mechanical and Manufacturing Engineering, Faculty of Engineering, Universiti Putra Malaysia, Serdang 43400, Malaysia
* Correspondence: saidur@sunway.edu.my

**Abstract:** Refrigeration systems have experienced massive technological changes in the past 50 years. Nanotechnology can lead to a promising technological leap in the refrigeration industry. Nano-refrigerant still remains unknown because of the complexity of the phase change process of the mixture including refrigerant, lubricant, and nanoparticle. In this study, the stability of $Al_2O_3$ nanofluid and the performance of a nano-refrigerant-based domestic refrigerator have been experimentally investigated, with the focus on the thermodynamic and energy approaches. It was found that by increasing the nanoparticle concentration, the stability of nano-lubricant was decreased and evaporator temperature gradient was increased. The average of the temperature gradient increment in the evaporator was 20.2% in case of using 0.1%-$Al_2O_3$. The results showed that the energy consumption of the refrigerator reduced around 2.69% when 0.1%-$Al_2O_3$ nanoparticle was added to the system.

**Keywords:** nano-refrigerant; nanofluid; refrigerator; energy efficiency; thermodynamic analysis; aluminum oxide

**Citation:** Sarrafzadeh Javadi, F.; Saidur, R. Thermodynamic and Energy Efficiency Analysis of a Domestic Refrigerator Using $Al_2O_3$ Nano-Refrigerant. *Sustainability* **2021**, *13*, 5659. https://doi.org/10.3390/su13105659

Academic Editors: Mehdi Seyedmahmoudian, Alex Stojcevski, Ben Horan and Saad Mekhilef

Received: 13 April 2021
Accepted: 5 May 2021
Published: 18 May 2021



## 1. Introduction

In the past 50 years, refrigeration systems have experienced massive technological changes. Many of the recent changes in refrigeration systems are due to the rapid changes in technology and environmental challenges. Beside the harmful effect of refrigerant on the environment, scientists have warned that the continuous release of refrigerants into the atmosphere will destroy the Earth's ozone layer. Ozone layer depletion might lead to global warming followed by natural disasters. HFC-134a (R134a) is the most common refrigerant used in domestic refrigerators due to zero ozone depletion potential (ODP), low global warming potential (GWP), being non-flammable, and favorable thermodynamic properties [1,2].

Application of nanomaterials in various fields of engineering has become an interesting topic and remained challenging in some aspects during the last two decades [3–7]. Since 1995, when the term nanofluid was introduced by Choi [8] to describe a new class of heat transfer fluids, many studies have been conducted in this field to discover advantages and disadvantages in order to overcome the application barriers [9–11]. The main objective has been to create a new type of coolant with higher heat transfer capability, which has been used in variety of products such as computers, power electronics, car engines, heat exchangers, and high-powered lasers. Enhancement in thermal properties such as thermal conductivity due to the presence of nanoparticles has attracted great interest of researchers, but instability of nano-size solid particles in the basefluid, especially at high concentrations, still remains a challenge in its application [12–15]. Many studies have been conducted on the fundamental properties, application, and characterization of different types of nanoparticles dispersed in different base fluids [16–19]. However, application of nanoparticle in a low temperature base fluid such as refrigerant remains unknown because of the difficulties during measuring the fundamental properties of the mixture and liquid-to-vapor phase change of the fluid. Therefore, the function of nano-refrigerant in a phase-change process including migration of nanoparticle during boiling, sedimentation, and thermophysical characteristics of different phases is very controversial.

Comparatively, few investigations have considered certain effects of nano refrigerant on the refrigeration system, and most of them have evaluated the fundamental properties of nano-refrigerants [20,21] or impacts of using nano-refrigerant on the heat transfer characteristics in a pipe [22–24]. In addition, it is essential to consider the applicability of nano-refrigerants in a real refrigeration system.

There are a limited number of investigations on nano-refrigerants available in the literature. Peng et al. [22] have investigated the heat transfer characteristics of the flow boiling of a refrigerant-based nanofluid inside a horizontal smooth tube. They have found that the heat transfer coefficient of the nanofluid is larger than that of the pure refrigerant. The nucleate pool boiling heat transfer enhancement of refrigerant-based nanofluid with low concentration of additives was reported by Peng et al. [23]. The experiment by Henderson et al. [24] on the flow-boiling of R-134a/polyolester mixture showed that the heat transfer coefficient increased more than 100% over the baseline by adding CuO nanoparticle into the mixture. In a fundamental aspect, several experimental articles investigated the characterization of refrigerant-based nanofluid [20,21,25,26]. There are only two articles available reporting the effect of refrigerant-based nanofluid in a refrigeration system. According to the work done by Bi et al. [25], R-134a/mineral oil with $TiO_2$ nanoparticles worked normally in the refrigerator and the performance of refrigerator was better than that of R-134a and POE oil. The improved refrigerator performance with $TiO_2$-R600a nano-refrigerant compared to pure R600a was achieved experimentally by Bi et al. [26] as well.

Influence of CuO nanoparticles on the boiling performance of R134a/polyolester lubricant oil mixture was experimentally investigated by Kedzierski et al. [27]. The experiment was done on a roughened, horizontal, and flat surface. They found 50% to 275% heat transfer improvement for R134a/polyolester by adding 0.5% mass fraction of CuO nanoparticle. The average boiling heat transfer enhancement was 19% and 12% in case of using 1% and 2% nanoparticle loading concentration, respectively. They also concluded that the thermal conductivity of the lubricant was increased around 20%. The result of the same investigation done by Kedzierski [28], showed 400% heat transfer improvement by adding 0.5% mass fraction of $Al_2O_3$ nanoparticles into the R134a/polyolester lubricant oil.

The feasibility of synthesis of refrigerant-based nanofluids, as well as characterizing their thermal behavior, must be taken into consideration.

In recent years, some studies have been reported on phase-change heat transfer of nanofluid. Most of them focused on pool boiling heat transfer. Two-phase flow heat transfer investigation was studied by Bartelt et al. [29]. The effect of CuO nanoparticle on the flow boiling of R134a/POE mixtures in a horizontal tube was examined. At least, 42% and 50% heat transfer enhancements were concluded as the effect of using 1% and 2% mass fraction of nanoparticles, respectively. No effect on the heat transfer coefficient was observed in case of using 0.5% mass fraction of CuO nanoparticle.

Household refrigerator freezer market is one of the major segments of the refrigeration industry. The widespread use of household refrigerator freezers provides an opportunity for sustainable energy saving, and the 100 million new units sold annually around the world represent a considerable potential of energy consumption in this field. Consequently, it can lead to huge amounts of energy saving by considering the energy consumption reduction methods.

This study investigates the performance of a domestic refrigerator using nano-refrigerant by focusing on the thermodynamic and energy approach. To analyze the thermodynamic and energy efficiency of the domestic refrigerator, experimental study was conducted, and the effect of different mass fractions of nanoparticles were investigated. This study can be a viable means of elucidating the effect of nanoparticles on the performance of the refrigeration systems.

## 2. Materials and Methods

The domestic refrigerator used in the experiment was a SR 30NMB type manufactured by Samsung Company Limited, which was a double-door, freezer and refrigerator, evaporator fin type, and natural convection condenser type. The picture of the refrigerator which was used as a test-rig and the schematic diagram of the experimental set-up and the charging mechanism, are shown in Figures 1 and 2, respectively. This refrigerator was designed to work with R-134a refrigerant and the technical specifications are shown in Table 1.

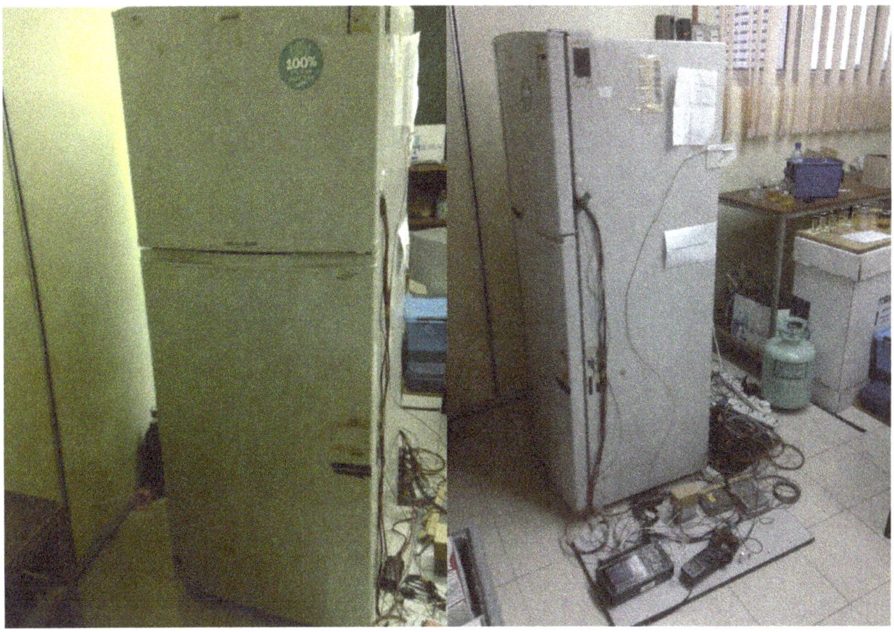

**Figure 1.** Domestic refrigerator test-rig.

**Table 1.** Technical specifications of refrigerator test unit.

| Item | | Specification |
|---|---|---|
| Model name | | SR 30 NMB |
| Type | | 2-Door Freezer/refrigerator |
| Power source | | 230~240 V/50 Hz |
| Net Capacity Lit (cu.ft.) | Freezer | 68 (2.4) |
| | Refrigerator | 186 (6.6) |
| | Total | 254 (9.0) |
| Refrigerant | | R 134a (140 g) |
| Compressor model | | SD162CL1U/T3 |

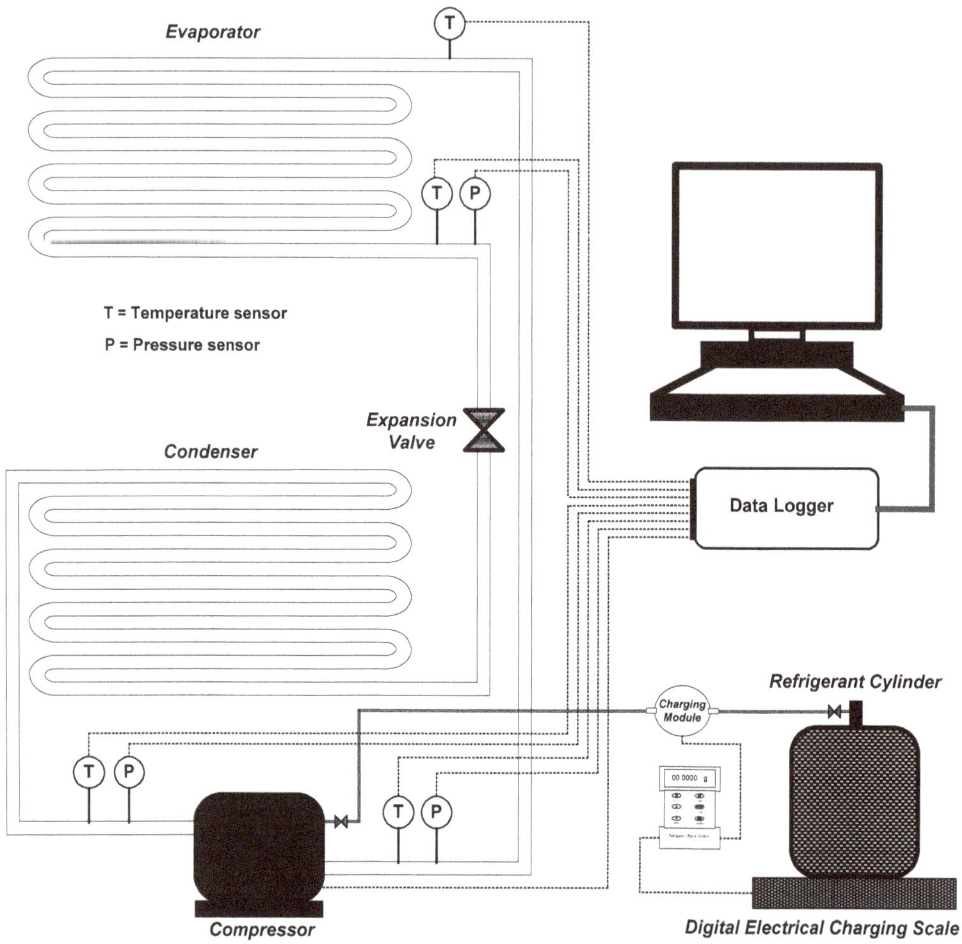


**Figure 2.** Schematic diagram of the experiment set-up and charging system.

The refrigerator's performance has been investigated with no load and closed door condition. The refrigerator was fitted with the thermocouples, pressure transducers, and power meter, and the other components remained intact.

### 2.1. Experiment Condition

In order to carry out the tests, the sequence of the clauses in international standard of household refrigerating standard, refrigerator–freezers characteristics and test methods were considered. The refrigerator was placed in a temperature controlled room and all experiments have been done in a steady-state operating condition. There was no ceiling fan and air conditioning system to force the movement of air inside the room. Therefore, heat transfer occurred by natural convection in the condenser and refrigerator walls [30].

### 2.2. Experiment Procedure

The same procedure was used for all experiments. Nanofluids were prepared based on the proposed method in the literature. Refrigerant and nanoparticle in this experiment were R134a and $Al_2O_3$. As the baseline experiment, the system was operated with R134a and polyolester oil (POE) as a lubricant. In this study, nano-refrigerant was a mixture of $Al_2O_3$ nanoparticles, POE lubricant oil, and R134a refrigerant. First, the required amount of $Al_2O_3$ nanoparticles and POE lubricant oil were mixed and sonicated for at least 30 min to achieve a homogenized mixture before being injected to the compressor. Finally, R134a was charged into the system. The thermocouples, pressure transducers, and power meter were installed on the system (Figure 2) and connected to the data logger to record the required data during all experiments. Vacuum and pressurize tests were performed before and after charging the refrigerant for at least 30 min to ensure that there was no leakage in the system.

To ensure that the previous working fluid was cleaned out of the passages, the fluid was driven out and the system was washed with based refrigerant and lubricant. It was done by charging the base lubricant into the compressor, then evacuating the system from air and moisture, followed by charging the system with refrigerant, and finally making it run for a few hours with the base fluid passing within the system. At the end, the system was made empty from refrigerant, and the oil was driven out from the compressor.

### 2.3. Refrigeration System Performance

Generally, heat travels from a hot space to a cold space due to a certain temperature difference. However, in refrigeration systems, it is done in the opposite direction as heat transfers from a lower temperature region to a higher temperature one. Figure 3, shows a schematic of a vapor compression refrigeration system.

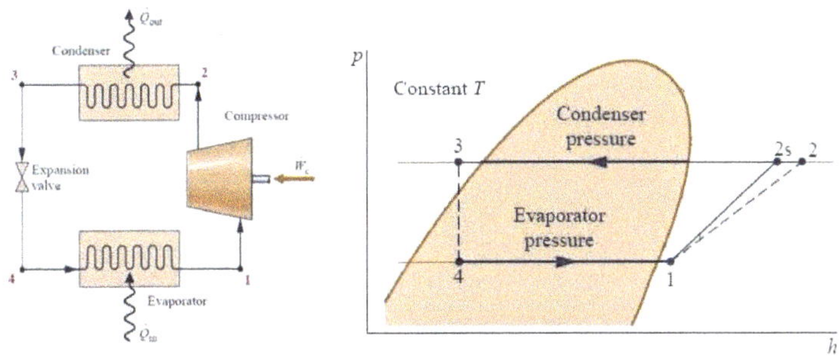


**Figure 3.** Schematic of a vapor compression system (Adapted from [31]).

The ratio of the cooling or refrigeration capacity (desired output) to the energy input into the system (required input) is called the coefficient of performance (COP), and expressed as:

$$COP = \frac{Cooling.capacity}{Work.input} = \frac{Q_{rf}}{W_{net,in}} \quad (1)$$

According to Figure 3, the refrigeration cycle is divided into four stages during which the properties of refrigerant change. Based on thermodynamics, heat transfer characteristics and thermal performance in each stage are explained on the following.

The compressor is the main power-consuming device in the refrigeration system. This energy is used to increase the pressure of the refrigerant vapor and to circulate it through the system. Consequently, the temperature of vapor refrigerant rises which in turn increases the enthalpy of the refrigerant at the outlet of the compressor.

The compressor work is determined with according to the *P-h* diagram and can be written as:

$$w = (h_2 - h_1) \quad (2)$$

where $h_2$ and $h_1$ are the enthalpies of refrigerant at the outlet and inlet of the compressor, respectively.

The compressor of the set-up refrigerator is a hermetic reciprocating type, (220 Volts, 50 Hz) that is thermally protected, and design for use with R-134a.

A condenser operates as a heat transfer device to release the heat from high-pressure superheated refrigerant vapor (discharged from the compressor) to the surrounding. As a result of heat rejection in the condenser, vapor refrigerant becomes a liquid at the outlet. Heat transfer rate in the condenser is calculated by:

$$q_{con} = (h_3 - h_2) \tag{3}$$

In this study, condenser is the natural convection cooling type.

Capillary tube operates as an expansion valve. The pressure of the liquid refrigerant is reduced by the capillary tube. The pressure of the liquid drops slightly in the first two-thirds of the length of the capillary tube. In current refrigerator, the capillary tube has 0.75 mm inside diameter and 3400 mm length.

The evaporator absorbs heat from its surroundings (inside refrigerator and freezer space) and transfers it to the refrigerant inside the evaporator. The refrigerant phase changes during the evaporation process from a liquid to a vapor, and at the evaporator exit is slightly superheated. This slight overdesign ensures that the refrigerant is completely vaporized when entering the compressor. The refrigeration effect is defined as the heat rejected by a unit mass of refrigerant during the evaporating process in the evaporator. It can be written as:

$$q_{rf} = h_1 - h_4 \tag{4}$$

where $h_1$ and $h_4$ are the refrigerant enthalpies at the outlet and inlet of the evaporator, respectively.

Refrigerating capacity, or cooling capacity,$Q_{rf}$, is the actual rate of heat which is removed by refrigerant in the evaporator, and can be calculated by [1]:

$$Q_{rf} = \dot{m}_r(h_1 - h_4) \tag{5}$$

### 2.4. Nano-Refrigerant Properties

According to the literatures, the physical and thermophysical properties of nanofluid can be calculated based on the following equations [32,33]. The specific heat of nanofluids and volume fraction of nanoparticle in the basefluid are expressed as:

$$C_{p,nf} = \frac{\phi(\rho c_p)_n + (1-\phi)(\rho c_p)_f}{\phi\rho_n + (1-\phi)\rho_f} \tag{6}$$

$$\phi = \frac{V_n}{V_n + V_f} \tag{7}$$

The density of nano-refrigerant, as a physical property of mixture, is introduced by:

$$\rho_{eff} = (1-\phi)\rho_f + \phi\rho_n \tag{8}$$

where subscripts $f$ and $n$ refer to fluid and nanoparticle, respectively. Dynamic viscosity of nanofluid is determined using the following equation:

$$\mu_{nf} = \frac{\mu_f}{(1-\phi)^{2.5}} \tag{9}$$

The viscosity of the mixture is related to the viscosity of refrigerant and the volume fraction of nanoparticles.

Enthalpy is a critical parameter in evaporation process for calculating the thermodynamic characteristics of the system and it depends on temperature, pressure, and state of the fluid (Liquid–Vapor). By assuming that the nanoparticles are volatile as basefluid, and that using the low nanoparticle concentration can lead to that the enthalpy difference of fluid in the case of using nanoparticle could be relevant to the temperature difference compared to the fluid without nanoparticle. Enthalpies of various fluids are available in different temperatures and pressures. As these data are empirical-based, there is no available data for nano-refrigerants that causes a major gap in the calculations related to application of nano-refrigerants. Therefore, future investigations including experimental and numerical methods are needed to obtain accurate thermophysical properties of nanofluids during evaporation and condensation.

### *2.5. Energy Consumption and Energy Efficiency of Refrigerator*

The purpose of the energy consumption test is to check the energy consumption of refrigerator under specific test conditions according to the international standard of household refrigerating standard, refrigerator–freezers characteristics and test methods.

The energy consumption was measured for a period of 24 h after stable operating conditions. Each test was repeated several times to ensure the reliability of the results.

The measurement of energy consumption was carried out under empty condition with all compartments simultaneously being in operation and has been expressed in kilowatt hours per 24 h (kWh/24 h) [30].

## **3. Results and Discussions**

All experiments were done in the same condition to avoid the effects of external parameters on the results. Figure 4, shows the ambient temperature was controlled during the experiments and it followed the same pattern in all experiments. Relative humidity also was controlled in the accepted range of the international standard.

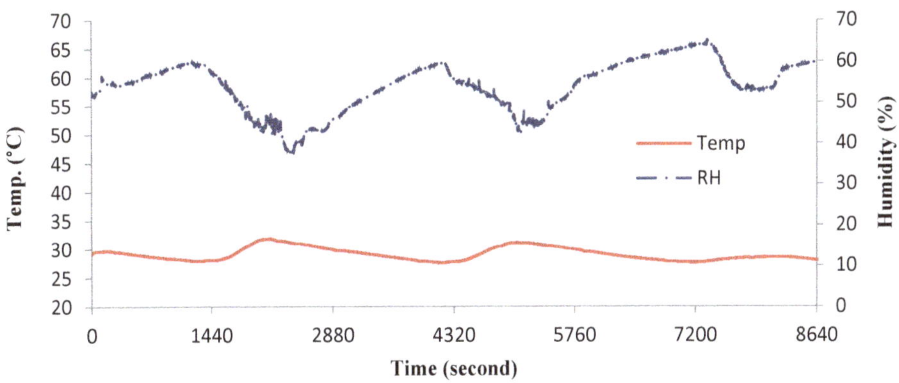


**Figure 4.** Ambient temperature and relative humidity of test room during experiments.

### *3.1. Stability of Nanofluid*

Stability of the prepared nanofluids was investigated while keeping the samples at similar conditions. Experiments show that the stability of nanofluid decreases with by increasing the concentration of nanoparticles. It may happen as a result of increasing agglomeration process. When the number of nanoparticle molecules increases in the constant volume, the interaction of molecules within the solution causes the formation of aggregates due to strong van der Walls forces [34].

Stability of $Al_2O_3$ and polyolester lubricant oil mixtures with 0.05%, 0.1%, and 0.3% mass fractions is shown in Figure 5. All mixtures were stable in the first hour after preparation. Nanofluid with 0.3%-$Al_2O_3$ started to sediment after some hours. This signifies the insufficient stability of the mixture to be used in the system. Sedimentation of mixture with 0.1%-$Al_2O_3$ increased at the end of day 1 and it almost completely sedimented on day 3. 0.05%-$Al_2O_3$ was stable even after 4 days, which proved the possibility for long-term stability for the lowest loading concentration of nanoparticles. Therefore, mixtures of 0.05% and 0.1%-$Al_2O_3$ nanoparticle and lubricant oil were chosen for this experiment.

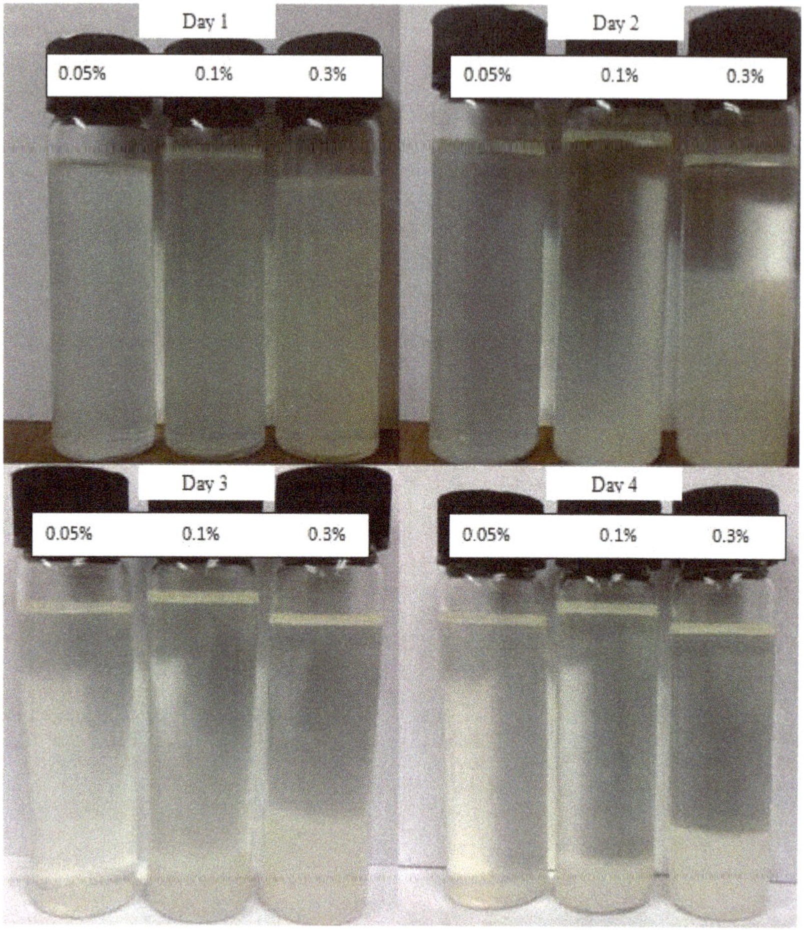


**Figure 5.** Stability of $Al_2O_3$-POE lubricant oil with different concentrations.

### 3.2. *Effect of Nano-Refrigerant on Evaporator Temperature Gradient*

Evaporator temperature is one of the most important parameters when investigating the heat transfer analysis in refrigeration systems. In the evaporation process, heat is transferred from the cold region into the refrigerant medium through three steps. Heat transfer to the liquid refrigerant before evaporation, during evaporation, and to the refrigerant vapor after completing the evaporation right before leaving the evaporator. These three steps, altogether, are known as refrigeration effect in the evaporator and can be calculated based on Equation (4). According to Equations (4) and (5), the heat transfer and cooling effect in the refrigeration system is explained based on the enthalpy difference of refrigerant fluid between outlet and inlet of the evaporator. Based on the explanation in the nano-refrigerant properties section regarding the enthalpy difference in refrigeration systems in the case of using nanoparticles compared to without nanoparticles, the evaporator performance can be analyzed based on temperature gradient of the evaporator. Higher temperature gradient causes an increase in the cooling effect and performance of the evaporator due to higher enthalpy difference between outlet and inlet.

Figure 6, shows the temperature gradient of refrigerant during one on–off cycle in the evaporator. As it can be seen, the temperature gradient of nanoparticle based refrigerant is higher than that of the pure refrigerant (R134a). It proves that the heat transfer is improved in the case of using nano-refrigerants. The average temperature gradient improvement in the evaporator was 1.9 °C (equal to 20.2%) when using 0.1%-$Al_2O_3$. The possible explanation for the increased heat transfer is the enhanced thermophysical properties of nano-refrigerants, such as thermal conductivity due to the existence of solid nanoparticles in the refrigerant. The experiments were conducted for 7 consecutive days in order to evaluate the stability of nano-refrigerant. The obtained results for last day are approximately similar to the acquired results in the initiation of the measurements which proves the stability of nano-refrigerant during the course of experiment. Although higher temperature gradient and consequently higher cooling effect can be achieved at higher nanoparticle concentrations, there is a limitation due to stability problem.

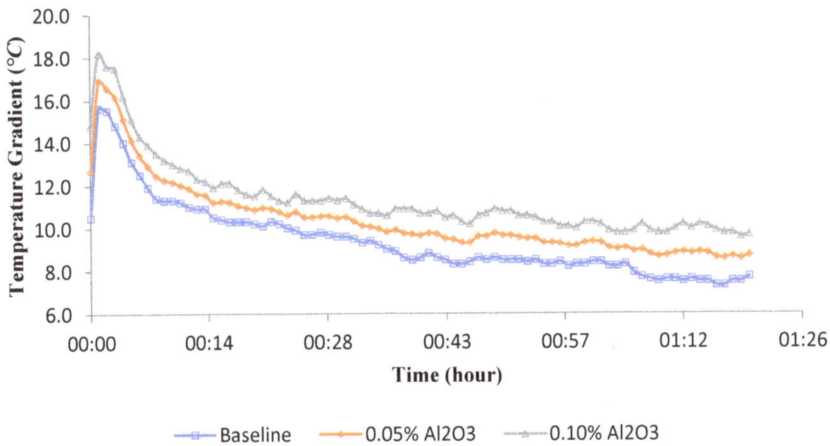


**Figure 6.** Temperature gradient in evaporator.

### 3.3. *Energy Consumption by the Compressor*

The energy consumption of each test is shown in Figures 7 and 8. Graphs show decrements in energy consumption after adding nanoparticles to the refrigerant. Every test was done at least three times under the same condition to ensure the repeatability of the results. The energy consumption of the refrigerator with normal refrigerant medium (R134a-POE) was 3.821 kWh/day. The maximum reduction in the energy consumption was around 2.69% in the case of using 0.1%-$Al_2O_3$ nanoparticles. The other nanoparticle concentration (0.05%-$Al_2O_3$) also led to reduction in the energy consumption by 1%. The results from previous experiments on the use of nanoparticle for refrigeration applications showed a similar behavior in terms of energy consumption reduction [25,26]. This can lead to significant long-term energy saving and emission reduction, as current refrigeration methods are a main part of environmental pollutants, when considered as a national policy [35].

Analysis of Figures 7 and 8, appeals that despite the equal off-cycles duration the on-cycles duration was decreased after using nano-refrigerant. This means that the cooling velocity of nano-refrigerant system was more quickly that the R134a system. Therefore, energy consumption of the refrigerator was reduced due to the decrease in the total on-cycles duration in a complete standard cycle.

The reason behind the decrement in the on-cycles time of the compressor can be the enhancement in the heat transfer characteristics of the evaporator as a result of improved thermophysical characteristics of the refrigerant after dispersing the nanoparticles in the refrigerant.

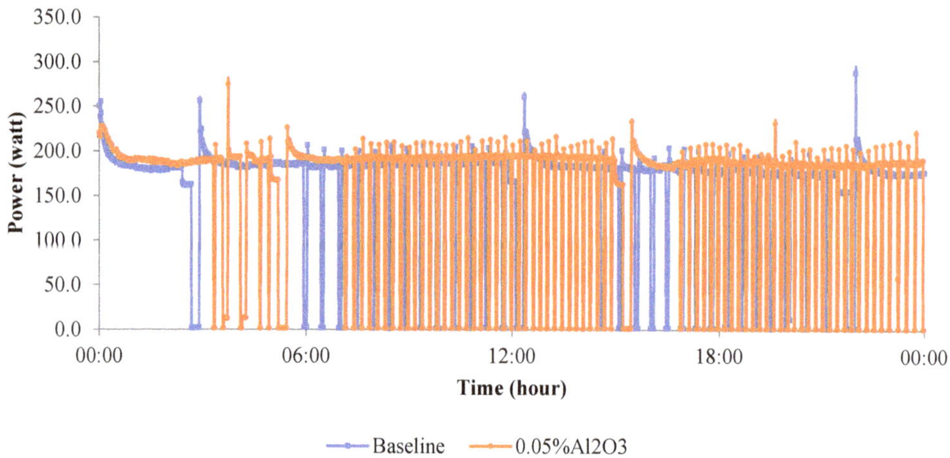


**Figure 7.** Energy consumption of refrigerator with 0.05%-$Al_2O_3$ and without nanoparticle.

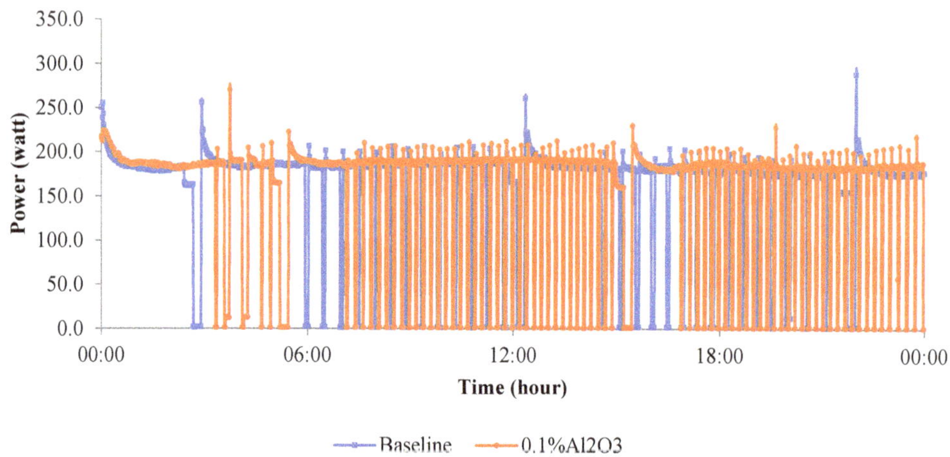


**Figure 8.** Energy consumption of refrigerator with 0.1%-$Al_2O_3$ and without nanoparticle.

*3.4. Compressor Discharge and Suction Pressure Analysis*

Figures 9 and 10, compare the compressor discharge and suction pressures of the test-rig over one on–off cycle, respectively. These figures show that both pressures were reduced for the case of nano-refrigerants.

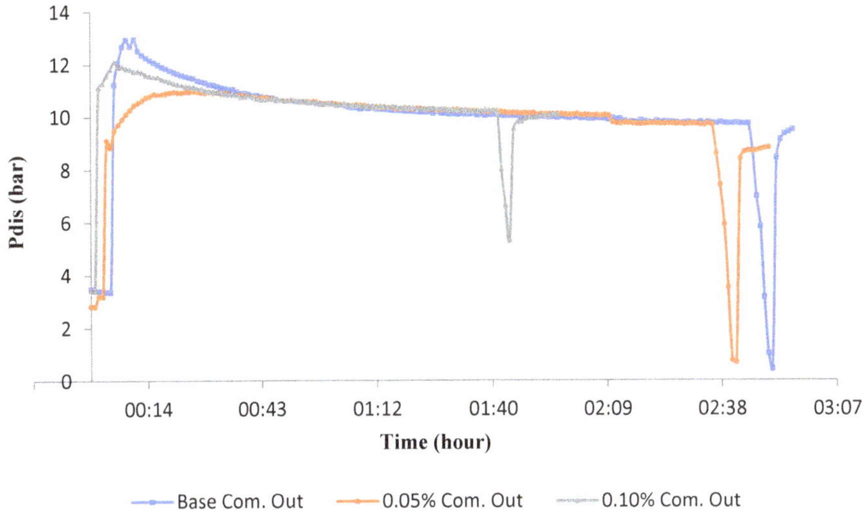


**Figure 9.** Compressor discharge pressure.

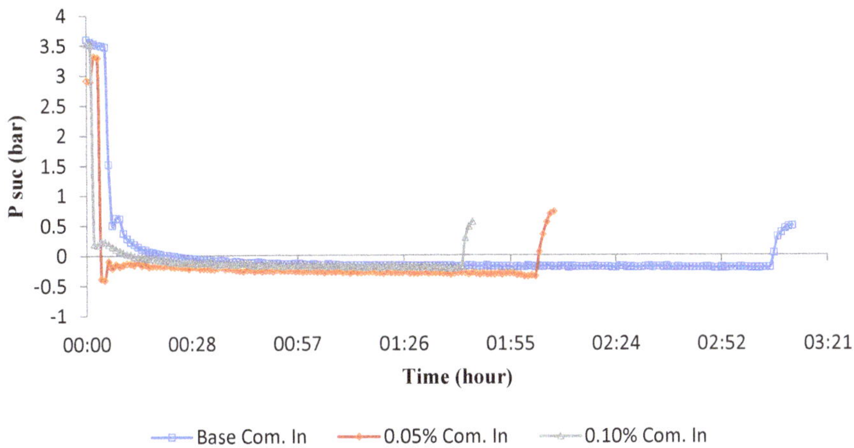


**Figure 10.** Compressor suction pressure.

For 0.05%-$Al_2O_3$ nanoparticles, the discharge pressure of the compressor was lower than that of the base fluid at the first minutes of compressor operating time, but it was almost the same as the base fluid after the pressure becomes stable. It shows that small amounts of nanoparticles do not have substantial effect on the discharge pressure, but causes the discharge pressure to decrease at the first minutes of the on-cycle time. Nano-refrigerant with 0.1% nanoparticles also follows the same trend as 0.05%, but at a higher discharge pressure between those for the base fluid and 0.05% nanofluid. On the other hand, both Figures 9 and 10 demonstrate that the operating time of compressor in cases of using nanofluids are less than base fluid which causes the energy consumption of the compressor to drop during a complete standard cycle. The reason might be that the nanoparticles improve the heat transfer capability of the refrigerant and then causes the working time of the compressor to decrease. Figure 10 also demonstrates that existence of nanoparticles in the refrigeration system caused the suction pressure and working time of the compressor to reduce. Decreasing the suction pressure of the compressor could be the result of two phenomena: the increase in the pressure drop in the system which is the result of increase in the viscosity of the working fluid, as well as the decrease in the discharge pressure of the compressor.

### 3.5. Pressure Drop in the System

Pressure drop is one of the important parameters in fluid systems, especially when the existing fluid is replaced by a new fluid. Figure 11 shows the pressure drop between the outlet and inlet of the compressor. This includes the summation of the pressure drop of the condenser, evaporator, capillary tube, and pipes in the system from inlet to the outlet of the compressor.

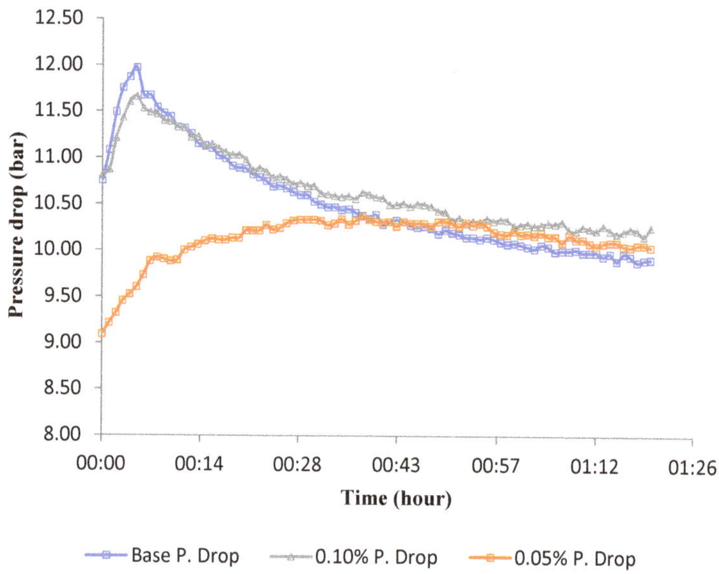


**Figure 11.** Total pressure drop in the system.

As can be seen, Figure 11 indicates the pressure drop in the system for base fluid and nanofluid with different nanoparticle concentration over one on-off cycle. It shows that base fluid and nano-refrigerant with 0.1% nanoparticle have almost the same pressure drops. 0.1% nano-refrigerant owns less pressure drop at the beginning of the cycle. However, its pressure drop increases slightly and reaches to well above that of the baseline at the end of the cycle. Nano-refrigerant with 0.05%-$Al_2O_3$ follows a different pattern. Apparently, the pressure drop in the beginning of the cycle is much less than those of the other samples. It increases slightly during the initial one-third of the cycle, reaches to the same pressure drop as other experiments, and follows the same pattern during the rest of the cycle. Domination of the lubricity characteristic of nanoparticle in low concentration to the increasing of the density and viscosity due to adding solid particles to the fluid could be the reason. Therefore, increasing the concentration of nanoparticle in the base fluid causes to increase pressure drop and more energy consumption in each on-cycle.

## 4. Conclusions

A test-rig was constructed in order to evaluate the performance of a household refrigerator-freezer using refrigerant R-134a as a working fluid. The test-rig included instrumentation to measure the thermodynamic properties of the system. The energy consumption test was used to evaluate energy consumption of the refrigerator. The system was tested by the normal working fluid, as well as $Al_2O_3$ nano-refrigerant with different nanoparticle concentrations. Finally, all data were compared to evaluate the effect of nano-refrigerant on the thermodynamic performance, heat transfer characteristics, and energy efficiency of the domestic refrigerator. After the successful investigation of the measured parameters, the following conclusions have been drawn:

- Stability of $Al_2O_3$ nano-lubricant oil decreases by increasing the nanoparticle concentration in the base fluid. Therefore, application of nanofluid with high nanoparticle concentration is limited and alternative preparation methods and using additives are needed to improve the stability. However, nanofluid with low nanoparticle concentration is stable for longtime;
- Evaporator temperature gradient is increased by using nano-refrigerant. It proves that thermodynamic behavior of fluid is improved. An increment of 20.2% occurred at the temperature gradient of the evaporator for 0.1%-$Al_2O_3$;
- It has been found that the electricity consumption of the refrigerator was 2.69% lower than that of the base fluid (R134a) when 0.1%-$Al_2O_3$ nanoparticle was added to the system. This value was 1% for the case of 0.05%-$Al_2O_3$;
- It is apparent from the data that the on-cycles duration was less for nano-refrigerants, but off-cycles duration was nearly the same for both baseline and nanofluid. It proves that the cooling velocity in nano-refrigerant system was happened quicker than the normal refrigerant. This can be a reason of reduction in energy consumption during a complete standard cycle;
- Suction and discharge pressures of the compressor decreased when using nano-refrigerant compared to the case of pure refrigerant. The result of increasing the pressure drop in the system due to adding nanoparticles to the fluid appears in the suction pressure and its effects on the discharge pressure;
- Effect of nano-refrigerant on the performance of the compressor could be a reason for decreasing the discharge pressure of the compressor;
- Finally, it can be concluded that using nanoparticles in a refrigeration system can improve thermodynamic characteristics and decrease energy consumption of a domestic refrigerator.

## 5. Recommendations

The present research investigated the performance of the domestic refrigerator using nano-refrigerant. In fact, the feasibility of using nano-refrigerant as a refrigerant medium in domestic refrigerator was taken into account. Utilization of nano-refrigerant requires a wide range of information about the properties of the nanofluid such as enthalpy, viscosity, thermal conductivity, and compatibility of the nano-refrigerant with the other material in the system. Compatibility is very important and should be examined for all parts of the system to ensure that there are no negative effects on the system components in the short and long run.

The following recommendations can be suggested for the future research on the application of nano-refrigerant: overcoming the stability problem, direct preparation method for nano-refrigerant, measuring the fundamental properties of nano-refrigerant especially in cases where the boiling temperature of refrigerant is lower than the ambient temperature, the effects of using solid nanoparticles in the refrigeration system, such as compatibility with the equipment, chemical reaction during longtime operation, solubility of the mixture of lubricant oil, refrigerant, and nanoparticle.

**Author Contributions:** F.S.J.: Conceptualization, Methodology, Investigation, Writing—Original Draft, Writing—Review & Editing. R.S.: Supervision, Validation, Project administration, Funding acquisition. All authors have read and agreed to the published version of the manuscript.

**Funding:** This research is supported by High Impact Research MoE Grant UM.C/625/1/HIR/MoE/ENG/40 from the Ministry of Education Malaysia.

**Institutional Review Board Statement:** Not applicable.

**Informed Consent Statement:** Not applicable.

**Data Availability Statement:** The data presented in this study are available on request from the corresponding author.

**Conflicts of Interest:** The authors declare no conflict of interest.

## References

1. Wang, S.K. *Handbook of Air Conditioning and Refrigeration*, 2nd ed.; McGraw Hill: New York, NY, USA, 2000.
2. Althouse, A.D.; Turnquist, C.H.; Bracciano, A.F.; Bracciano, D.C.; Bracciano, G.M. *Modern Refrigeration and Air Conditioning*, 9th ed.; Goodheart-Willcox: Tinley Park, IL, USA, 2004.
3. Aslfattahi, N.; Saidur, R.; Arifutzzaman, A.; Sadri, R.; Bimbo, N.; Mohd Sabri, M.F.; Maughan, P.A.; Bouscarrat, L.; Dawson, R.J.; Mohd Said, S.; et al. Experimental investigation of energy storage properties and thermal conductivity of a novel organic phase change material/MXene as A new class of nanocomposites. *J. Energy Storage* **2020**, *27*, 101115. [CrossRef]
4. Abdelrazik, A.; Tan, K.H.; Aslfattahi, N.; Arifutzzaman, A.; Saidur, R.; Al-Sulaiman, F.A. Optical, stability and energy performance of water-based MXene nanofluids in hybrid PV/thermal solar systems. *Solar Energy* **2020**, *204*, 32–47. [CrossRef]
5. Aslfattahi, N.; Samylingam, L.; Abdelrazik, A.S.; Arifutzzaman, A.; Saidur, R. MXene based new class of silicone oil nanofluids for the performance improvement of concentrated photovoltaic thermal collector. *Sol. Energy Mater. Sol. Cells* **2020**, *211*, 110526. [CrossRef]
6. Aslfattahi, N.; Saidur, R.; Sabri, M.F.M.; Arifutzzaman, A. Experimental investigation on thermal stability and enthalpy of eutectic alkali metal solar salt dispersed with MgO nanoparticles. In Proceedings of the 4th International Tropical Renewable Energy Conference, Bali, Indonesia, 14–16 August 2019.
7. Das, L.; Habib, K.; Saidur, R.; Aslfattahi, N.; Yahya, S.M.; Rubbi, F. Improved thermophysical properties and energy efficiency of aqueous ionic liquid/MXene nanofluid in a hybrid PV/T solar system. *Nanomaterials* **2020**, *10*, 1372. [CrossRef] [PubMed]
8. Das, S.; Choi, S.U.; Yu, W.; Pradeep, T. *Nanofluids Science and Technology*, 1st ed.; Wiley: New Jersey, NJ, USA, 2008.
9. Rubbi, F.; Habib, K.; Saidur, R.; Aslfattahi, N.; Yahya, S.M.; Das, L. Performance optimization of a hybrid PV/T solar system using Soybean oil/MXene nanofluids as A new class of heat transfer fluids. *Solar Energy* **2020**, *208*, 124–138. [CrossRef]
10. Aslfattahi, N.; Saidur, R.; Sidik, N.A.C.; Sabri, M.F.M.; Zahir, M.H. Experimental assessment of a novel eutectic binary molten salt-based hexagonal boron nitride nanocomposite as a promising PCM with enhanced specific heat capacity. *J. Adv. Res. Fluid Mech. Therm. Sci.* **2020**, *68*, 73–85. [CrossRef]
11. Samylingam, L.; Aslfattahi, N.; Saidur, R.; Yahya, S.M.; Afzal, A.; Arifutzzaman, A.; Tan, K.H.; Kadirgama, K. Thermal and energy performance improvement of hybrid PV/T system by using olein palm oil with MXene as a new class of heat transfer fluid. *Sol. Energy Mater. Sol. Cells* **2020**, *218*, 110754. [CrossRef]
12. Aslfattahi, N.; Saidur, R.; Arifutzzaman, A.; Abdelrazik, A.S.; Samylingam, L.; Sabri, M.F.M.; Sidik, N.A.C. Improved thermophysical properties and energy efficiency of hybrid PCM/graphene-silver nanocomposite in a hybrid CPV/thermal solar system. *J. Therm. Anal. Calorim.* **2020**, *142*, 1–18.
13. Aslfattahi, N.; Saidur, R.; Sabri, M.F.M.; Arifutzzaman, A. Thermal conductivity and rheological investigation of aqueous poly (ethylene) glycol/MXene as a novel heat transfer fluid. In *AIP Conference Proceedings*; AIP Publishing LLC.: New York, NY, USA, 2021.
14. Parashar, N.; Aslfattahi, N.; Yahya, S.M.; Saidur, R. ANN Modeling of Thermal Conductivity and Viscosity of MXene-Based Aqueous IoNanofluid. *Int. J. Thermophys.* **2021**, *42*, 1–24. [CrossRef]
15. Parashar, N.; Aslfattahi, N.; Yahya, S.M.; Saidur, R. *Prediction of the Dynamic Viscosity of MXene/Palm Oil Nanofluid Using Support Vector Regression*; AIP Publishing LLC.: New York, NY, USA, 2021.
16. Abdullah, N.; Saidur, R.; Zainoodin, A.M.; Aslfattahi, N. Optimization of electrocatalyst performance of platinum–ruthenium induced with MXene by response surface methodology for clean energy application. *J. Clean. Prod.* **2020**, *277*, 123395. [CrossRef]
17. Abdelrazik, A.; Tan, K.H.; Aslfattahi, N.; Saidur, R.; Al-Sulaiman, F.A. Optical properties and stability of water-based nanofluids mixed with reduced graphene oxide decorated with silver and energy performance investigation in hybrid photovoltaic/thermal solar systems. *Int. J. Energy Res.* **2020**, *44*, 11487–11508. [CrossRef]
18. Krishna, Y.; Saidur, R.; Aslfattahi, N.; Faizal, M.; Ng, K.C. Enhancing the thermal properties of organic phase change material (palmitic acid) by doping MXene nanoflakes. In *AIP Conference Proceedings*; AIP Publishing LLC.: New York, NY, USA, 2020.
19. Krishna, Y.; Aslfattahi, N.; Saidur, R.; Faizal, M.; Ng, K.C. Fatty acid/metal ion composite as thermal energy storage materials. *SN Appl. Sci.* **2020**, *2*, 1–10. [CrossRef]
20. Mahbubul, I.M.; Saidur, R.; Amalina, M.A. Pressure Drop Characteristics of $TiO_2$–R123 Nanorefrigerant in a Circular Tube. *Eng. e-Trans* **2011**, *6*, 124–130.
21. Mahbubul, I.M.; Saidur, R.; Amalina, M.A. Investigation of Viscosity of R123-$TiO_2$ Nanorefrigerant. *Int. J. Mech. Mater. Eng.* **2012**, *7*, 146–151.
22. Peng, H.; Ding, G.; Jiang, W.; Hu, H.; Gao, Y. Heat transfer characteristics of refrigerant-based nanofluid flow boiling inside a horizontal smooth tube. *Int. J. Refrig.* **2009**, *32*, 1259–1270. [CrossRef]
23. Peng, H.; Ding, G.; Hu, H. Effect of surfactant additives on nucleate pool boiling heat transfer of refrigerant-based nanofluid. *Exp. Therm. Fluid Sci.* **2011**, *35*, 960–970. [CrossRef]
24. Henderson, K.; Park, Y.G.; Liu, L.; Jacobi, A.M. Flow-boiling heat transfer of R-134a-based nanofluids in a horizontal tube. *Int. J. Heat Mass Transf.* **2010**, *53*, 944–951. [CrossRef]

25. Ding, G.; Peng, H.; Jiang, W.; Gao, Y. The migration characteristics of nanoparticles in the pool boiling process of nanorefrigerant and nanorefrigerant–oil mixture. *Int. J. Refrig.* **2009**, *32*, 114–123. [CrossRef]
26. Saidur, R.; Kazi, S.N.; Hossain, M.S.; Rahman, M.M.; Mohammed, H.A. A review on the performance of nanoparticles suspended with refrigerants and lubricating oils in refrigeration systems. *Renew. Sustain. Energy Rev.* **2011**, *15*, 310–323. [CrossRef]
27. Bi, S.; Shi, L.; Zhang, L.L. Application of nanoparticles in domestic refrigerators. *Appl. Therm. Eng.* **2008**, *28*, 1834–1843. [CrossRef]
28. Bi, S.; Guo, K.; Liu, Z.; Wu, J. Performance of a domestic refrigerator using $TiO_2$-R600a nano-refrigerant as working fluid. *Energy Convers. Manag.* **2011**, *52*, 733–737. [CrossRef]
29. Kedzierski, M.A.; Gong, M. Effect of CuO nanolubricant on R134a pool boiling heat transfer. *Int. J. Refrig.* **2009**, *32*, 791–799. [CrossRef]
30. Kedzierski, M.A. Effect of $Al_2O_3$ nanolubricant on R134a pool boiling heat transfer. *Int. J. Refrig.* **2011**, *34*, 498–508. [CrossRef]
31. Bartelt, K.; Park, Y.; Liu, L.; Jacobi, A. Flow-Boiling of R-134a/POE/CuO Nanofluids in a Horizontal Tube. In Proceedings of the International Refrigeration and Air Conditioning Conference, Champaign and Urbana, West Lafayette, IN, USA, 14–17 July 2008.
32. Standard, BS EN ISO 15502:2005: Household refrigerating appliances. In *Characteristics and Test Methods*; British Standard/European Standard/International Organization for Standardization, BSI Standards Publication: London, UK, 2005.
33. Micael, J.M.; Shapiro, H.N.; Boettner, D.D.; Bailey, M.B. *Fundamentals of Engineering Thermodynamics*, 7th ed.; Wiley: Hoboken, NJ, USA, 2011.
34. Pak, B.C.; Cho, Y.I. Hydrodynamic and Heat Transfer Study of Dispersed Fluids with Submicron Metallic Oxide Particles. *Exp. Heat Transf.* **1998**, *11*, 151–170. [CrossRef]
35. Brinkman, H.C. The Viscosity of Concentrated Suspensions and Solutions. *J. Chem. Phys.* **1952**, *20*, 571. [CrossRef]

sustainability

Article

# Modeling Renewable Energy Systems by a Self-Evolving Nonlinear Consequent Part Recurrent Type-2 Fuzzy System for Power Prediction

Jafar Tavoosi [1,*], Amir Abolfazl Suratgar [2], Mohammad Bagher Menhaj [2], Amir Mosavi [3,4,*], Ardashir Mohammadzadeh [5] and Ehsan Ranjbar [2]

1 Department of Electrical Engineering, Faculty of Engineering, Ilam University, Ilam, Iran
2 Center of Excellence on Control and Robotics, Department of Electrical Engineering, Amirkabir University of Technology, 424 Hafez Ave, Tehran, Iran; a-suratgar@aut.ac.ir (A.A.S.); menhaj@aut.ac.ir (M.B.M.); en.ranjbar.eeng@gmail.com (E.R.)
3 Faculty of Civil Engineering, Technische Universität Dresden, 01069 Dresden, Germany
4 John von Neumann Faculty of Informatics, Obuda University, 1034 Budapest, Hungary
5 Department of Electrical Engineering, University of Bonab, Bonab 551761167, Iran; a.mzadeh@ubonab.ac.ir
* Correspondence: j.tavoosi@ilam.ac.ir (J.T.); amir.mosavi@mailbox.tu-dresden.de (A.M.)

**Abstract:** A novel Nonlinear Consequent Part Recurrent Type-2 Fuzzy System (NCPRT2FS) is presented for the modeling of renewable energy systems. Not only does this paper present a new architecture of the type-2 fuzzy system (T2FS) for identification and behavior prognostication of an experimental solar cell set and a wind turbine, but also, it introduces an exquisite technique to acquire an optimal number of membership functions (MFs) and their corresponding rules. Using nonlinear functions in the "Then" part of fuzzy rules, introducing a new mechanism in structure learning, using an adaptive learning rate and performing convergence analysis of the learning algorithm are the innovations of this paper. Another novel innovation is using optimization techniques (including pruning fuzzy rules, initial adjustment of MFs). Next, a solar photovoltaic cell and a wind turbine are deemed as case studies. The experimental data are exploited and the consequent yields emerge as convincing. The root-mean-square-error (RMSE) is less than 0.006 and the number of fuzzy rules is equal to or less than four rules, which indicates the very good performance of the presented fuzzy neural network. Finally, the obtained model is used for the first time for a geographical area to examine the feasibility of renewable energies.

**Keywords:** self-evolving; nonlinear consequent part; convergence analysis; renewable energy; type-2 fuzzy; artificial intelligence; machine learning; big data; data science; fuzzy logic; energy

**Citation:** Tavoosi, J.; Suratgar, A.A.; Menhaj, M.B.; Mosavi, A.; Mohammadzadeh, A.; Ranjbar, E. Modeling Renewable Energy Systems by a Self-Evolving Nonlinear Consequent Part Recurrent Type-2 Fuzzy System for Power Prediction. *Sustainability* **2021**, *13*, 3301. https://doi.org/10.3390/su13063301

Academic Editor: Mehdi Seyedmahmoudian

Received: 1 February 2021
Accepted: 7 March 2021
Published: 17 March 2021



## 1. Introduction

Renewable energy is expanding rapidly around the world. There are two main reasons for this: one is the issue of fossil fuel pollution and the other is the high cost of fossil fuels. Therefore, research in this field should be developed and supported. One of the powerful tools in data analysis and inference is computational intelligence. Neural networks share lots of significant benefits such as landmark computation ability, parallel processing and adaptation. The fuzzy systems are able to utilize the expert knowledge entitled "if-then rules" and possess actual parameter concepts. As is well known, mathematical modeling is a substantial preliminary step in many control issues. On the other hand, prediction, simulation and modeling of complicated systems established upon physical and chemical principles appear industrious in such a way that they will not yield consolidated mathematical forms [1]. One may suggest system identification as a solution to cope with this problematic issue. This method puts the mathematical equations at the access point, utilizing input-to-output data analysis to increase the efficiency of dynamic process calculations [2]. Computational intelligence lies among the most efficient methods with excellent fulfillment. Many papers have recently been published on fuzzy modeling and identification. Nonlinear system identification, founded on fuzzy and neuro-fuzzy models, was surveyed [3]. Computational intelligence becomes extremely feasible in the area of renewable energy [4]. For design MPPT control [5], solar water heater selection [6], photovoltaic system failure diagnosis [7] and solar power plant location alternatives [8], computational intelligence has been used. Neural networks were also used by Grahovac et al. [9] in order to model and anticipate bio-ethanol generation from the intermediates and byproducts yielded in the beet-to-sugar procedure. The productivity of the neuro-fuzzy controller in extraction of the maximum yield by flow and energy optimization was demonstrated by Khiareddine et al. [10] in comparison with fuzzy and algorithm controllers. It was asserted that the neuro-fuzzy control system is worthy of being executed in an experimental setup in Tunisia. Ocario et al. [11] testified wind power forecasts in the Portuguese system, exploiting a novel hybrid evolutionary–adaptive methodology. Etemadi et al. [12] predicted the wind power produced by data-driven fuzzy modeling.

Type-2 fuzzy (T2F) logic, which appears more capable and flexible in comparison to type-1, has been investigated for the last ten years. A novel method was suggested for general T2F clustering by Doostparast et al. [13]. Some other applications of T2F sets can be found in textile engineering [14] and aerospace engineering [15]. Fuzzy c-means clustering and high order cognitive map were exerted by Lu in order to model and predict time series by T1FS [16]. T2FS identification has engrossed many researchers [17–23]. Abiyev et al. [17] took advantage of T2F clustering to organize construction of a wavelet TSK-based T2FS. They brought forth an adaptive law to update the parameters of the antecedent part and ultimately, they employed a gradient learning algorithm to bring parameters of the descendant part up to date. T2FSs were applied for elicitation of fuzzy rules and casting derogatory features off [24]. The proposed mechanism took advantage of the self-evolution capability in such a way that identification of the integral structure of the network would become efficient and there would be no requirement for initial start-up of the network structure. The antecedent part and modulation parameters are trained in order to hold parameter learning in the network true, utilizing back-propagation errors. Tuning parameters of the resultant part, the rule-ordered Kalman filter algorithm assists in network sharpness amelioration. The genetic algorithm [25] and PSO [26] are among the learning mechanism of T2F neural networks which have been conversed and scrutinized so far. Research development on T2F systems has brought about their vast usages in various fields such as time-series prediction [27], DC motor control [28], clinical practice guideline encryption [29], pattern recognition [30], robot control [31] and control of nonlinear systems [32,33]. A new smart type of reduction is held forth in [34]. A T2FS is optimized by its type-1 counterpart in [35]. The learning process was held true, merging and extending the type-1 membership functions. Henceforth, the novel constructed T2FS went under implementation on a programmable chip.

It is worth noting that most of the control engineers and system analyzers consider actual systems represented in nonlinear dynamics; not only do these system outputs momentarily turn dependent upon the input, but also, they appear reliant on the delayed inputs/outputs. This leads to a responsible consideration of both external and internal dynamics as a non-negligible essential remark in system modeling. Delayed inputs/outputs have to be used in external dynamics. Another feedback, denoted as "recurrent neuron", has to be exerted in internal dynamics. Wu et al. [36] presented the solution of recurrent FSs for problematic classification. Not only does this paper contribute to minimization of the cost function utilizing a recurrent fuzzy neural network, but it also proposes maximization of the discriminability of adopting a novel approach. Some modern recurrent fuzzy systems are presented in [37]. This special kind of neural network in the resultant part functions input variables in a nonlinear manner. There have hardly been any studies on recurrent T2F systems so far. Some of them are surveyed in the following. A contributive recurrent interval T2FS is presented in order to identify nonlinear systems in [30]. The novel technique requires initial information about plant order and input numbers as well.

Furthermore, the convergence issue in the learning algorithm is not taken into consideration and conversed even theoretically. Juang et al. [15] put forth another contributive recurrent T2F neural network to model dynamical systems. There is not any rule pruning, which leads to extremely overlapped fuzzy sets. Soft switching of the nonlinear model is superior to the linear one in order to identify nonlinear systems [1]. Consequently, our suggested technique is established upon the nonlinear resultant part in fuzzy rules. Rarely may one find comprehensive works on nonlinear consequent parts in fuzzy systems; however, some of the studies in this arena are shortly surveyed in the following. A reduction in the number of rules was carried out by Moodi in a fuzzy system using the TSK fuzzy model accompanied by a nonlinear consequent part [38]. The result of a rule is supposed to comprise a linear term and a nonlinear one. In their attempts, the numerous rules decrease and model precision simultaneously shows an increase at the cost of complication abundance in the fuzzy model. The NFNN was constructed applying fuzzy rules which merge nonlinear functions. The linear consequent part requires more rules to achieve the desired precision during the modeling of complicated nonlinear processes. The increasing number of rules represents the increasing number of neurons [39]. Some recent works on T2F neural networks can be seen in many applications such as 2DOF robot control [40], 3 parallel robots control [41], PMSM control [42], water temperature control [43,44], environmental temperature control [45] and UAV control [46]. Tavoosi and Badamchizadeh [47] proposed a T2S with linear "then part" for dynamic modeling. Their pivotal contribution was rule pruning in such a way that an increase in learning speed would be targeted to attain a reduction in the parameters in both MF parameters and descendant parts. Tavoosi et al. [48,49] have made another contribution to the issue, bringing forth a novel technique for analyzing the stability of one class of T2F systems. Another analysis method for stability was also suggested by Jahangiri et al. [50]. Suratgar and Nikravesh [51] proposed a modern technique of fuzzy linguistic modeling as well as integral stability analysis. In [52], a fuzzy neural network has been used for wind speed forecasting. In [53], a comparison between ANFIS and an autoregressive method for wind speed/power prediction has been performed. In [54], a fuzzy control on the basis of a predictive technique for a governing system has been presented. In [55], a multilayer perceptron is combined with an adaptive fuzzy system to forecast the performance of a wind turbine. Some disadvantages and shortcomings of the works studied above are: lack of convergence proof, long training time (not usable in online applications), high complexity of the model, lack of proper accuracy. On the other hand, so far, no applied research has been conducted to use renewable energies in the Ilam region. Unfortunately, there are no wind turbines in this area, and solar cells have also not been used on a large scale to supply electricity to a neighborhood or even several houses. Due to this issue, the main innovation of this paper is the feasibility study of new energy use in the Ilam region.

Therefore, this paper proposes NCPRT2FS for nonlinear system identification. The nonlinear systems here are the same as solar cells and wind turbines. The objective of identifying the system is to use it to specify the efficiency of the renewable energy system in the Ilam region. The innovations of this article are as follows: (1) Using a nonlinear consequent part in the rules. (2) Introducing a new mechanism in structure learning. (3) Using an adaptive learning rate (different from the other studies in the literature). (4) Convergence analysis of the T2F neural network learning algorithm. Finally, (5) New optimization techniques (including pruning fuzzy rules, initial adjustment of MFs, etc.). The paper is divided into six sections. Section 2 presents a short surveying of T2F logic. Section 3 entails an inspection of the structure of NCPRT2FS. The learning convergence of NCPRT2FS is subsumed relying upon Lyapunov theory in Appendix A. Section 4 presents simulative identification studies, taking into account a solar photovoltaic cell and a wind turbine as the case studies and utilizing their experimental data.

## 2. A Review on T2FSs

Firstly, Zadeh brought forward type-1 fuzzy logic, and introduced the T2F logic in order to provide solutions to some problems of type-1 ten years later. He deemed a fuzzy set where its MF was fuzzy and entitled a "type-2 fuzzy set". T2F sets may typically be exploited when the determination of accurate membership function becomes arduous. For instance, some time series predictions lie among problematic cases, which necessitate the usage of T2F sets. Hence, exploiting T2F sets emerges as advantageous in order to describe some system behaviors.

Certain defects with type-1 fuzzy sets were scrutinized by Castro et al. [56]. Research on T2F systems was limited before the years of 1998. Critical and controversial questions and debate on T2F logic and its usage commenced after publication of a book which contained the solidarity and intersection of T2F sets [57]. Extensive information on T2FS computation, such as defuzzification and type reduction, was suggested by Mendel [58]. A general T2F set, $\widetilde{A}$, may be specified by (1):

$$\widetilde{A} = \int_{x \in X} \mu_{\widetilde{A}}(x)/x = \frac{\int_{x \in X}\left[\int_{\mu \in J_x} \frac{f_x(\mu)}{\mu}\right]}{x} \quad (1)$$

where $\mu_{\widetilde{A}}(x)$ is a secondary MF; $J_x$ represents the primary membership of $x \in X$, with $\mu \in J_x$; $f_x(\mu) \in [0,1]$ denotes a secondary membership. The primary and secondary MFs in Gaussian form are illustrated in Figure 1.

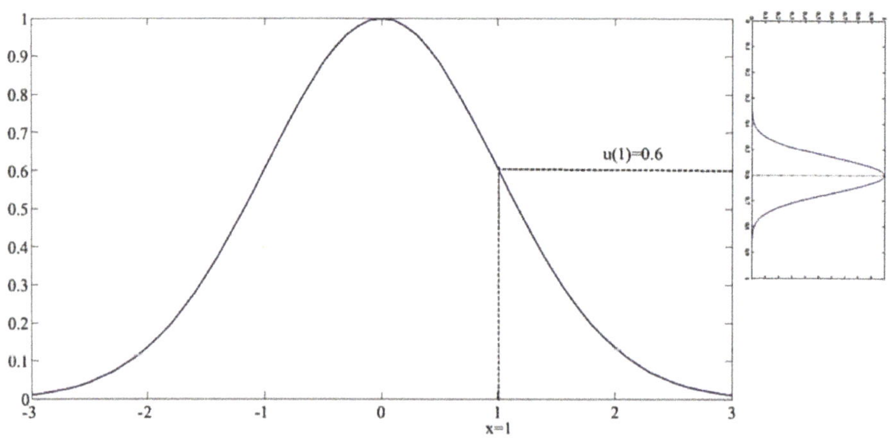


**Figure 1.** Primary and secondary membership functions (MFs).

Note that the secondary MFs lead to interval T2F ones, while $f_x(\mu) = 1$, $\forall \mu \in J_x \subseteq [0,1]$. For more explanation, a crisp number would be fuzzified in two stages supposing that Gaussian MF was exerted to attain a T2F number. First,

$$\mu_1 = \exp\left(-0.5 \cdot \frac{(x-M)^2}{{\sigma_x}^2}\right) \quad (2)$$

where $\mu_1$ is the primary membership and $M$ and $\sigma_x$ are the primary mean and spread of Gaussian MF, respectively; then,

$$\mu_2(x, \mu_1) = \exp\left(-0.5 \cdot \frac{(a-\mu_1(x))^2}{{\sigma_m}^2}\right) \quad (3)$$

where $\mu_2(x, \mu_1)$ is the secondary degree, $a \in [0, 1]$ is the domain of the secondary MF for each $x$, and $\sigma_m$ is the secondary spread of the Gaussian MF.

Simple and special kinds of general T2F sets change the same as the interval T2F one. Figure 2 depicts two interval T2F sets. A fuzzy set specified by a Gaussian MF by mean/width $m/[\sigma_1, \sigma_2]$ is demonstrated in Figure 2a. Two T2F sets are given in Figure 2. Figure 2b illustrates a fuzzy set with an MF of Gaussian form encompassing a distinct standard deviation of σ. However, the mean value is quite uncertain and adopts values in the interval of $[m_1, m_2]$.

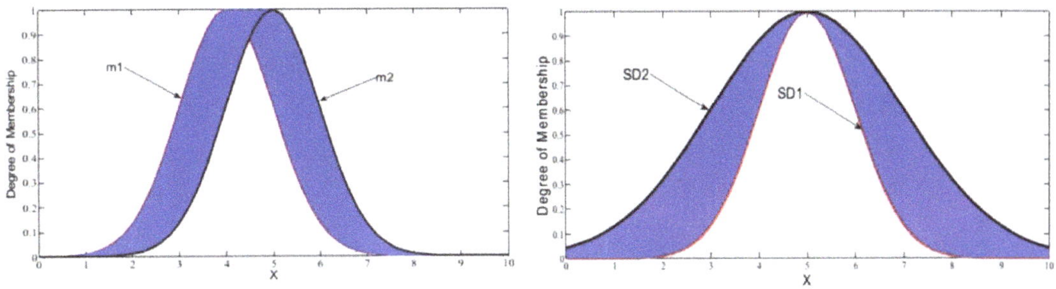


(**a**) (**b**)

**Figure 2.** (**a**) Uncertainty in width and (**b**) uncertainty in center.

An MF of Gaussian form with determined σ and uncertain $m$, as seen in Figure 2a, is applied through all of this paper.

*Type-2 Fuzzy Systems*

One may gain a certain number by defuzzifying a T1FS [59], whereas T2FS yields a T2F set. This is the reason one has to endeavor to succeed in the reduction in fuzzy set type from two to one in a process entitled "Type Reduction". The process is a challenging issue of high significance in T2F systems [60]. Figure 3 displays the structure of a T2F system.

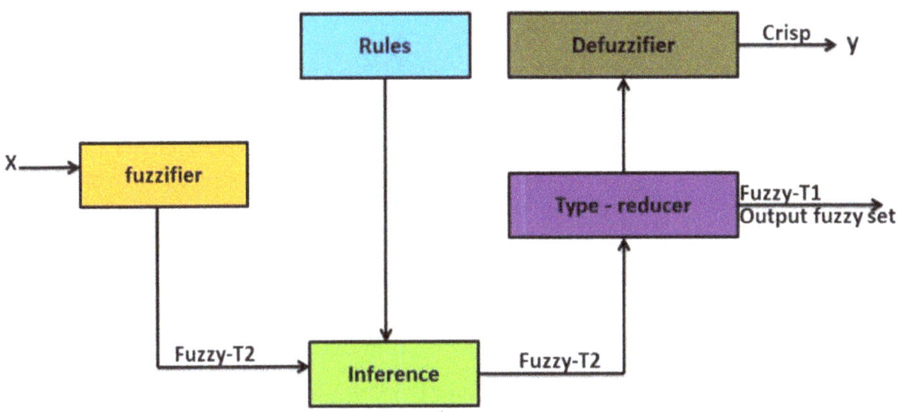


**Figure 3.** The structure of a T2F system.

As can be easily grasped through Figure 4, construction of the T2FS will be the same as the organization of type-1 if the "Type-Reduction" block is neglected.

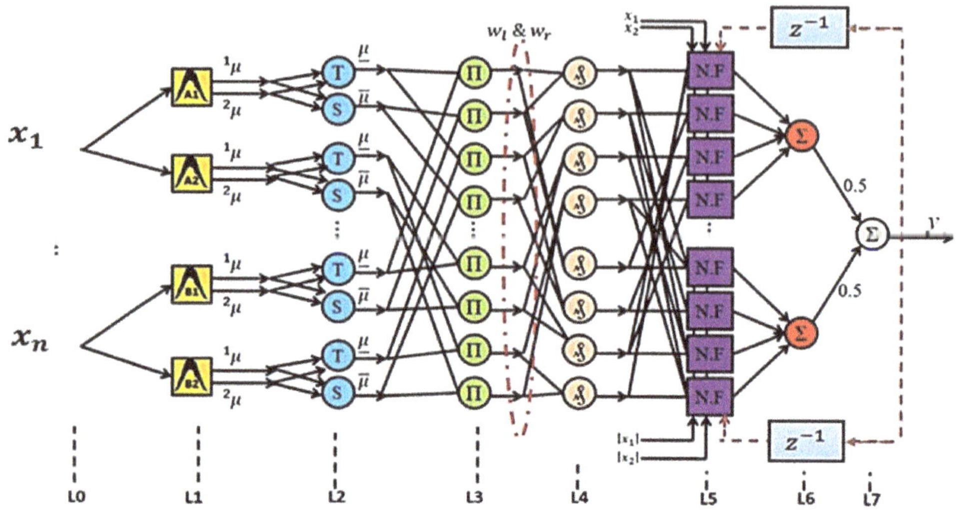


**Figure 4.** The structure of the proposed NCPRT2FS.

## 3. The Proposed NCPRT2FS

Section 3 tries to consolidate the nonlinear descendant or resultant part of recurrent T2F systems into a formula. Taking into account two informative and useful points that are mentioned later, the descriptive equation of (1) establishes the kth rule:

1) TSK-based T2FSs, usually yield a polynomial constructive of the inputs;
2) The outputs are represented by T1F sets [61].

This study recommends a novel NCPRT2FS, of which its total construction is illustrated in Figure 4. As one may see, the system clearly embodies seven layers. Generally speaking, the kth rule would be demonstrated in the following terms in a first-order T2FS with a TSK model by $M$ rules and $n$ inputs:

$$R^k: if\ x_1\ is\ \widetilde{A}_1^k\ and \ldots and\ x_n\ is\ \widetilde{A}_n^k\ then\ \widetilde{y}_k = C_{k,0} + C_{k,1}x_1 + \ldots + C_{k,n}x_n$$

where $k = 1, \ldots, M$ is the number of rules, $x_i (i = 1, \ldots, n)$ are inputs, and $\widetilde{y}_k$ is the output of the $kth$ rule. $\widetilde{y}_k$ is an interval T1F set and $\widetilde{A}_i^k$ are antecedent sets; $C_{k,i} \in [c_{k,i} - s_{k,i}, c_{k,i} + s_{k,i}]$ represent consequent sets, where $c_{k,i}$ represents the center of $C_{k,i}$ and $s_{k,i}$ is the spread of $C_{k,i}$.

In this paper, the nonlinear consequent part is taken into account. The resulting kth rule in NCPRT2FS, which has two antecedent variables and three outputs with delayed time shift ranging from one unit to three in the descendant part, is demonstrated in (2):

$$\begin{aligned} &R^k: if\ x_1\ is\ \widetilde{A}_1^k\ and\ x_2\ is\ \widetilde{A}_2^k\ then \\ \widetilde{y}_k = {} & C_{k,0} + C_{k,1}x_1 + C_{k,2}x_2 + C_{k,3}y(t-1) + C_{k,4}x_1x_2 + C_{k,5}x_1y(t-1) + C_{k,6}x_2y(t-1) \\ & + C_{k,7}x_1^2 + C_{k,8}x_2^2 + C_{k,9}y^2(t-1) + C_{k,10}x_1x_2y(t-1) \end{aligned} \tag{4}$$

One may make an extension to fuzzy rule (2) considering $n$ antecedent variables and time-delayed outputs in the descendant part with a delaying shift in time ranging from one unit to m units. $n$ may be designed remarking nonlinearity degree and complexity of the unknown system, which is going to be identified next.

The layers' details are as:

**Layer 0**: This layer denotes the inputs.

**Layer 1**: The outputs of fuzzification are written as:

$$ {}^{1}\mu_{k,i}\left(x_i, \left[\sigma_{k,i}, {}^{1}m_{k,i}\right]\right) = e^{-0.5\left(\frac{x_i - {}^{1}m_{k,i}}{\sigma_{k,i}}\right)^2} \quad (5) $$

$$ {}^{2}\mu_{k,i}\left(x_i, \left[\sigma_{k,i}, {}^{2}m_{k,i}\right]\right) = e^{-0.5\left(\frac{x_i - {}^{2}m_{k,i}}{\sigma_{k,i}}\right)^2} \quad (6) $$

where $m_{k,i} \in \left[{}^{1}m_{k,i}, {}^{2}m_{k,i}\right]$ and $\sigma_{k,i}$ are the uncertain mean and spread for $kth$ rule and $ith$ input.

**Layer 2**: The T-norm and S-norm are computed as:

$$ \underline{\mu}_{k,i}(x_i) = {}^{1}\mu_{k,i}(x_i).{}^{2}\mu_{k,i}(x_i), \quad k = 1,2,\ldots,M, \; i = 1,2,\ldots,n \quad (7) $$

$$ \overline{\mu}_{k,i}(x_i) = {}^{1}\mu_{k,i}(x_i) + {}^{2}\mu_{k,i}(x_i) - \underline{\mu}_{k,i}(x_i) \quad (8) $$

**Layer 3**: The rule firings ($\underline{f}^k$ and $\overline{f}^k$) are:

$$ \underline{f}^k = \prod_{i=1}^{n} \underline{\mu}_{k,i} \; ; \; \overline{f}^k = \prod_{i=1}^{n} \overline{\mu}_{k,i} \quad (9) $$

**Layer 4**: The left-most/right-most firing are obtained as:

$$ f_l^k = \frac{\overline{w}_l^k \overline{f}^k + \underline{w}_l^k \underline{f}^k}{\overline{w}_l^k + \underline{w}_l^k} \; ; \; f_r^k = \frac{\overline{w}_r^k \overline{f}^k + \underline{w}_r^k \underline{f}^k}{\overline{w}_r^k + \underline{w}_r^k} \quad (10) $$

where $w$ are adjustable weights.

**Layer 5**: The rule left/right firings are:

$$ \begin{aligned} y_l^k = {} & c_{k,0} + c_{k,1}x_1 + c_{k,2}x_2 + c_{k,3}y(t-1) + c_{k,4}x_1x_2 + c_{k,5}x_1y(t-1) \\ & + c_{k,6}x_2y(t-1) + c_{k,7}x_1^2 + c_{k,8}x_2^2 + c_{k,9}y^2(t-1) \\ & + c_{k,10}x_1x_2y(t-1) - s_{k,0} - s_{k,1}|x_1| - s_{k,2}|x_2| - s_{k,3}|y(t-1)| \\ & - s_{k,4}|x_1x_2| - s_{k,5}|x_1y(t-1)| - s_{k,6}|x_2y(t-1)| - s_{k,7}x_1^2 \\ & - s_{k,8}x_2^2 - s_{k,9}y^2(t-1) - s_{k,10}x_1x_2y(t-1) \end{aligned} \quad (11) $$

$$ \begin{aligned} y_r^k = {} & c_{k,0} + c_{k,1}x_1 + c_{k,2}x_2 + c_{k,3}y(t-1) + c_{k,4}x_1x_2 + c_{k,5}x_1y(t-1) \\ & + c_{k,6}x_2y(t-1) + c_{k,7}x_1^2 + c_{k,8}x_2^2 + c_{k,9}y^2(t-1) \\ & + c_{k,10}x_1x_2y(t-1) + s_{k,0} + s_{k,1}|x_1| + s_{k,2}|x_2| + s_{k,3}|y(t-1)| \\ & + s_{k,4}|x_1x_2| + s_{k,5}|x_1y(t-1)| + s_{k,6}|x_2y(t-1)| + s_{k,7}x_1^2 \\ & + s_{k,8}x_2^2 + s_{k,9}y^2(t-1) + s_{k,10}x_1x_2y(t-1) \end{aligned} \quad (12) $$

**Layer 6**: $\hat{y}_l$ and $\hat{y}_r$ are:

$$ \hat{y}_l = \frac{\sum_{k=1}^{M} f_l^k y_l^k}{\sum_{k=1}^{M} f_l^k} \quad (13) $$

$$ \hat{y}_r = \frac{\sum_{k=1}^{M} f_r^k y_r^k}{\sum_{k=1}^{M} f_r^k} \quad (14) $$

**Layer 7**: The output is:

$$ \hat{y} = \frac{\hat{y}_l + \hat{y}_r}{2} \quad (15) $$

In this article, structure learning is realized by exploiting T2F clustering. As one knows, an efficacious method is suggested to procreate fuzzy rules in real-time and decrease computations in antecedent part in structure optimization [62]. Structure learning appears as a great assistance in the simplification of T2FS, taking advantage of the reduction in fuzzy rules. Scrutinizing more, its duty is not only the production of novel membership but also pruning additional MFs and rules. In the input layer, a rule geometrically represents a cluster. Its firing degree could be taken into account as the degree of input data that belongs to a cluster. The center of the firing degree in NCPRT2FS is calculated by (16) since it is an interval.

$$f_k = \frac{\underline{f}^k + \overline{f}^k}{2} \tag{16}$$

Additionally, for generation of a new MF, find:

$$\mu_{\tilde{A}_i^k} = \frac{\underline{\mu}_{\tilde{A}_i^k} + \overline{\mu}_{\tilde{A}_i^k}}{2} \quad , \quad i = 1, 2, \ldots, n \tag{17}$$

For every incoming data $\vec{x} = \{x_1, \ldots, x_n\}$, calculate:

$$I = \arg \max_{1 \le k \le M(t)} f_k \tag{18}$$

For newly generated rules:

$$I_i = \arg \max_{1 \le k \le k_i(t)} \mu_{\tilde{A}_i^k} \quad , \quad i = 1, 2, \ldots, n \tag{19}$$

where $M(t)$ is the existing number of rules at time $t$. If $I \le \varnothing_{th}$, the system generates a new rule, where $\varnothing_{th} \in \ (0\ 1)$ is a threshold that is defined [63]. If $I_i > \rho$, where $\rho \in \ [0\ 1]$ is a previously defined threshold, then use the existing fuzzy set $\tilde{A}_i^{I_i}$ as the antecedent part of the new rule in input variable $i$. Otherwise, one could produce a novel MF in input variable $i$ and hold the equation, $k_i(t+1) \ = k_i(t) + 1$, true. The number of MFs is defined by the parameter $\rho$ in each input variable. Fuzzy clustering is a technique to structure a fuzzy model [64]. A new T2F clustering technique, which is a development of Krishnapuram and Keller Possibilistic C-Mean (PCM) [65], is suggested and described by:

$$J_m(x, \tilde{\mu}, c) = min \left[ \sum_{i=1}^{c} \sum_{j=1}^{N} \tilde{\mu}_{ij}^m D_{ij} + \sum_{i=1}^{c} \eta_i \sum_{j=1}^{N} (1 - \tilde{\mu}_{ij})^m \right] \tag{20}$$

$$S.T: \begin{cases} 0 < \sum_{j=1}^{N} \tilde{\mu}_{ij} < N \\ \tilde{\mu}_{ij} \in [0, 1] \quad \forall i, j \\ \max \tilde{\mu}_{ij} > 0 \quad \forall j \end{cases} \tag{21}$$

where $\tilde{\mu}_{ij}$ is type-2 MF in the $j^{th}$ data for the $i^{th}$ cluster. Moreover, the symbols $D_{ij}$, c, and N are the Euclidean distance of the $j^{th}$ data in the $i^{th}$ cluster center, clusters and input data numbers, respectively. $\eta_i$ is also a positive number. $D_{ij}$ has to be as small as possible as the first term. On the other hand, the memberships in a cluster have to be greater as much as possible. They have to stay in the interval of [0, 1] and their sum is confined to become smaller than the number of input data. Equation (21) appears as the descriptive term. That $\eta_i$ corresponds to ith cluster, and is of the order of $D_{ij}$, is greatly welcomed [65]. The distance to the cluster's center must be as low as possible (first term). It is desirable that $\eta_i$ relate to $i^{th}$ cluster and be of the order of $D_{ij}$ [63].

$$\eta_i = \frac{\sum_{j=1}^{N} \tilde{\mu}_{ij}^m D_{ij}}{\sum_{j=1}^{N} \tilde{\mu}_{ij}^m} \quad \forall i = 1, \ldots, c$$

Using (20), the optimal values of the centers of the clusters are achieved. The initial $m_{k,i}$ and $\sigma_{k,i}$ for the $k_i(t+1)$ *th* interval T2F set are:

$$m_{k,i} \in \left[v_i - 0.1v_i,\ v_i + 0.1v_i\right]$$

$$\sigma_{k_i(t+1)i} = \beta \left| v_i - \frac{{}^1m_{I_i,i} + {}^2m_{I_i,i}}{2} \right|$$

where $v_i$ is the optimal value of the cluster's center; $\beta > 0$ denotes the degree of overlap between 2 fuzzy sets. In this study, $\beta$ is considered to be 0.5 [61]. The parameters of the consequent part are initialized as:

$$\left[c_{k,0} - s_{k,0}, c_{k,0} + s_{k,0}\right] = \left[yd - 0.1,\ yd + 0.1\right] \quad , \quad k = 1, 2, \ldots, M \tag{22}$$

where $yd$ is the target signal for input $\vec{x} = \{x_1, \ldots, x_n\}$. All the other consequent parameters are zero.

By repeating the above process for each training dataset, new rules are created one after the other until NCPRT2FS is finally complete. The network output is calculated for each input applied. The calculated output is then compared to the target to obtain an error. Assume that the input–output data pair $\{(x_p : t_p)\}\ \forall p = 1, \ldots, q$, where $p$ represents the data numbers and *x/t* is the input/output, respectively. The NCPRT2FS output error can be expressed as follows:

$$e_p = t_p - \hat{y}_p, \tag{23}$$

$$E_p = \frac{1}{2}e_p^2 = \frac{1}{2}\left(t_p - \hat{y}_p\right)^2 \tag{24}$$

$$E = \sum_{p=1}^{q} E_p \tag{25}$$

The gradient-based learning algorithm is used for updating the parameters. The mathematical relation of the gradient-based update algorithm is as follows.

$$W_{new} = W_{old} - \eta \frac{\partial E}{\partial W}$$

See Appendix A for more details on the parameter update formulation. We choose the initial $\eta$ as:

$$\eta = \frac{1}{max\left|\frac{\partial \hat{y}(k)}{\partial W}\right|^2}$$

After all the data have been applied, the variable learning rate is determined by the following form.

$$\begin{cases} if \quad \frac{RMSE\ (l)}{RMSE\ (l-1)} < 1 & \rightarrow \quad \eta(l) = \eta(l-1) \\ if \quad \frac{RMSE\ (l)}{RMSE\ (l-1)} \geq 1 & \rightarrow \quad \eta(l) = 0.9 \times \eta(l-1) \end{cases}$$

where $l$ is the number of iterations. The *RMSE* formula is as follows:

$$RMSE = \sqrt{\frac{1}{N}\sum_{p=1}^{N}\left(t_p - \hat{y}_p\right)^2}$$

where $t_p$ and $\hat{y}_p$ are actual and model (NCPRT2FS) outputs at $p$ moment, respectively. The total number of data is denoted by $N$.

## 4. Simulation Results

Two real renewable energy systems are used for identification. The structure is shown in Figure 5.

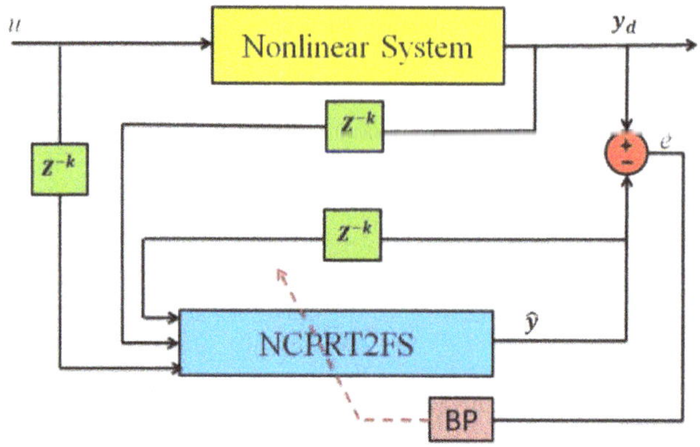


**Figure 5.** The structure of the system and the NCPRT2FS-based identifier.

The inputs to the NCPRT2FS-based identifier are the main input and delayed system output. The parameters of the NCPRT2FS structure should be adjusted to minimize plant output yd and identification yield $\hat{y}$ for all input values of x.

**Example 1:** Real data of a 660kw wind turbine (see Figure 6) have been taken from the Iran Renewable Energy Organization (SUNA) (http://www.suna.org.ir/en/home/ 1 March 2021). The model of the wind turbine is S47-660kw, made by VESTAS (Denmark), and information is given in Table 1.

**Figure 6.** Manjil and Rudbar Wind Farm.

In this example, $u(k)$, $k = 1, \ldots, 365$ is wind speed that is fed to the wind turbine system and obtains the 365 samples of $y(k)$, which is the output power of the wind turbine. The other conditions are the same as example 1. Figure 7 exhibits the identification performance of the NCPRT2FS. Here, the output (solid line) and the NCPRT2FS identifier output (dashed line) are shown.

**Table 1.** Information for Example 1.

| Cut-in wind speed: | | 4 m/s | |
|---|---|---|---|
| Survival wind speed: | | 60 m/s | |
| Rated wind speed: | | 15 m/s | |
| Cut-out wind speed: | | 25 m/s | |
| **Rotor:** | | **Generator:** | |
| Number of blades: | 3 | Type: | Asynchronous |
| Swept area: | 1.735 $m^2$ | Number: | 1.0 |
| Type: | 22.90 | Grid connection: | Thyristor |
| Rotor speed, max: | 28.50 U/min | | |
| Tipspeed: | 70.10 m/s | Voltage: | 400 V |
| Diameter: | 47 m | Speed, max: | 1.650 U/min |
| Material: | GFK | Grid frequency: | 50 Hz |

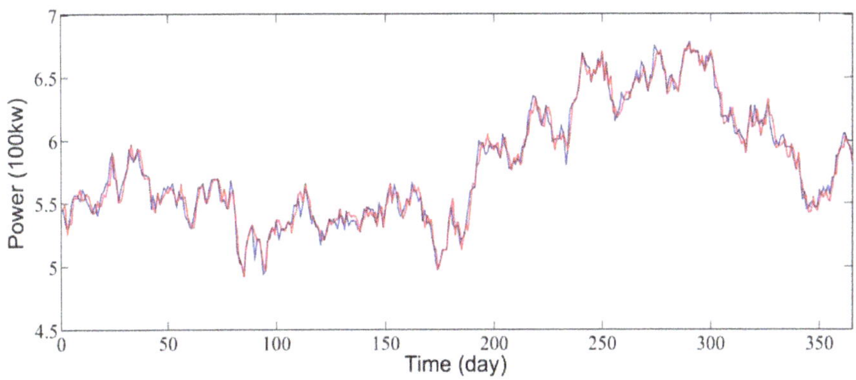


**Figure 7.** Identification performance of the NCPRT2FS for wind turbine.

The trained NCPRT2FS is used to calculate wind power in a place called Ilam (A city in the west of the Islamic Republic of Iran). Figure 8 shows the wind speed of Ilam for a year. Figure 9 shows the predicted wind power in Ilam.

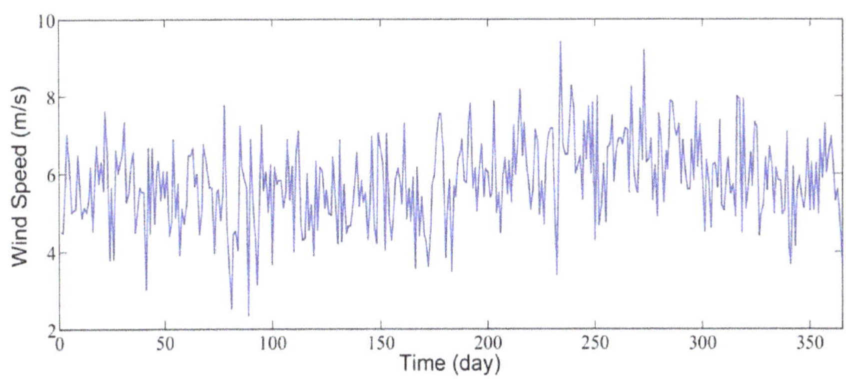


**Figure 8.** Wind speed of a place in Ilam for a year.

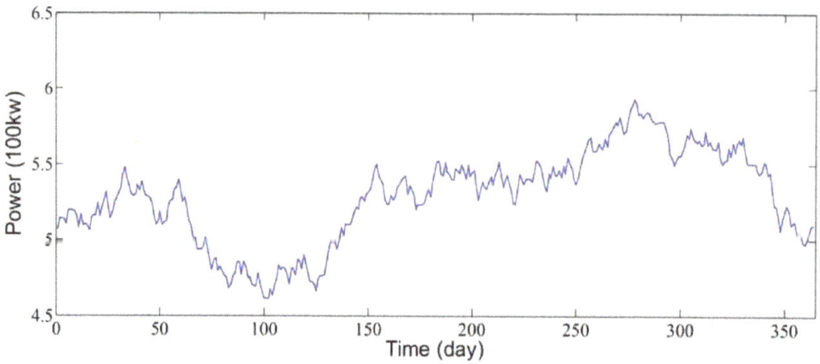


**Figure 9.** Predicted wind power of a place in Ilam for a year.

The final values of the parameters of NCPRT2FS are shown in Table 2.

**Table 2.** The final values of NCPRT2FS parameters.

| | | | $^1m_{ij}$ | | $^2m_{ij}$ | | $\sigma_{ij}$ | |
|---|---|---|---|---|---|---|---|---|
| Antecedent parameters | **u(k)** | | $^1m_{11} = 3.62$<br>$^1m_{21} = 6.13$<br>$^1m_{31} = 8.19$ | | $^2m_{11} = 4.32$<br>$^2m_{21} = 7.02$<br>$^2m_{31} = 9.51$ | | $\sigma_{11} = 0.38$<br>$\sigma_{21} = 1.10$<br>$\sigma_{31} = 0.89$ | |
| | **y(k-1)** | | $^1m_{12} = 4.93$<br>$^1m_{22} = 5.34$<br>$^1m_{32} = 5.81$<br>$^1m_{42} = 6.11$ | | $^2m_{12} = 5.12$<br>$^2m_{22} = 5.66$<br>$^2m_{32} = 5.98$<br>$^2m_{42} = 6.48$ | | $\sigma_{12} = 0.21$<br>$\sigma_{22} = 0.09$<br>$\sigma_{32} = 0.36$<br>$\sigma_{42} = 0.18$ | |
| fourth layer adaptive weights | $\overline{w}_r^1 = 1.92$ | | $\underline{w}_r^1 = 1.50$ | | $\overline{w}_l^1 = 1.00$ | | $\underline{w}_l^1 = 0.63$ | |
| | $\overline{w}_r^2 = 1.66$ | | $\underline{w}_r^2 = 0.92$ | | $\overline{w}_l^2 = 0.71$ | | $\underline{w}_l^2 = 0.06$ | |
| | $\overline{w}_r^3 = 0.80$ | | $\underline{w}_r^3 = 0.70$ | | $\overline{w}_l^3 = 0.56$ | | $\underline{w}_l^3 = 0.43$ | |
| | $\overline{w}_r^4 = 1.87$ | | $\underline{w}_r^4 = 0.94$ | | $\overline{w}_l^4 = 0.85$ | | $\underline{w}_l^4 = 0.77$ | |
| consequent parameters | **Rule 1** | **Rule 2** | **Rule 3** | **Rule 4** | **Rule 1** | **Rule 2** | **Rule 3** | **Rule 4** |
| | $s_{1,0} = 0.40$ | $s_{2,0} = 0.33$ | $s_{3,0} = 0.27$ | $s_{4,0} = 0.52$ | $c_{1,0} = 1.00$ | $c_{2,0} = 1.40$ | $c_{3,0} = 1.00$ | $c_{4,0} = 1.40$ |
| | $s_{1,1} = 0.55$ | $s_{2,1} = 0.39$ | $s_{3,1} = 0.48$ | $s_{4,1} = 0.43$ | $c_{1,1} = 1.10$ | $c_{2,1} = 1.00$ | $c_{3,1} = 1.00$ | $c_{4,1} = 1.00$ |
| | $s_{1,2} = 1.00$ | $s_{2,2} = 1.00$ | $s_{3,2} = 1.00$ | $s_{4,2} = 1.00$ | $c_{1,2} = 1.00$ | $c_{2,2} = 1.32$ | $c_{3,2} = 0.81$ | $c_{4,2} = 0.93$ |
| | $s_{1,3} = 0.43$ | $s_{2,3} = 0.39$ | $s_{3,3} = 0.65$ | $s_{4,3} = 0.90$ | $c_{1,3} = 1.00$ | $c_{2,3} = 1.00$ | $c_{3,3} = 1.65$ | $c_{4,3} = 1.82$ |
| | $s_{1,4} = 0.62$ | $s_{2,4} = 1.00$ | $s_{3,4} = 1.00$ | $s_{4,4} = 1.00$ | $c_{1,4} = 1.00$ | $c_{2,4} = 1.09$ | $c_{3,4} = 1.00$ | $c_{4,4} = 1.00$ |
| | $s_{1,5} = 0.87$ | $s_{2,5} = 0.10$ | $s_{3,5} = 1.00$ | $s_{4,5} = 1.00$ | $c_{1,5} = 1.10$ | $c_{2,5} = 1.00$ | $c_{3,5} = 1.55$ | $c_{4,5} = 1.90$ |
| | $s_{1,6} = 1.00$ | $s_{2,6} = 1.00$ | $s_{3,6} = 1.00$ | $s_{4,6} = 1.00$ | $c_{1,6} = 1.00$ | $c_{2,6} = 1.00$ | $c_{3,6} = 1.00$ | $c_{4,6} = 1.00$ |
| | $s_{1,7} = 0.69$ | $s_{2,7} = 0.66$ | $s_{3,7} = 0.31$ | $s_{4,7} = 0.06$ | $c_{1,7} = 0.80$ | $c_{2,7} = 0.72$ | $c_{3,7} = 0.67$ | $c_{4,7} = 0.81$ |
| | $s_{1,8} = 0.96$ | $s_{2,8} = 0.11$ | $s_{3,8} = 0.54$ | $s_{4,8} = 0.21$ | $c_{1,8} = 1.10$ | $c_{2,8} = 1.00$ | $c_{3,8} = 0.92$ | $c_{4,8} = 0.59$ |
| | $s_{1,9} = 0.30$ | $s_{2,9} = 0.32$ | $s_{3,9} = 0.36$ | $s_{4,9} = 0.98$ | $c_{1,9} = 0.95$ | $c_{2,9} = 0.77$ | $c_{3,9} = 1.00$ | $c_{4,9} = 1.00$ |
| | $s_{1,10} = 0.35$ | $s_{2,10} = 0.31$ | $s_{3,10} = 0.54$ | $s_{4,10} = 0.50$ | $c_{1,10} = 1.00$ | $c_{2,10} = 0.44$ | $c_{3,10} = 0.64$ | $c_{4,10} = 0.89$ |

**Example 2**: A real solar cell system is shown in Figure 10.

In this example, $u(k)$, $k = 1, \ldots, 600$ is solar radiation that is fed to the real solar cell system and 600 samples of $y(k)$ are obtained. The other conditions are the same as in examples 1 and 2. Figure 11 shows the identification performance of the NCPRT2FS for three solar radiations. Here, the plant output (solid line) and the NCPRT2FS identifier output (dashed line) are shown.

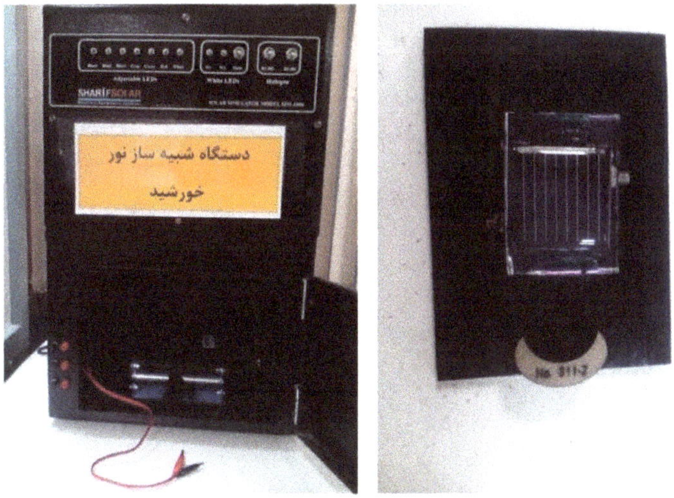


(**a**) (**b**)

**Figure 10.** Experimental solar cell testing system (**a**) and a solar cell (**b**).

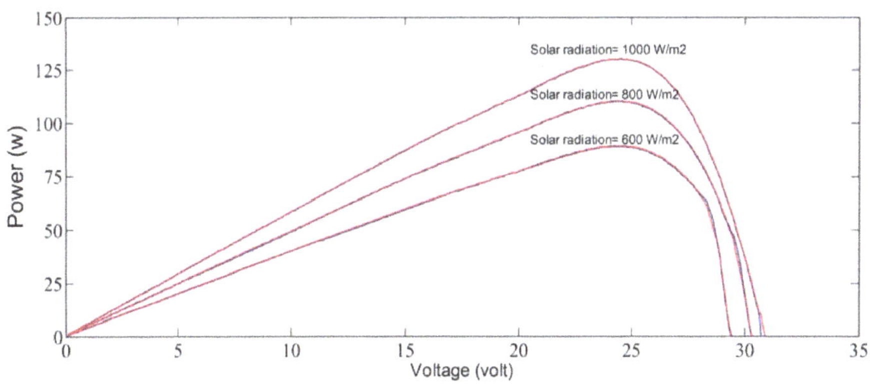


**Figure 11.** Identification results of the NCPRT2FS for solar cell.

After structure learning, for NCPRT2FS, three rules are generated and the RMSE value for the NCPRT2FS and IT2-TSK-FNN for the training and test are shown in Table 3. The final parameters are given in Table 3.

**Table 3.** The final values of NCPRT2FS parameters.

| | | $^1m_{ij}$ | $^2m_{ij}$ | $\sigma_{ij}$ |
|---|---|---|---|---|
| Antecedent parameters | **u(k)** | $^1m_{11} = 251$<br>$^1m_{21} = 598$<br>$^1m_{31} = 798$ | $^2m_{11} = 332$<br>$^2m_{21} = 615$<br>$^2m_{31} = 949$ | $\sigma_{11} = 43$<br>$\sigma_{21} = 12$<br>$\sigma_{31} = 211$ |
| | **y(k-1)** | $^1m_{12} = 69$<br>$^1m_{22} = 82$<br>$^1m_{32} = 93$ | $^2m_{12} = 75$<br>$^2m_{22} = 89$<br>$^2m_{32} = 97$ | $\sigma_{12} = 11$<br>$\sigma_{22} = 5$<br>$\sigma_{32} = 3$ |
| fourth layer adaptive weights | $\overline{w}_r^1 = 0.20$ | $\underline{w}_r^1 = 0.06$ | $\overline{w}_l^1 = 0.12$ | $\underline{w}_l^1 = 0.09$ |
| | $\overline{w}_r^2 = 1.80$ | $\underline{w}_r^2 = 1.00$ | $\overline{w}_l^2 = 1.42$ | $\underline{w}_l^2 = 0.98$ |
| | $\overline{w}_r^3 = 0.57$ | $\underline{w}_r^3 = 0.21$ | $\overline{w}_l^3 = 1.93$ | $\underline{w}_l^3 = 1.10$ |

**Table 3.** *Cont.*

| | Rule 1 | Rule 2 | Rule 3 | Rule 1 | Rule 2 | Rule 3 |
|---|---|---|---|---|---|---|
| consequent parameters | $s_{1,0} = 0.10$ | $s_{2,0} = 0.84$ | $s_{3,0} = 1.00$ | $c_{1,0} = 0.56$ | $c_{2,0} = 1.00$ | $c_{3,0} = 1.22$ |
| | $s_{1,1} = 0.32$ | $s_{2,1} = 0.39$ | $s_{3,1} = 0.37$ | $c_{1,1} = 0.94$ | $c_{2,1} = 1.60$ | $c_{3,1} = 1.00$ |
| | $s_{1,2} = 1.00$ | $s_{2,2} = 1.00$ | $s_{3,2} = 0.61$ | $c_{1,2} = 1.00$ | $c_{2,2} = 1.00$ | $c_{3,2} = 1.00$ |
| | $s_{1,3} = 0.22$ | $s_{2,3} = 1.20$ | $s_{3,3} = 0.50$ | $c_{1,3} = 1.00$ | $c_{2,3} = 1.77$ | $c_{3,3} = 1.20$ |
| | $s_{1,4} = 0.10$ | $s_{2,4} = 0.42$ | $s_{3,4} = 1.00$ | $c_{1,4} = 1.61$ | $c_{2,4} = 0.60$ | $c_{3,4} = 1.63$ |
| | $s_{1,5} = 0.47$ | $s_{2,5} = 1.00$ | $s_{3,5} = 1.00$ | $c_{1,5} = 1.30$ | $c_{2,5} = 1.00$ | $c_{3,5} = 2.00$ |
| | $s_{1,6} = 0.10$ | $s_{2,6} = 1.00$ | $s_{3,6} = 1.00$ | $c_{1,6} = 1.00$ | $c_{2,6} = 1.11$ | $c_{3,6} = 1.00$ |
| | $s_{1,7} = 1.20$ | $s_{2,7} = 1.00$ | $s_{3,7} = 0.19$ | $c_{1,7} = 1.10$ | $c_{2,7} = 1.50$ | $c_{3,7} = 0.88$ |
| | $s_{1,8} = 1.00$ | $s_{2,8} = 0.36$ | $s_{3,8} = 0.69$ | $c_{1,8} = 1.60$ | $c_{2,8} = 0.89$ | $c_{3,8} = 0.91$ |
| | $s_{1,9} = 1.00$ | $s_{2,9} = 0.28$ | $s_{3,9} = 0.11$ | $c_{1,9} = 1.53$ | $c_{2,9} = 0.95$ | $c_{3,9} = 0.48$ |
| | $s_{1,10} = 0.55$ | $s_{2,10} = 0.35$ | $s_{3,10} = 0.50$ | $c_{1,10} = 0.88$ | $c_{2,10} = 1.00$ | $c_{3,10} = 1.00$ |

The trained NCPRT2FS is used to calculate the solar power of Ilam. Figure 12 shows the solar radiation of Ilam for a year. Figure 13 shows the predicted solar power in Ilam.

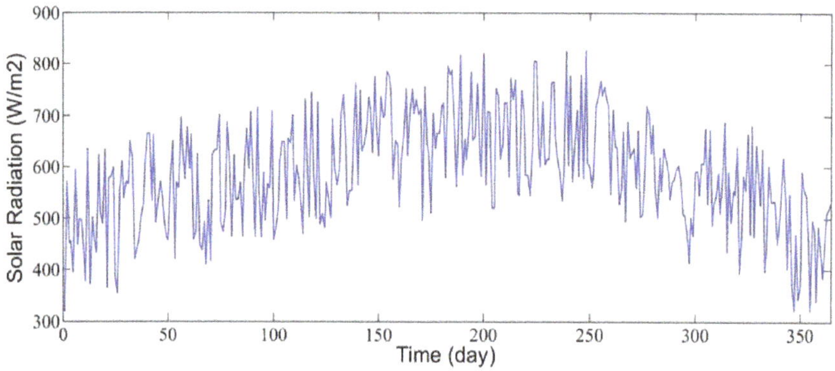


**Figure 12.** Solar radiation of Ilam.

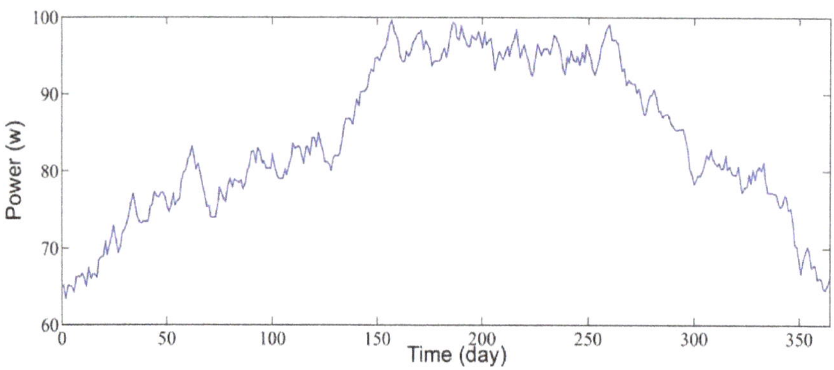


**Figure 13.** Predicted solar power in Ilam for a year.

Table 4. presents the comparison of our proposed method with another method (method of [46]).

**Table 4.** Comparison between results of the proposed method and the method of [46].

| Example | Method of [46] | | | | Proposed NCPRT2FS | | | |
|---|---|---|---|---|---|---|---|---|
| | Rules | Epochs | Run Time (s) | RMSE | Rules | Epochs | Run Time (s) | RMSE |
| 1 | 4 | 34 | 4 | 0.0159 | 4 | 31 | 6 | 0.0057 |
| 2 | 5 | 27 | 4 | 0.00759 | 3 | 39 | 7 | 0.0013 |

Simulations verify that the presented NCPRT2FS has high performances in function approximation and system identification. Table 4 shows that the number of rules of the proposed NCPRT2FS is almost less than the method of [53]; accuracy of identification is better than [53], but the training time in 10 runs (MATLAB 2011a; Dual CPU T3200 @ 2.00; RAM: 2.00 GB; GHz 2.00 GHz) is more than [53]. The references [23,46] presented two different T2F neural structures. They have also been used and evaluated only to identify some theory systems. In the present paper, however, the T2F neural network structure is different from references [23] and [53] and several experimental energy systems have been used for modeling.

## 5. Conclusions

In this paper, a novel Nonlinear Consequent Part Recurrent T2FS (NCPRT2FS) for identification and prediction of renewable energy systems was proposed. The nonlinear consequent part helps to better model highly nonlinear systems. Recurrent structure is a useful choice for the identification of dynamical systems. The self-evolving structure helps to obtain a simpler structure of the NCPRT2FS by ending up with a minimum number of fuzzy sets and fuzzy rules in the end. Simulations showed that the NCPRT2FS based on the backpropagation algorithm and adaptive optimization rate performs better than IT2-TSK-FNN [53] in identification. An S47-660 kw wind turbine (VESTAS company Denmark) and a solar cell were selected as case studies. After data gathering, the proposed method was finally used with the experimental data for the purpose of identification. The RMSE was less than 0.006 and the number of fuzzy rules was equal and less than 4 rules; therefore, the results easily demonstrated the remarkable capability of the NCPRT2FS developed in the paper. In order to continue the work and look to the future, we can use the evolutionary algorithms as a complement to the proposed method for the development of the fuzzy neural network (to increase accuracy, increase convergence, etc.). Different case studies (types of solar cells, types of wind turbines, etc.) should be identified and the appropriate renewable system can be extracted for each geographical location.

**Author Contributions:** Writing—Original draft preparation, J.T., A.A.S., M.B.M., A.M. (Ardashir Mohammadzadeh), E.R., A.M. (Amir Mosavi); Software, J.T., A.A.S., M.B.M., A.M. (Ardashir Mohammadzadeh), E.R., A.M. (Amir Mosavi); Validation, J.T., A.A.S., M.B.M., A.M. (Ardashir Mohammadzadeh), E.R., A.M. (Amir Mosavi); Supervision, J.T., A.M. (Amir Mosavi). All authors have read and agreed to the published version of the manuscript.

**Funding:** This research received Open Access Funding by the Publication Fund of the TU Dresden.

**Institutional Review Board Statement:** Not applicable.

**Informed Consent Statement:** Not applicable.

**Data Availability Statement:** The data are available from via, (j.tavoosi@ilam.ac.ir), (accessed on 1 March 2021), upon reasonable request.

**Conflicts of Interest:** The authors declare that they have no conflict of interest.

## Appendix A

To update the consequent part parameters, Equations (A1)–(A20) are used.

$$^{new}c_{k,0} = {}^{old}c_{k,0} + \eta{\cdot}0.5{\cdot}e_p{\cdot}\left[\frac{f_l^k}{\sum_{k=1}^M f_l^k} + \frac{f_r^k}{\sum_{k=1}^M f_r^k}\right] \tag{A1}$$

$$^{new}c_{k,i} = {}^{old}c_{k,i} + \eta{\cdot}0.5{\cdot}e_p{\cdot}\left[\frac{f_l^k}{\sum_{k=1}^M f_l^k} + \frac{f_r^k}{\sum_{k=1}^M f_r^k}\right]\cdot x_i \quad i = 1,2 \tag{A2}$$

$$^{new}c_{k,3} = {}^{old}c_{k,3} + \eta{\cdot}0.5{\cdot}e_p\cdot\left[\frac{f_l^k}{\sum_{k=1}^M f_l^k} + \frac{f_r^k}{\sum_{k=1}^M f_r^k}\right]\cdot y(t-1) \tag{A3}$$

$$^{new}c_{k,4} = {}^{old}c_{k,4} + \eta{\cdot}0.5{\cdot}e_p\cdot\left[\frac{f_l^k}{\sum_{k=1}^M f_l^k} + \frac{f_r^k}{\sum_{k=1}^M f_r^k}\right]\cdot x_1{\cdot}x_2 \tag{A4}$$

$$^{new}c_{k,5} = {}^{old}c_{k,5} + \eta{\cdot}0.5{\cdot}e_p\cdot\left[\frac{f_l^k}{\sum_{k=1}^M f_l^k} + \frac{f_r^k}{\sum_{k=1}^M f_r^k}\right]\cdot x_1{\cdot}y(t-1) \tag{A5}$$

$$^{new}c_{k,6} = {}^{old}c_{k,6} + \eta{\cdot}0.5{\cdot}e_p\cdot\left[\frac{f_l^k}{\sum_{k=1}^M f_l^k} + \frac{f_r^k}{\sum_{k=1}^M f_r^k}\right]\cdot x_2{\cdot}y(t-1) \tag{A6}$$

$$^{new}c_{k,7} = {}^{old}c_{k,7} + \eta{\cdot}0.5{\cdot}e_p\cdot\left[\frac{f_l^k}{\sum_{k=1}^M f_l^k} + \frac{f_r^k}{\sum_{k=1}^M f_r^k}\right]\cdot x_1^2 \tag{A7}$$

$$^{new}c_{k,8} = {}^{old}c_{k,8} + \eta{\cdot}0.5{\cdot}e_p\cdot\left[\frac{f_l^k}{\sum_{k=1}^M f_l^k} + \frac{f_r^k}{\sum_{k=1}^M f_r^k}\right]\cdot x_2^2 \tag{A8}$$

$$^{new}c_{k,9} = {}^{old}c_{k,9} + \eta{\cdot}0.5{\cdot}e_p\cdot\left[\frac{f_l^k}{\sum_{k=1}^M f_l^k} + \frac{f_r^k}{\sum_{k=1}^M f_r^k}\right]\cdot y^2(t-1) \tag{A9}$$

$$^{new}c_{k,10} = {}^{old}c_{k,10} + \eta{\cdot}0.5{\cdot}e_p\cdot\left[\frac{f_l^k}{\sum_{k=1}^M f_l^k} + \frac{f_r^k}{\sum_{k=1}^M f_r^k}\right]\cdot x_1{\cdot}x_2{\cdot}y(t-1) \tag{A10}$$

$$^{new}s_{k,0} = {}^{old}s_{k,0} + \eta{\cdot}0.5{\cdot}e_p\cdot\left[\frac{f_l^k}{\sum_{k=1}^M f_l^k} - \frac{f_r^k}{\sum_{k=1}^M f_r^k}\right] \tag{A11}$$

$$^{new}s_{k,i} = {}^{old}s_{k,i} + \eta{\cdot}0.5{\cdot}e_p\cdot\left[\frac{f_l^k}{\sum_{k=1}^M f_l^k} - \frac{f_r^k}{\sum_{k=1}^M f_r^k}\right]\cdot |x_i| \quad i = 1,2 \tag{A12}$$

$$^{new}s_{k,3} = {}^{old}s_{k,3} + \eta{\cdot}0.5{\cdot}e_p\cdot\left[\frac{f_l^k}{\sum_{k=1}^M f_l^k} - \frac{f_r^k}{\sum_{k=1}^M f_r^k}\right]\cdot |y(t-1)| \tag{A13}$$

$$^{new}s_{k,4} = {}^{old}s_{k,4} + \eta{\cdot}0.5{\cdot}e_p\cdot\left[\frac{f_l^k}{\sum_{k=1}^M f_l^k} - \frac{f_r^k}{\sum_{k=1}^M f_r^k}\right]\cdot |x_1 x_2| \tag{A14}$$

$$^{new}s_{k,5} = {}^{old}s_{k,5} + \eta{\cdot}0.5{\cdot}e_p\cdot\left[\frac{f_l^k}{\sum_{k=1}^M f_l^k} - \frac{f_r^k}{\sum_{k=1}^M f_r^k}\right]\cdot |x_1{\cdot}y(t-1)| \tag{A15}$$

$$^{new}s_{k,6} = {}^{old}s_{k,6} + \eta{\cdot}0.5{\cdot}e_p\cdot\left[\frac{f_l^k}{\sum_{k=1}^M f_l^k} - \frac{f_r^k}{\sum_{k=1}^M f_r^k}\right]\cdot |x_2{\cdot}y(t-1)| \tag{A16}$$

$$^{new}s_{k,7} = {}^{old}s_{k,7} + \eta{\cdot}0.5{\cdot}e_p\cdot\left[\frac{f_l^k}{\sum_{k=1}^M f_l^k} - \frac{f_r^k}{\sum_{k=1}^M f_r^k}\right]\cdot x_1^2 \tag{A17}$$

$$ {}^{new}s_{k,8} = {}^{old}s_{k,8} + \eta \cdot 0.5 \cdot e_p \cdot \left[ \frac{f_l^k}{\sum_{k=1}^{M} f_l^k} - \frac{f_r^k}{\sum_{k=1}^{M} f_r^k} \right] \cdot x_2^2 \tag{A18} $$

$$ {}^{new}s_{k,9} = {}^{old}s_{k,9} + \eta \cdot 0.5 \cdot e_p \cdot \left[ \frac{f_l^k}{\sum_{k=1}^{M} f_l^k} - \frac{f_r^k}{\sum_{k=1}^{M} f_r^k} \right] \cdot y^2(t-1) \tag{A19} $$

$$ {}^{new}s_{k,10} = {}^{old}s_{k,10} + \eta \cdot 0.5 \cdot e_p \cdot \left[ \frac{f_l^k}{\sum_{k=1}^{M} f_l^k} - \frac{f_r^k}{\sum_{k=1}^{M} f_r^k} \right] \cdot |x_1 \cdot x_2 \cdot y(t-1)| \tag{A20} $$

The learning rate is indicated by $\eta$.

To update the left and right weights, Equations (A21)–(A24) are used.

$$ {}^{new}\underline{w}_l^k = {}^{old}\underline{w}_l^k + \eta \cdot 0.5 \cdot e_p \cdot \frac{y_l^k - \hat{y}_l}{\sum_{j=1}^{M} f_l^j} \cdot \frac{\underline{f}^k - f_l^k}{\overline{w}_l^k + \underline{w}_l^k} \tag{A21} $$

$$ {}^{new}\overline{w}_l^k = {}^{old}\overline{w}_l^k + \eta \cdot 0.5 \cdot e_p \cdot \frac{y_l^k - \hat{y}_l}{\sum_{j=1}^{M} f_l^j} \cdot \frac{\overline{f}^k - f_l^k}{\overline{w}_l^k + \underline{w}_l^k} \tag{A22} $$

$$ {}^{new}\underline{w}_r^k = {}^{old}\underline{w}_r^k + \eta \cdot 0.5 \cdot e_p \cdot \frac{y_r^k - \hat{y}_r}{\sum_{j=1}^{M} f_r^j} \cdot \frac{\underline{f}^k - f_r^k}{\overline{w}_r^k + \underline{w}_r^k} \tag{A23} $$

$$ {}^{new}\overline{w}_r^k = {}^{old}\overline{w}_r^k + \eta \cdot 0.5 \cdot e_p \cdot \frac{y_r^k - \hat{y}_r}{\sum_{j=1}^{M} f_r^j} \cdot \frac{\overline{f}^k - f_r^k}{\overline{w}_r^k + \underline{w}_r^k} \tag{A24} $$

Finally, the equations for updating the antecedent parameters can be described as follows:

$$ {}^{1}m_{k,i}^{new} = {}^{1}m_{k,i}^{old} + \eta \cdot 0.5 \cdot e_p \cdot \left[ \frac{y_l^k - \hat{y}_l}{\sum_{j=1}^{M} f_l^j} \cdot \frac{\partial f_l^k}{\partial^1 m_{k,i}} + \frac{y_r^k - \hat{y}_r}{\sum_{j=1}^{M} f_r^j} \cdot \frac{\partial f_r^k}{\partial^1 m_{k,i}} \right] \tag{A25} $$

$$ {}^{2}m_{k,i}^{new} = {}^{2}m_{k,i}^{old} + \eta \cdot 0.5 \cdot e_p \cdot \left[ \frac{y_l^k - \hat{y}_l}{\sum_{j=1}^{M} f_l^j} \cdot \frac{\partial f_l^k}{\partial^2 m_{k,i}} + \frac{y_r^k - \hat{y}_r}{\sum_{j=1}^{M} f_r^j} \cdot \frac{\partial f_r^k}{\partial^2 m_{k,i}} \right] \tag{A26} $$

$$ \sigma_{k \cdot i}^{new} = \sigma_{k \cdot i}^{old} + \eta \cdot 0.5 \cdot e_p \cdot \left[ \frac{y_l^k - \hat{y}_l}{\sum_{j=1}^{M} f_l^j} \cdot \frac{\partial f_l^k}{\partial \sigma_{k,i}} + \frac{y_r^k - \hat{y}_r}{\sum_{j=1}^{M} f_r^j} \cdot \frac{\partial f_r^k}{\partial \sigma_{k,i}} \right] \tag{A27} $$

where

$$ \frac{\partial f_l^k}{\partial^1 m_{k,i}} = \frac{\overline{w}_l^k \cdot \left[ \overline{f}^k - {}^{2}\mu_{k,i} \cdot \prod_{l=1, l \neq i}^{n} \left( \overline{\mu}_{k,l} \right) \right] + \underline{w}_l^k \cdot \underline{f}^k}{\overline{w}_l^k + \underline{w}_l^k} \cdot \frac{x_i - {}^{1}m_{k,i}}{(\sigma_{k,i})^2}, \tag{A28} $$

$$ \frac{\partial f_l^k}{\partial^2 m_{k,i}} = \frac{\overline{w}_l^k \cdot \left[ \overline{f}^k - {}^{1}\mu_{k,i} \cdot \prod_{l=1, l \neq i}^{n} \left( \overline{\mu}_{k,l} \right) \right] + \underline{w}_l^k \cdot \underline{f}^k}{\overline{w}_l^k + \underline{w}_l^k} \cdot \frac{x_i - {}^{2}m_{k,i}}{(\sigma_{k,i})^2} \tag{A29} $$

$$ \frac{\partial f_l^k}{\partial \sigma_{k,i}} = \frac{\overline{w}_l^k \cdot \left[ \left( \overline{f}^k - {}^{2}\mu_{k,i} \cdot \prod_{l=1, l \neq i}^{n} \left( \overline{\mu}_{k,l} \right) \right) \cdot \frac{(x_i - {}^{1}m_{k,i})^2}{(\sigma_{k,i})^3} \right]}{\overline{w}_l^k + \underline{w}_l^k} + \frac{\overline{w}_l^k \cdot \left[ \left( \overline{f}^k - {}^{1}\mu_{k,i} \cdot \prod_{l=1, l \neq i}^{n} \left( \overline{\mu}_{k,l} \right) \right) \cdot \frac{(x_i - {}^{2}m_{k,i})^2}{(\sigma_{k,i})^3} \right]}{\overline{w}_l^k + \underline{w}_l^k} $$

$$+\frac{\underline{w}_l^k\cdot\underline{f}^k\cdot\left[\frac{\left(x_i-{}^1m_{k,i}\right)^2+\left(x_i-{}^2m_{k,i}\right)^2}{\left(\sigma_{k,i}\right)^3}\right]}{\overline{w}_l^k+\underline{w}_l^k} \tag{A30}$$

$$\frac{\partial f_r^k}{\partial^1 m_{k,i}}=\frac{\overline{w}_r^k\cdot\left[\overline{f}^k-{}^2\mu_{k,i}\cdot\prod_{l=1,l\neq i}^n\left(\overline{\mu}_{k,l}\right)\right]+\underline{w}_r^k\cdot\underline{f}^k}{\overline{w}_r^k+\underline{w}_r^k}\cdot\frac{x_i-{}^1m_{k,i}}{\left(\sigma_{k,i}\right)^2} \tag{A31}$$

$$\frac{\partial f_r^k}{\partial^2 m_{k,i}}=\frac{\overline{w}_r^k\cdot\left[\overline{f}^k-{}^1\mu_{k,i}\cdot\prod_{l=1,l\neq i}^n\left(\overline{\mu}_{k,l}\right)\right]+\underline{w}_r^k\cdot\underline{f}^k}{\overline{w}_r^k+\underline{w}_r^k}\cdot\frac{x_i-{}^2m_{k,i}}{\left(\sigma_{k,i}\right)^2} \tag{A32}$$

$$\frac{\partial f_r^k}{\partial\sigma_{k,i}}=\frac{\overline{w}_r^k\cdot\left[\left(\overline{f}^k-{}^2\mu_{k,i}\cdot\prod_{l=1,l\neq i}^n\left(\overline{\mu}_{k,l}\right)\right)\cdot\frac{\left(x_i-{}^1m_{k,i}\right)^2}{\left(\sigma_{k,i}\right)^3}\right]}{\overline{w}_r^k+\underline{w}_r^k}$$
$$+\frac{\overline{w}_r^k\cdot\left[\left(\overline{f}^k-{}^1\mu_{k,i}\cdot\prod_{l=1,l\neq i}^n\left(\overline{\mu}_{k,l}\right)\right)\cdot\frac{\left(x_i-{}^2m_{k,i}\right)^2}{\left(\sigma_{k,i}\right)^3}\right]}{\overline{w}_r^k+\underline{w}_r^k}$$
$$+\frac{\underline{w}_r^k\cdot\underline{f}^k\cdot\left[\frac{\left(x_i-{}^1m_{k,i}\right)^2+\left(x_i-{}^2m_{k,i}\right)^2}{\left(\sigma_{k,i}\right)^3}\right]}{\overline{w}_r^k+\underline{w}_r^k} \tag{A33}$$

*Convergence Analysis of Learning Algorithm*

The Lyapunov function is used to guarantee learning algorithm convergence. The Lyapunov function is defined as

$$V_p(k)=E_p(k)=\frac{1}{2}e_p^2(k)=\frac{1}{2}\left(t_p(k)-\hat{y}_p(k)\right)^2 \tag{A34}$$

Equation (A35) shows the Lyapunov function changes.

$$\Delta V_p(k)=V_p(k+1)-V_p(k)=\frac{1}{2}\left(e_p^2(k+1)-e_p^2(k)\right) \tag{A35}$$

Next, the moment error is calculated from Equation (A36).

$$e_p(k+1)=e_p(k)+\Delta e_p(k)\cong e_p(k)+\left[\frac{\partial\,e_p(k)}{\partial W}\right]^T\Delta W \tag{A36}$$

In Equation (A36), $\Delta W$ is parameter changing, where $W=\left[\sigma_{k,i},{}^1m_{k,i},{}^2m_{k,i},c_{k,i},s_{k,i}\right]$. In Equation (A37), the general form of gradient-based updating is presented.

$$W(k+1)=W(k)+\Delta W(k)=W(k)+\eta\cdot\left(-\frac{\partial E_p(k)}{\partial W}\right) \tag{A37}$$

where

$$\frac{\partial E_p(k)}{\partial W}=-e_p(k)\cdot\frac{\partial\hat{y}}{\partial W} \tag{A38}$$

Equation (A35) can be rewritten as Equation (A39).

$$\Delta V_p(k)=\frac{1}{2}\left(e_p^2(k+1)-e_p^2(k)\right) \tag{A39}$$
$$=\frac{1}{2}\left[\left(e_p(k+1)-e_p(k)\right)\right]\cdot\left[\left(e_p(k+1)+e_p(k)\right)\right]$$
$$=\frac{1}{2}\Delta e_p(k)\left[2\left(e_p(k)\right)+\Delta e_p(k)\right]$$

$$= \frac{1}{2}\Delta e_p(k)\ \left[2(e_p(k))\ +\ \Delta e_p(k)\right]$$

$$= \left[\frac{\partial e_p(k)}{\partial W}\right]^T \cdot\eta\cdot e_p(k)\cdot\frac{\partial\hat{y}(k)}{\partial W}\cdot\left\{e_p(k)+\frac{1}{2}\left[\frac{\partial e_p(k)}{\partial W}\right]^T\cdot\eta\cdot e_p(k)\cdot\frac{\partial\hat{y}(k)}{\partial W}\right\}$$

$$= -\left[\frac{\partial\hat{y}(k)}{\partial W}\right]^T \cdot\eta\cdot e_p(k)\cdot\frac{\partial\hat{y}(k)}{\partial W}\cdot\left\{e_p(k)\ -\ \frac{1}{2}\left[\frac{\partial\hat{y}(k)}{\partial W}\right]^T\cdot\eta\cdot e_p(k)\cdot\frac{\partial\hat{y}(k)}{\partial W}\right\}$$

$$= -\eta\cdot\left(e_p(k)\right)^2\left|\frac{\partial\hat{y}(k)}{\partial W}\right|^2\cdot\left[1-\frac{1}{2}\eta\cdot\left|\frac{\partial\hat{y}(k)}{\partial W}\right|^2\right]$$

In order for $\Delta V_p(k) < 0$, then:

$$0 < \eta < \frac{2}{max\left|\frac{\partial\hat{y}(k)}{\partial W}\right|^2} \tag{A40}$$

If (A40) holds for every parameter $W = \left[\sigma_{k,i}, {}^1m_{k,i}, {}^2m_{k,i}, c_{k,i}, s_{k,i}\right]$, then the algorithm is definitely convergent.

## References


1. Quaranta, G.; Lacarbonara, W.; Masri, S.F. A review on computational intelligence for identification of nonlinear dynamical systems. *Nonlinear Dyn.* **2020**, *99*, 1709–1761. [CrossRef]
2. Andrukhiv, A.; Sokil, M.; Fedushko, S.; Syerov, Y.; Kalambet, Y.; Peracek, T. Methodology for increasing the efficiency of dynamic process calculations in elastic elements of complex engineering constructions. *Electronics* **2021**, *10*, 40. [CrossRef]
3. Shekhovtsov, A.; Kołodziejczyk, J.; Sałabun, W. Fuzzy Model identification using monolithic and structured approaches in decision problems with partially incomplete data. *Symmetry* **2020**, *12*, 1541. [CrossRef]
4. Mombeini, H.; Yazdani-Chamzini, A.; Streimikiene, D.; Zavadskas, E.K. New fuzzy logic approach for the capability assessment of renewable energy technologies: Case of Iran. *Energy Environ.* **2018**, *29*, 511–532. [CrossRef]
5. Sakthivel, K.; Devaraj, B.; Banu, D.; Narmatha Selvi, R.; Agnes Idhaya, V. A hybrid wind-solar energy system with ANFIS based MPPT controller. *J. Intell. Fuzzy Syst.* **2018**, *35*, 1579–1595.
6. Shuli, L.; Xinwang, L.; Dongwei, L. A prospect theory based MADM method for solar water heater selection problems. *J. Intell. Fuzzy Syst.* **2017**, *32*, 1855–1865.
7. Kaid, I.E.; Hafaifa, A.; Guemana, M.; Hadroug, N.; Kouzou, A.; Mazouz, L. Photovoltaic system failure diagnosis based on adaptive neuro fuzzy inference approach: South Algeria solar power plant. *J. Clean. Prod.* **2018**, *204*, 169–182. [CrossRef]
8. Samanlioglu, F.; Ayağ, Z. A fuzzy AHP-PROMETHEE II approach for evaluation of solar power plant location alternatives in Turkey. *J. Intell. Fuzzy Syst.* **2017**, *33*, 859–871. [CrossRef]
9. Grahovac, J.; Jokic, A.; Dodic, J.; Vucurovic, D.; Dodic, S. Modelling and prediction of bioethanol production from intermediates and byproduct of sugar beet processing using neural networks. *Renew. Energy* **2016**, *85*, 953–958. [CrossRef]
10. Khiareddine, A.; Salah, C.B.; Mimouni, M.F. Power management of a photovoltaic/battery pumping system in agricultural experiment station. *Sol. Energy* **2015**, *112*, 319–338. [CrossRef]
11. Osorio, G.J.; Matias, J.C.O.; Catalao, J.P.S. Short-term wind power forecasting using adaptive neuro-fuzzy inference system combined with evolutionary particle swarm optimization, wavelet transform and mutual information. *Renew. Energy* **2015**, *75*, 301–307. [CrossRef]
12. Etemadi, M.; Abdollahi, A.; Rashidinejad, M.; Aalami, H.A. Wind turbine output power prediction in a probabilistic framework based on fuzzy intervals. *Iran J. Sci. Technol. Trans. Electr. Eng.* **2021**, *45*, 131–139. [CrossRef]
13. Torshizi, A.D.; Fazel Zarandi, M.H. A new cluster validity measure based on general type-2 fuzzy sets: Application in gene expression data clustering. *Knowl. Based Syst.* **2014**, *64*, 81–93. [CrossRef]
14. Hesarian, M.S.; Tavoosi, J.; Hosseini, S.H. Neuro-fuzzy modelling and experimental study of the physiological comfort of green cotton fabric based on yarn properties. *Int. J. Eng.* **2020**, *33*, 2443–2449. [CrossRef]
15. Tavoosi, J. A new type-2 fuzzy sliding mode control for longitudinal aerodynamic parameters of a commercial aircraft. *J. Eur. Des. Systèmes Autom.* **2020**, *53*, 479–485. [CrossRef]
16. Lu, W.; Yang, J.; Liu, X.; Pedrycz, W. The modeling and prediction of time series based on synergy of high-order fuzzy cognitive map and fuzzy c-means clustering. *Knowl. Based Syst.* **2014**, *70*, 242–255. [CrossRef]
17. Abiyev, R.H.; Kaynak, O.; Kayacan, E. A type-2 fuzzy wavelet neural network for system identification and control. *J. Frankl. Inst.* **2013**, *350*, 1658–1685. [CrossRef]
18. Wang, H.; Luo, C.; Wang, X. Synchronization and identification of nonlinear systems by using a novel self-evolving interval type-2 fuzzy LSTM-neural network. *Eng. Appl. Artif. Intell.* **2019**, *81*, 79–93. [CrossRef]

19. Luo, C.; Tan, C.; Wang, X.; Zheng, Y. An evolving recurrent interval type-2 intuitionistic fuzzy neural network for online learning and time series prediction. *Appl. Soft Comput.* **2019**, *78*, 150–163. [CrossRef]
20. El-Nagar, A. Nonlinear dynamic systems identification using recurrent interval type-2 TSK fuzzy neural network—A novel structure. *ISA Trans.* **2018**, *72*, 205–217. [CrossRef]
21. Tavoosi, J.; Suratgar, A.A.; Menhaj, M.B. Nonlinear system identification based on a self-organizing type-2 fuzzy RBFN. *Eng. Appl. Artif. Intell.* **2016**, *54*, 26–38. [CrossRef]
22. Anh, N.; Suresh, S.; Pratama, M.; Srikanth, N. Interval prediction of wave energy characteristics using meta-cognitive interval type-2 fuzzy inference system. *Knowl. Based Syst.* **2019**, *169*, 28–38. [CrossRef]
23. Tavoosi, J.; Suratgar, A.A.; Menhaj, M.B. Stable ANFIS2 for Nonlinear System Identification. *Neurocomputing* **2016**, *182*, 235–246. [CrossRef]
24. Ali, F.; Kim, E.K.; Kim, Y.G. Type-2 fuzzy ontology-based semantic knowledge for collision avoidance of autonomous underwater vehicles. *Inf. Sci.* **2015**, *295*, 441–464. [CrossRef]
25. Sun, Z.; Wang, N.; Bi, Y. Type-1/type-2 fuzzy logic systems optimization with RNA genetic algorithm for double inverted pendulum. *Appl. Math. Model.* **2015**, *39*, 70–85. [CrossRef]
26. Khooban, M.H.; Alfi, A.; Abadi, D.N.M. Control of a class of non-linear uncertain chaotic systems via an optimal Type-2 fuzzy proportional integral derivative controller. *IET Sci. Meas. Technol.* **2013**, *7*, 50–58. [CrossRef]
27. Lee, C.H.; Chang, F.Y.; Lin, C.M. An efficient interval type-2 fuzzy CMAC for chaos time-series prediction and synchronization. *IEEE Trans. Cybern.* **2014**, *44*, 329–341. [CrossRef] [PubMed]
28. Yu, W.S.; Chen, H.S. Interval type-2 fuzzy adaptive tracking control design for PMDC motor with the sector dead-zones. *Inf. Sci.* **2014**, *288*, 108–134. [CrossRef]
29. Esposito, M.; Pietro, G.D. Interval type-2 fuzzy logic for encoding clinical practice guidelines. *Knowl. Based Syst.* **2013**, *54*, 329–341. [CrossRef]
30. Lin, Y.Y.; Chang, J.Y.; Pal, N.R.; Lin, C.T. A Mutually recurrent interval Type-2 neural fuzzy system (MRIT2NFS) with self-evolving structure and parameters. *IEEE Trans. Fuzzy Syst.* **2013**, *21*, 492–509. [CrossRef]
31. Tavoosi, J. A new Type-2 fuzzy systems for flexible-joint robot arm control. *Aut J. Model. Simul.* **2019**, *51*. [CrossRef]
32. Asad, Y.P.; Shamsi, A.; Ivani, H.; Tavoosi, J. Adaptive intelligent inverse control of nonlinear systems with regard to sensor noise and parameter uncertainty (magnetic ball levitation system case study). *Int. J. Smart Sens. Intell. Syst.* **2016**, *9*, 148–169.
33. Tavoosi, J.; Badamchizadeh, M.A.; Ghaemi, S. Adaptive inverse control of nonlinear dynamical system using Type-2 fuzzy neural networks. *J. Control* **2011**, *5*, 52–60.
34. Fazlya, M.; Pedram, M.Z.; Salarieh, H.; Alasty, A. Parameter estimation and interval type-2 fuzzy sliding mode control of a z-axis MEMS gyroscope. *ISA Trans.* **2013**, *52*, 900–911. [CrossRef] [PubMed]
35. Melin, P.; Castillo, O. A review on type-2 fuzzy logic applications in clustering, classification and pattern recognition. *Appl. Soft Comput.* **2014**, *21*, 568–577. [CrossRef]
36. Wu, G.D.; Zhu, Z.W. An enhanced discriminability recurrent fuzzy neural network for temporal classification problems. *Fuzzy Sets Syst.* **2014**, *237*, 47–62. [CrossRef]
37. Tavoosi, J.; Mohammadzadeh, A. A new recurrent radial basis function network-based model predictive control for a power plant boiler temperature control. *Int. J. Eng.* **2021**, *34*, 667–675.
38. Moodi, H.; Farrokhi, M. Robust observer design for Sugeno systems with incremental quadratic nonlinearity in the consequent. *Int. J. Appl. Math. Comput. Sci.* **2013**, *23*, 711–723. [CrossRef]
39. Tavoosi, J. A novel recurrent Type-2 fuzzy neural network stepper motor control. *Mechatron. Syst. Control* **2021**, *49*. [CrossRef]
40. Tavoosi, J.; Mohammadi, F. A new Type-II fuzzy system for flexible-joint robot arm control. In Proceedings of the 6th International Conference on Control, Instrumentation and Automation (ICCIA), Sanandaj, Iran, 30–31 October 2019; pp. 1–4. [CrossRef]
41. Tavoosi, J.; Mohammadi, F. A 3-PRS parallel robot control based on Fuzzy-PID controller. In Proceedings of the 6th International Conference on Control, Instrumentation and Automation (ICCIA), Sanandaj, Iran, 30–31 October 2019; pp. 1–4. [CrossRef]
42. Tavoosi, J.; Azami, R. A New Method for controlling the speed of a surface permanent magnet synchronous motor using fuzzy comparative controller with hybrid learning. *Comput. Intell. Electr. Eng.* **2019**, *10*, 57–68.
43. Tavoosi, J. An experimental study on inverse adaptive neural fuzzy control for nonlinear systems, International. *J. Knowl. Based Intell. Eng. Syst.* **2020**, *24*, 135–143. [CrossRef]
44. Tavoosi, J. Stable backstepping sliding mode control Based on ANFIS2 for a class of nonlinear systems. *Jordan J. Electr. Eng.* **2020**, *6*, 49–62. [CrossRef]
45. Tavoosi, J. Sliding mode control of a class of nonlinear systems based on recurrent type-2 fuzzy RBFN. *Int. J. Mechatron. Autom.* **2020**, *7*, 230–240.
46. Tavoosi, J. Hybrid intelligent adaptive controller for tiltrotor UAV. *Int. J. Intell. Unmanned Syst.* **2020**, in press. [CrossRef]
47. Tavoosi, J.; Badamchizadeh, M.A. A class of type-2 fuzzy neural networks for nonlinear dynamical system identification. *Neural Comput. Appl.* **2013**, *23*, 707–717. [CrossRef]
48. Tavoosi, J.; Suratgar, A.A.; Menhaj, M.B. Stability analysis of recurrent type-2 TSK fuzzy systems with nonlinear consequent part. *Neural Comput. Appl.* **2017**, *28*, 47–56. [CrossRef]
49. Tavoosi, J.; Suratgar, A.A.; Menhaj, M.B. Stability analysis of a class of MIMO recurrent type-2 fuzzy systems. *Int. J. Fuzzy Syst.* **2017**, *19*, 895–908. [CrossRef]

50. Jahangiri, F.; Doustmohammadi, A.; Menhaj, M.B. An adaptive wavelet differential neural networks based identifier and its stability analysis. *Neurocomputing* **2012**, *77*, 12–19. [CrossRef]
51. Suratgar, A.A.; Nikravesh, S.K. A new method for linguistic modeling with stability analysis and applications. *Intell. Autom. Soft Comput.* **2009**, *15*, 329–342. [CrossRef]
52. Li, C.; Wang, L.; Zhang, G.; Wang, H.; Shang, F. Functional-type single-input-rule-modules connected neural fuzzy system for wind speed prediction. *J. Autom. Sin.* **2017**, *4*, 751–762. [CrossRef]
53. Karakuş, O.; Kuruoğlu, E.E.; Altınkaya, M.A. One-day ahead wind speed/power prediction based on polynomial autoregressive model. *IET Renew. Power Gener.* **2017**, *11*, 1430–1439. [CrossRef]
54. Tian, Y.; Wang, B.; Zhu, D.; Wu, F. Takagi–Sugeno fuzzy generalised predictive control of a time-delay non-linear hydro-turbine governing system. *IET Renew. Power Gener.* **2019**, *13*, 2338–2345. [CrossRef]
55. Morshedizadeh, M.; Kordestani, M.; Carriveau, R.; Ting, D.S.; Saif, M. Power production prediction of wind turbines using a fusion of MLP and ANFIS networks. *IET Renew. Power Gener.* **2018**, *12*, 1025–1033. [CrossRef]
56. Castro, J.R.; Castillo, O.; Martínez, L.G. Interval Type-2 Fuzzy Logic Toolbox. *Eng. Lett.* **2007**, *15*, 1.
57. Mendel, J.M. *Uncertain Rule-Based Fuzzy Logic Systems: Introduction and New Directions*; Prentice-Hall: Upper New Jersey River, NJ, USA, 2001.
58. Mendel, J.M. Advances in type-2 fuzzy sets and systems. *Inf. Sci.* **2007**, *177*, 84–110. [CrossRef]
59. Kizielewicz, B.; Sałabun, W. A new approach to identifying a multi-criteria decision model based on stochastic optimization techniques. *Symmetry* **2020**, *12*, 1551. [CrossRef]
60. Karnik, N.N.; Mendel, J.M.; Liang, Q. Type-2 fuzzy logic systems. *IEEE Trans. Fuzzy Syst.* **1999**, *7*, 643–658. [CrossRef]
61. Singh, M.; Srivastava, S.; Hanmandlu, M.; Gupta, J.R.P. Type-2 fuzzy wavelet networks (T2FWN) for system identification using fuzzy differential and Lyapunov stability algorithm. *Appl. Soft Comput.* **2009**, *9*, 977–989. [CrossRef]
62. Juang, C.F.; Lin, Y.Y.; Tu, C.C. A recurrent self-evolving fuzzy neural network with local feedbacks and its application to dynamic system processing. *Fuzzy Sets Syst.* **2010**, *161*, 2552–2568. [CrossRef]
63. Lin, T.C.; Kuo, C.H.; Balas, V.E. Real-time fuzzy system identification using uncertainty bounds. *Neurocomputing* **2014**, *125*, 195–216. [CrossRef]
64. Pedrycz, W.; Izakian, H. Cluster–centric fuzzy modeling. *IEEE Trans. Fuzzy Syst.* **2014**, *22*, 1585–1597. [CrossRef]
65. Krishnapuram, R.; Keller, J.M. A possibilistic approach to clustering. *IEEE Trans. Fuzzy Syst.* **1993**, *1*, 98–110. [CrossRef]

*sustainability*

MDPI

*Article*

# Synthesizing Multi-Layer Perceptron Network with Ant Lion Biogeography-Based Dragonfly Algorithm Evolutionary Strategy Invasive Weed and League Champion Optimization Hybrid Algorithms in Predicting Heating Load in Residential Buildings

**Hossein Moayedi [1,2] and Amir Mosavi [3,4,5,6,*]**

1 Institute of Research and Development, Duy Tan University, Da Nang 550000, Vietnam; hosseinmoayedi@duytan.edu.vn
2 Faculty of Civil Engineering, Duy Tan University, Da Nang 550000, Vietnam
3 Thuringian Institute of Sustainability and Climate Protection, 07743 Jena, Germany
4 School of Economics and Business, Norwegian University of Life Sciences, 1430 Ås, Norway
5 John von Neumann Faculty of Informatics, Obuda University, 1034 Budapest, Hungary
6 School of the Built Environment, Oxford Brookes University, Oxford OX3 0BP, UK
* Correspondence: amir.mosavi@mailbox.tu-dresden.de

**Citation:** Moayedi, H.; Mosavi, A. Synthesizing Multi-Layer Perceptron Network with Ant Lion Biogeography-Based Dragonfly Algorithm Evolutionary Strategy Invasive Weed and League Champion Optimization Hybrid Algorithms in Predicting Heating Load in Residential Buildings. *Sustainability* **2021**, *13*, 3198. https://doi.org/10.3390/su13063198

Academic Editor: Mehdi Seyedmahmoudian

Received: 11 January 2021
Accepted: 17 February 2021
Published: 15 March 2021

**Publisher's Note:** MDPI stays neutral with regard to jurisdictional claims in published maps and institutional affiliations.



**Abstract:** The significance of accurate heating load (HL) approximation is the primary motivation of this research to distinguish the most efficient predictive model among several neural-metaheuristic models. The proposed models are formulated through synthesizing a multi-layer perceptron network (MLP) with ant lion optimization (ALO), biogeography-based optimization (BBO), the dragonfly algorithm (DA), evolutionary strategy (ES), invasive weed optimization (IWO), and league champion optimization (LCA) hybrid algorithms. Each ensemble is optimized in terms of the operating population. Accordingly, the ALO-MLP, BBO-MLP, DA-MLP, ES-MLP, IWO-MLP, and LCA-MLP presented their best performance for population sizes of 350, 400, 200, 500, 50, and 300, respectively. The comparison was carried out by implementing a ranking system. Based on the obtained overall scores (OSs), the BBO (OS = 36) featured as the most capable optimization technique, followed by ALO (OS = 27) and ES (OS = 20). Due to the efficient performance of these algorithms, the corresponding MLPs can be promising substitutes for traditional methods used for HL analysis.

**Keywords:** energy-efficient building; heating load; neural computing; biogeography-based optimization; big data; machine learning; artificial intelligence; deep learning; building energy; smart buildings, IoT; smart city

## 1. Introduction

Energy consumption analysis of buildings is a very significant task, due to the high rate of energy consumed in this sector [1]. Heating, ventilating, and air conditioning (HVAC) [2] is a state-of-the-art system that controls the heating load (HL) and cooling load (CL) in the buildings. Considering the crucial importance of the subject, the approaches such as regression analysis [3,4] and time series [5] cannot satisfy the accuracy required for estimating these parameters. As well as this, other difficulties such as the non-linearity of the problem have driven many scholars to improve the flexibility of intelligent models. As discussed by many scholars, along with well-known models (e.g., decision-making [6–9]), the artificial intelligence techniques have provided a high capability in the estimation of non-linear and intricate parameters [10–12]. Plenty of scientific efforts (e.g., concerning environmental subjects [13–23], gas consumption modeling [24,25], sustainable developments [26], pan evaporation and soil precipitation simulation [26–31], energy-related estimations [32–39], water supply assessment [16,40–49], computer vision and visual processing [50–57], building and structural design analysis [8,58–61], behavior of structural components [60,62–64], measurement techniques [43,50,65,66], climatic-related calculations [64], and analysis that deals with feature selection [64,67–72]) have been associated with these computational technologies. In an artificial neural network (ANN), for example, a mapping process between the input and target parameters is carried out by mimicking the neural-based method established in the human brain [73–76]. Different structures (and consequently diverse types) of ANNs have been designed for specific objectives (e.g., multi-layer perceptron (MLP) [77–79]). Going into deep processors such as ANN, a so-called method "deep learning" emerges, which has successfully modeled various phenomena and parameters [8,80–82]. Diagnostic problem and medical sciences, for instance, are two subjects which have been nicely solved by extreme machine learning approaches [83–86].

Up to now, diverse notions of soft computing techniques (e.g., support vector machine (SVM) and artificial neural network (ANN)) have been effectively used for energy consumption modeling [87–91]. Roy, et al. [92] proposed multivariate adaptive regression splines (MARS) coupled with an extreme learning machine (ELM) for predicting the HL and CL. They used the first model to perform importance analysis of the parameters to feed the second model. Likewise, Sholahudin and Han [93] used an ANN along with the Taguchi method for investigating the effect of the input factors on the HL. The feasibility of a random forest predictive method was investigated by Tsanas and Xifara [94] and Gao et al. [95] for both HL and CL factors. The latter reference is a comprehensive comparative study that compares the simulation capability of sixteen machine learning models (e.g., elastic net, radial basis function regression). This study also confirmed the high efficiency of alternating model tree and rules decision table models. Chou and Bui [91] proposed the combination of ANN and SVM as a proper model for new designs of energy-conserving buildings. The applicability of the neuro-fuzzy approach (ANFIS) for predicting the HL and CL was explored by Nilashi et al. [96]. They used expectation-maximization and principal component analysis along with the ANFIS, respectively, for clustering objective and removing noise. Referring to obtained values of mean absolute error (MAE) (0.16 and 0.52 for the HL and CL predictions, respectively), they concluded that the proposed model is accurate enough for this aim.

In addition, studies in different fields have shown that utilizing metaheuristic algorithms is an effective idea for improving the accuracy of typical predictors [97,98]. For energy-efficient buildings, Moayedi et al. [99] improved the ANN for forecasting the CL by benefiting from the foraging/social behavior of ants, Harris hawks, and elephant (i.e., the EHO algorithm). The results (e.g., the correlation values over 85%) show that the applied algorithms can satisfactorily handle the optimization task. An EHO-based CL predictive formula was also presented. Amar and Zeraibi [100] used the firefly algorithm to optimize the SVM (parameters) for HL modeling in district heating systems. Their model outperformed genetic programming and ANN. Moayedi et al. [99] employed a grasshopper optimization algorithm (GOA) and grey wolf optimization (GWO) algorithms for enhancing the HL prediction of ANN. A significant decrease in the MEA calculated for the ANN (from 2.0830 to 1.7373 and 1.6514, respectively, by incorporation of the GOA and GWO) means that the algorithms can build a more reliable ANN network compared to the typical back-propagation one. In addition, other studies such as [26] outlined the competency of such algorithms in the same fields. As a visible gap of knowledge, despite the variety of studies that have mainly focused on broadly used metaheuristic techniques [101], there are still some algorithms that need to be evaluated. Therefore, assessing the performance of six novel optimization techniques, namely ant lion optimization (ALO), biogeography-based optimization (BBO), many-objective sizing optimization [102–104], data-driven robust optimization [35,105], the dragonfly algorithm (DA), evolutionary strategy (ES), invasive weed optimization (IWO), and league champion optimization (LCA), is the central aim of the present paper.

## 2. Data Provision and Analysis

Providing a reliable dataset is an essential step in intelligent model implementation. These data are used in two stages. Firstly, the significant share is analyzed by the models to infer the relationship between the intended factors and independent variables. The rests are then used to represent unseen conditions of the problem and the performance of the model for stranger data.

In this article, the used dataset was downloaded from a freely available data repository (http://archive.ics.uci.edu/mL/datasets/Energy+efficiency, accessed on 20 December 2020) based on a study by Tsanas and Xifara [94]. They analyzed 768 residential buildings with different geometries using Ecotect software [106] to obtain the HL and CL as the outputs. They set the information of eight independent factors, namely relative compactness (RC), overall height (OH), surface area (SA), orientation, wall area (WA), glazing area (GA), roof area (RA), and glazing area distribution (GAD). Figure 1 shows the distribution of these factors versus the HL, which we aim to predict in this study. Based on plenty of previous studies [97], a random division process was carried out to specify 538 samples (i.e., 70% of the whole) and 230 rows (i.e., 30% of the whole) to the training and testing sets, respectively.

## 3. Methodology

The overall methodology used in this study is shown in Figure 2.

### 3.1. Artificial Neural Network

ANNs are popular data mining techniques based on the biological mechanism of the neural network [107]. ANNs are able to deal with highly complicated engineering simulations because of the non-linear analysis option [108,109]. This approach distinguishes itself by different notions including multi-layer perceptron (MLP) [110], radial basis function [111], and general regression [112]. In this study, an MLP network was selected as the basic method. Figure 3 depicts the MLP general structure predicting $M$ output variables by taking into consideration $L$ input factors. It is important to note that in an MLP, more than one hidden layer can be sandwiched between two other layers. However, theoretical studies have demonstrated the efficiency of unique hidden layer MLPs for any problem.

ANNs normally benefit from the training scheme of Levenberg–Marquardt (LM), an approximation to the method of Newton [113] (Equation (1)). The LM is known to be quicker and enjoy more power compared to conventional gradient descent technique [114,115].

$$\Delta\underline{x} = -\left[\nabla^2 V(\underline{x})\right]^{-1} \nabla V(\underline{x}) \tag{1}$$

where $\nabla V(\underline{x})$ and $\nabla^2 V(\underline{x})$ are the gradient and the Hessian matrix, respectively. The following equation expresses $V(\underline{x})$ as a sum of squares function:

$$V(\underline{x}) = \sum_{i=1}^{N} e_i^2(\mathrm{x}) \tag{2}$$

Next, let $J(\mathrm{x})$ be the Jacobean matrix, then it can be written:

$$\begin{gathered} \nabla V(\underline{x}) = J(\underline{x})\, \mathrm{e}(\underline{x}) \\ \nabla^2 V(\underline{x}) = J^T(\underline{x}) J(\underline{x}) + S(\underline{x}), \\ S(\underline{x}) = \sum_{i=1}^{N} \mathrm{e}_i\, \nabla^2 \mathrm{e}_i(\underline{x}) \end{gathered} \tag{3}$$

Equation (1) can be written as follows when $S(x) \approx 0$:

$$\Delta\underline{x} = \left[J^T(\underline{x}) J(\underline{x})\right]^{-1} J^T(\underline{x})\, \underline{\mathrm{e}}(\underline{x}) \tag{4}$$

Lastly, Equation (5) presents the central equation of the LM, based on the Gauss–Newton method.

$$\Delta \underline{x} = \left[ J^T(\underline{x}) J(\underline{x}) + \mu I \right]^{-1} J^T(\underline{x})\, \underline{e}(\underline{x}) \quad (5)$$

Remarkably, high and low values of $\mu$ turn this algorithm to steepest descent (with step $1/\mu$) and Gauss–Newton, respectively.

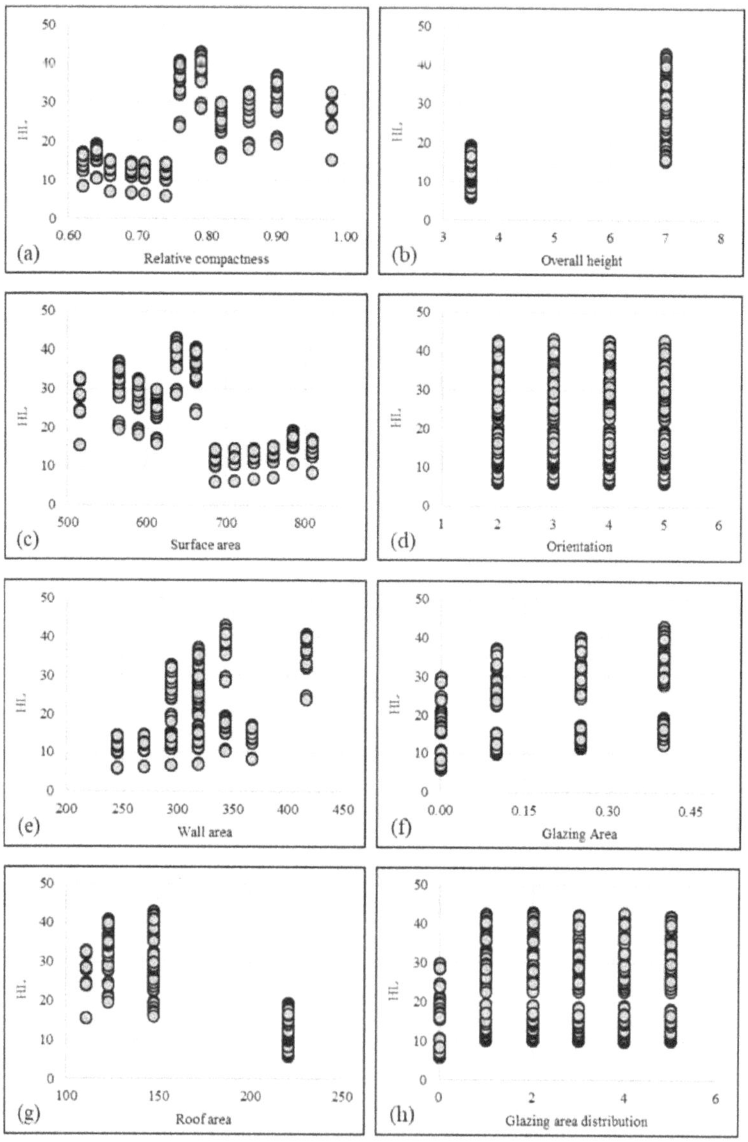


**Figure 1.** The distribution of the heating load (HL) versus environmental factors: (**a**) relative compactness, (**b**) overall height, (**c**) surface area, (**d**) orientation, (**e**) wall area, (**f**) glazing area, (**g**) roof area, and (**h**) glazing area distribution.

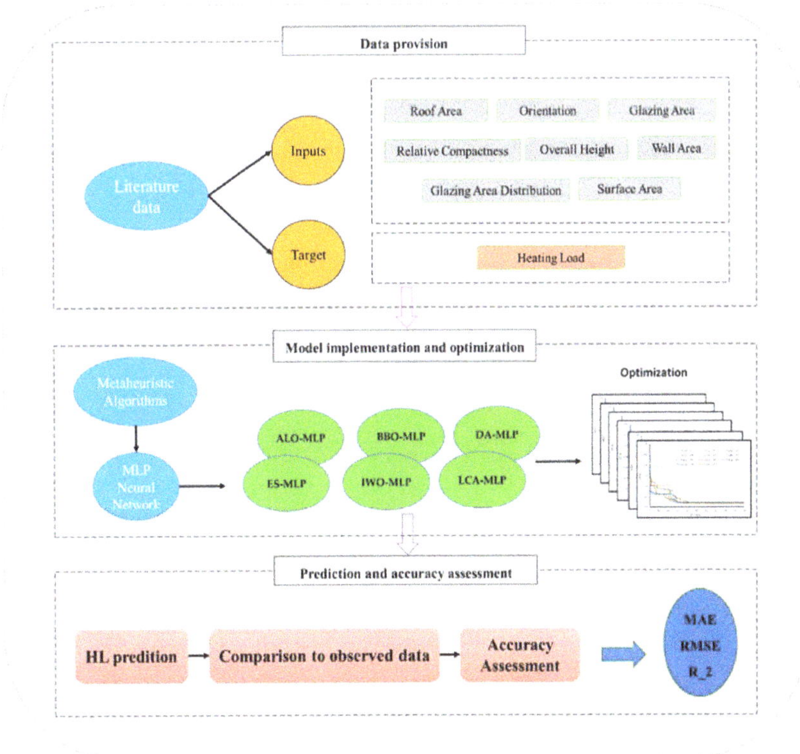


**Figure 2.** The general path of the study.

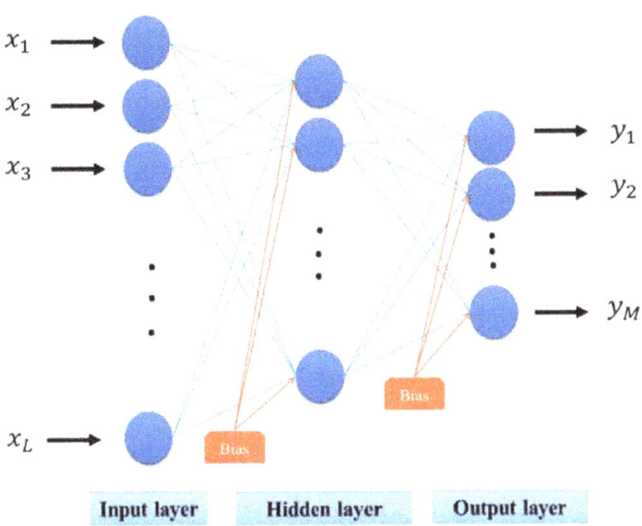


**Figure 3.** Multi-layer perceptron (MLP) general structure predicting M output variables.

### 3.2. *Swarm-Based Metaheuristic Ideas*

Optimization algorithms which have recently been very popular for enhancing the performance of predictive models (e.g., ANNs) are based on swarm functioning of a group of corresponding individuals. They are mostly inspired by nature and seek an optimal global solution for a defined problem by analyzing the relationship between the existing parameters. Coupled with an ANN, these optimizers seek to adjust the biases and weights. This process is better explained in the next section. Here, the overall idea of the intended algorithms is briefly described.

Ant lion optimization (Mirjalili [116]) is a recently-developed hybrid model that mimics the herding behavior of ant lions. It comprises different stages in which the prey (usually an ant) gets trapped and hunted in a hole by a random walk. The capability of the individuals is evaluated by a "roulette wheel selection" function. Biogeography-based optimization is based on two items: (a) the information concerning biogeography and (b) the way different species are distributed. This algorithm was designed by Simon [117] and was used by Mirjalili, et al. [118] to train an MLP network. In the BBO, there are migration and mutation steps and the population is made up of "habits". Note that these habits are evaluated by two indices called the habitat suitability index and suitability index variable. The dragonfly algorithm is another population-based optimization technique proposed by Mirjalili [119]. Based on the Reynolds swarm intelligence, the DA draws on three stages, namely separation, alignment, and cohesion. The name evolutionary strategy implies a stochastic search approach proposed by Schwefel [120]. In the ES, two operators of selection and mutation act during the evolution and adaption stages. The population is produced with offspring variables and the offspring's modality is compared to that of the parents. Inspired by the colonizing behavior of weeds, invasive weed optimization was presented by Mehrabian and Lucas [121]. The optimal solution of this algorithm is the most suitable site for the plants to grow and reproduce. The algorithm begins with the initialization and after reproducing, it runs the stages called spatial dispersal and competitive exclusion, and gets stopped after meeting with the termination measures. Last but not least, league champion optimization is suggested by Kashan [122], mimicking sporting competitions in leagues. The LCA tries to find the best-fitted solution to the problem by implementing an artificial league including schedule programming and determining the winner/looser teams. More information about the mentioned algorithms (e.g., mathematical relationships) was detailed in previous studies (for the ALO [123,124], BBO [125], DA [126], ES [127], IWO [128], and LCA [129,130]).

### 3.3. Hybridization Process and Sensitivity Analysis

In order to develop the proposed neural-metaheuristic ensembles, the algorithms should be hybridized with the ANN. To this end, utilizing the provided data, the general equation of an MLP neural network is yielded to the ALO, BBO, DA, ES, IWO, and LCA as the problem function. But before that, it is required to determine the most suitable structure (i.e., the number of neurons) of it. As explained previously, the number of neurons in the first and the last layers is equal to the number of input and output variables, respectively. Hence, only the number of hidden neurons can be varied. Based on a trial-and-error process, it was set to five. Therefore, the network architecture was distinguished as $8 \times 5 \times 1$.

Each ensemble was executed within 1000 repetitions, where the mean square error (MSE) was defined to measure the performance error during them (objective function = MSE). For greater reliability of the results, a sensitivity analysis was carried out in this part. Eleven different population sizes, including 25, 50, 100, 150, 200, 250, 300, 350, 400, 450, and 500, were tested for each model, and the best-fitted complexity was used to predict the HL in the following. The convergence curves belonging to elite networks of each model are presented in Figure 4. According to these charts, for all algorithms, the error is chiefly reduced within the first half of the iterations. Test best population sizes are determined 350, 400, 200, 500, 50, and 300 for the ALO-MLP, BBO-MLP, DA-MLP, ES-MLP, IWO-MLP, and LCA-MLP, respectively.

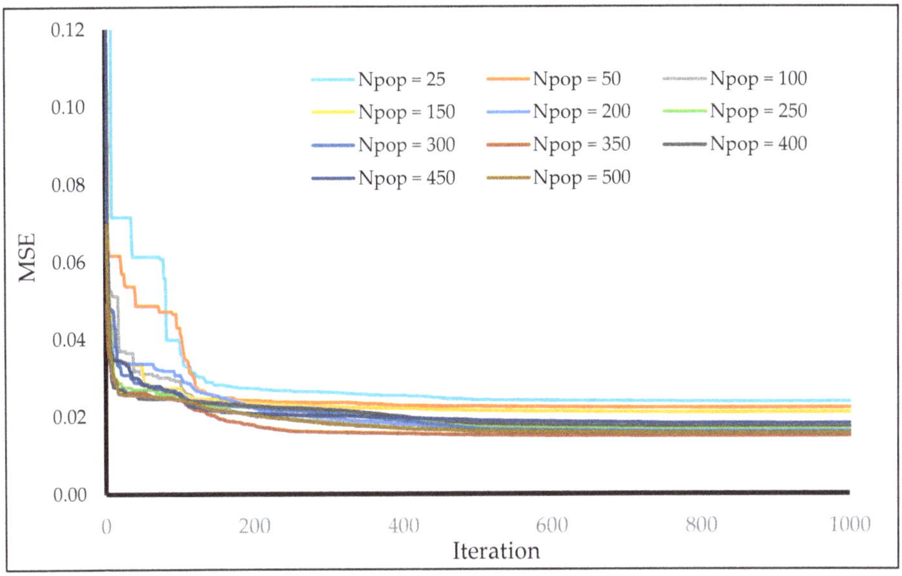


(a)

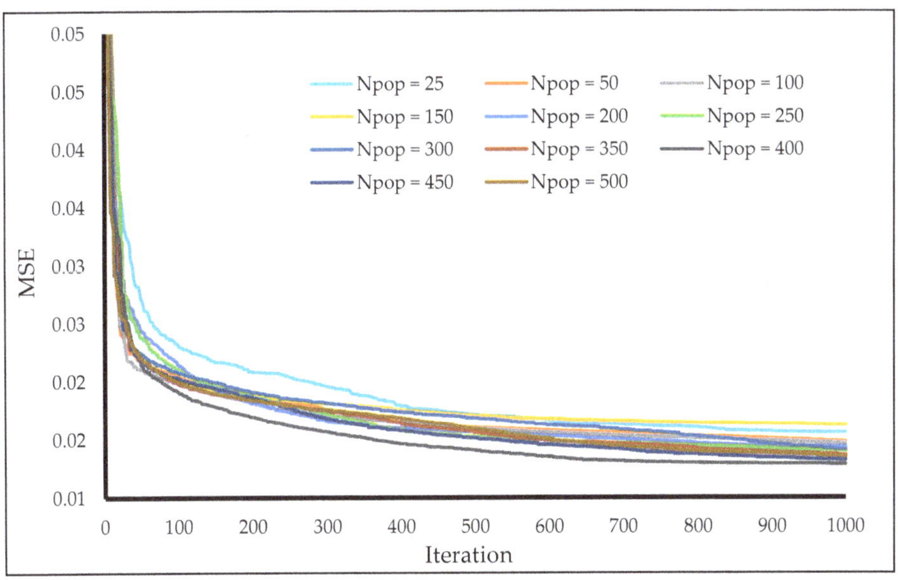


(b)

**Figure 4.** *Cont.*

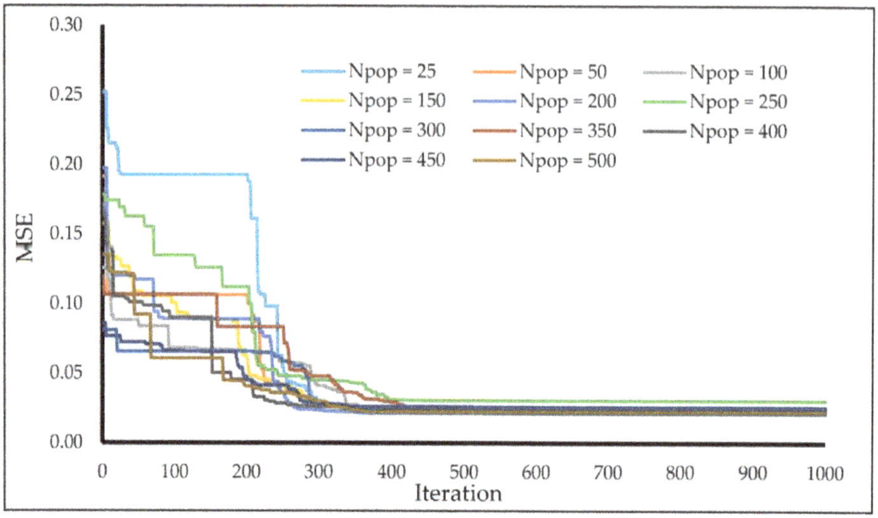


(**c**)

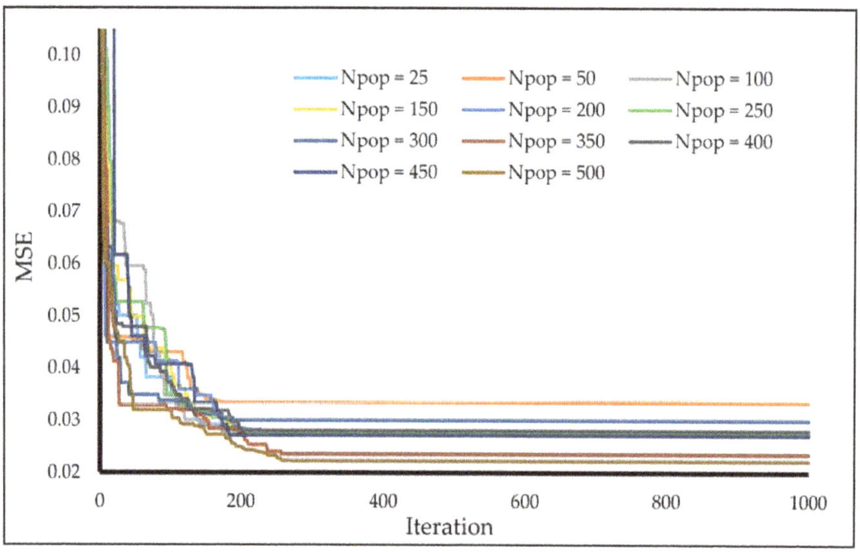


(**d**)

**Figure 4.** *Cont.*

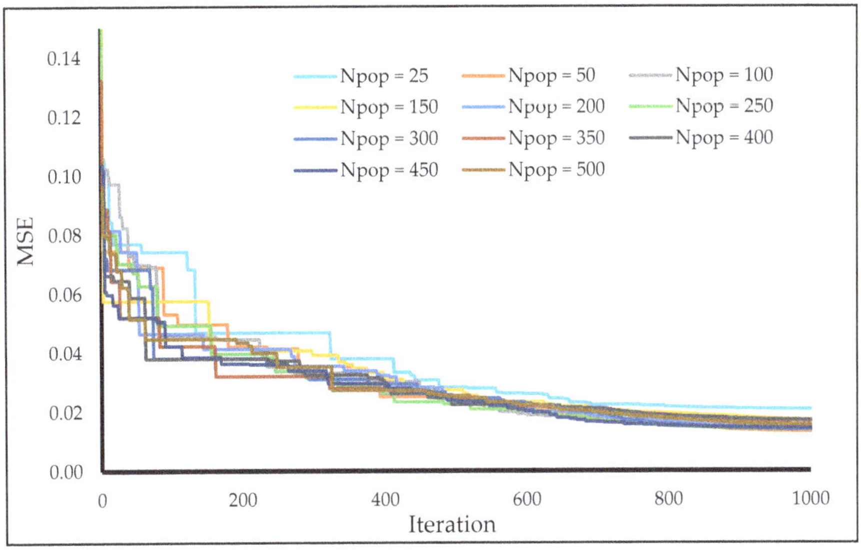


(e)

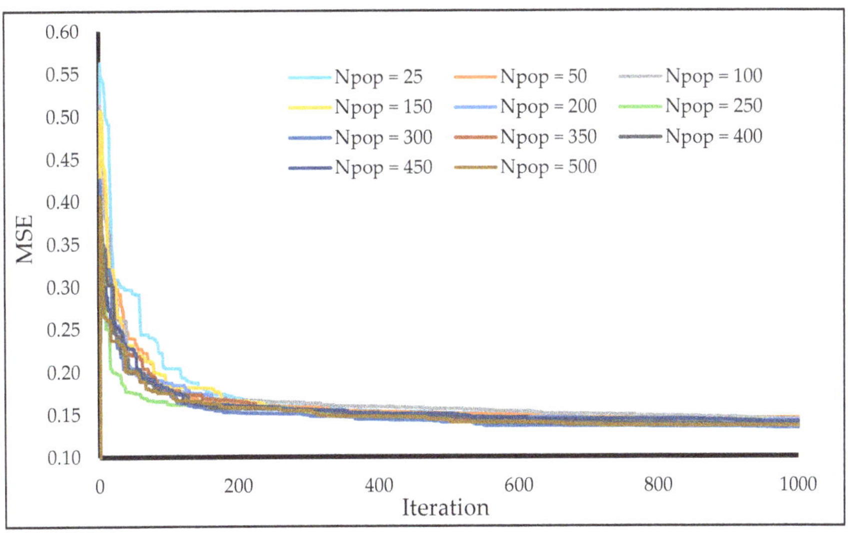


(f)

**Figure 4.** The sensitivity analysis accomplished for determining the best population size of the (**a**) ant lion optimization (ALO)-MLP, (**b**) biogeography-based optimization (BBO)-MLP, (**c**) dragonfly algorithm (DA)-MLP, (**d**) evolutionary strategy (ES)-MLP, (**e**) invasive weed optimization (IWO)-MLP, and (**f**) league champion optimization (LCA)-MLP.

## 4. Results and Discussion

### 4.1. Statistical Accuracy Assessment

Three broadly used criteria are applied to measure the prediction accuracy of the implemented models by reporting the error and correlation of the results. For this purpose, *MAE* (along with the *RMSE*) and the coefficient of determination ($R^2$) are used. These criteria are applied to the data belonging to the training and testing groups to demonstrate the qualities of learning and prediction, respectively. Assuming $G$ as the total number of samples, and $J_{i\ observed,}$ and $J_{i\ predicted}$ as the real and forecasted HL values, Equations (6)–(8) formulate the *RMSE*, *MAE*, and $R^2$.

$$RMSE = \sqrt{\frac{1}{G}\sum_{i=1}^{G}\left[\left(J_{i_{observed}} - J_{i_{predicted}}\right)\right]^2} \quad (6)$$

$$MAE = \frac{1}{G}\sum_{i=1}^{G}\left|J_{i_{observed}} - J_{i_{predicted}}\right| \quad (7)$$

$$R^2 = 1 - \frac{\sum_{i=1}^{G}\left(J_{i_{predicted}} - J_{i_{observed}}\right)^2}{\sum_{i=1}^{G}\left(J_{i_{observed}} - \overline{J}_{observed}\right)^2} \quad (8)$$

where $\overline{J}_{observed}$ denotes the mean of $J_{i\ observed}$ values.

### 4.2. Training Results

The results of elite structures of each model are evaluated in this section. Figure 5 shows the training results. In this regard, the error (=real HL − forecasted HL) is calculated and marked for all 538 samples. In this phase, the maximum and minimum of the (positive) error values were 0.0136 and 6.4455, 0.0018 and 6.0681, 0.0019 and 9.2773, 0.0248 and 7.3006, 0.0184 and 6.3776, and 0.0715 and 8.4620, respectively, for the leaning process of ALO-MLP, BBO-MLP, DA-MLP, ES-MLP, IWO-MLP, and LCA-MLP ensembles.

Referring to the calculated RMSEs (2.6054, 2.5359, 3.4314, 2.7146, 3.2506, and 3.8297), all six models achieved a reliable performance in understanding the non-linear relationship of the HL and eight influential factors. Another piece of evidence that supports this claim is the *MAE* index (2.0992, 2.0846, 2.9402, 2.0848, 2.8709, and 3.4091). Furthermore, the correlation between the expected and real HLs is higher than 92% in all models. In detail, the values of $R^2$ are 0.9539, 0.9596, 0.9222, 0.9357, 0.9547, and 0.9386.

### 4.3. Validation Results

The developed models are then applied to the second group of data to assess the generalization capability of them. Figure 6 depicts the correlation between the expected HLs and networks' products. As is seen, all obtained $R^2$s (0.9406, 0.9516, 0.9340, 0.9318, 0.9131, and 0.9400) reflect higher than 93% accuracy for all models. In this phase, the errors range between −5.5792 and 6.9349, −5.6311 and 6.3000, −9.3137 and 6.8288, −7.0282 and 7.0647, −6.2505 and 5.8823, and −8.2384 and 6.1992, respectively.

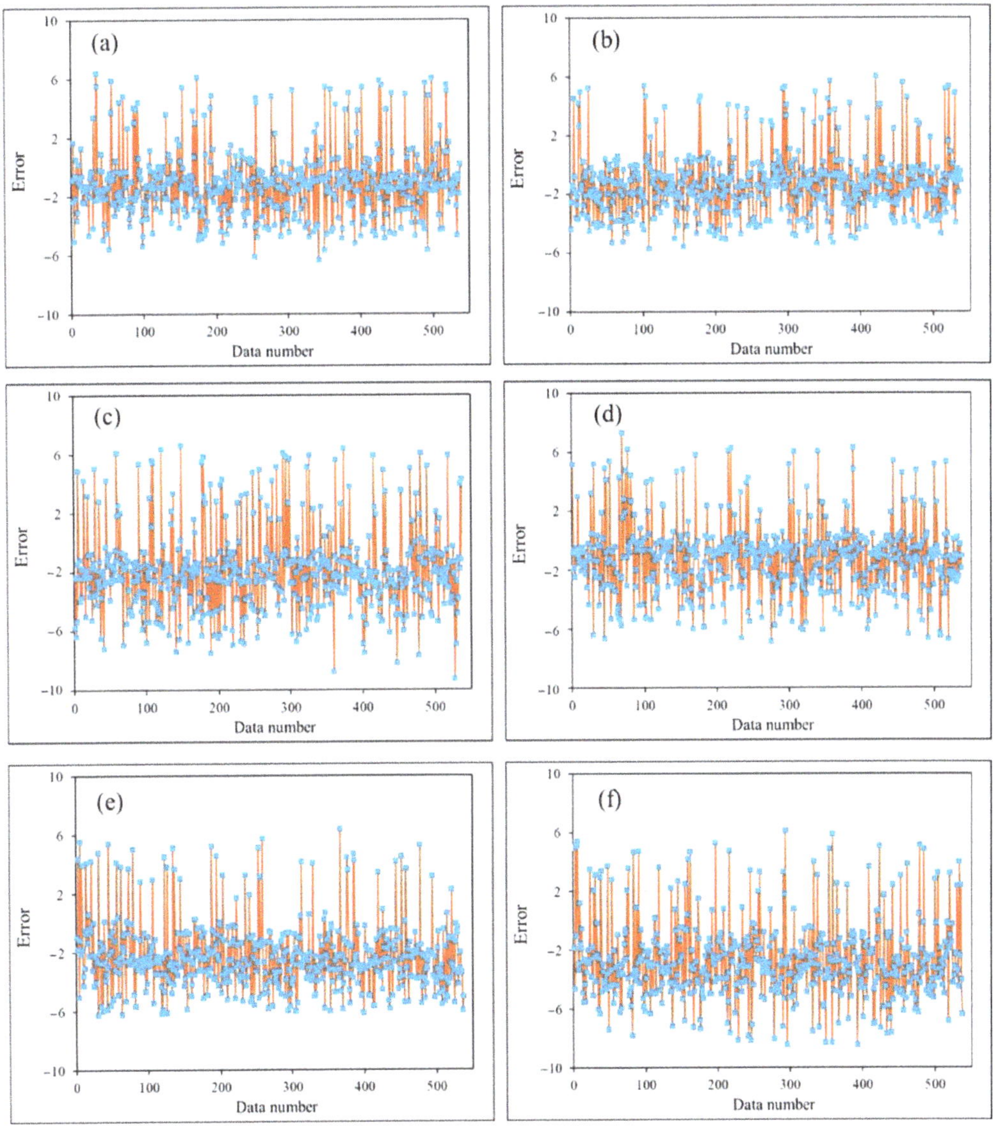


**Figure 5.** The training errors calculated for the (**a**) ALO-MLP, (**b**) BBO-MLP, (**c**) DA-MLP, (**d**) ES-MLP, (**e**) IWO-MLP, and (**f**) LCA-MLP prediction.

Considering the computed RMSE (2.7162, 2.4807, 3.3998, 3.0958, 3.3524, and 3.2954) as well as the MAE (2.1865, 1.8284, 2.8713, 2.5072, 2.9702, and 2.7807) error criteria, it can be deduced that the networks' prediction for unseen environmental conditions has a good level of accuracy. More clearly, the values of mean absolute percentage error were 10.01%, 9.78%, 13.59%, 12.63%, 13.01%, and 13.01%, respectively.

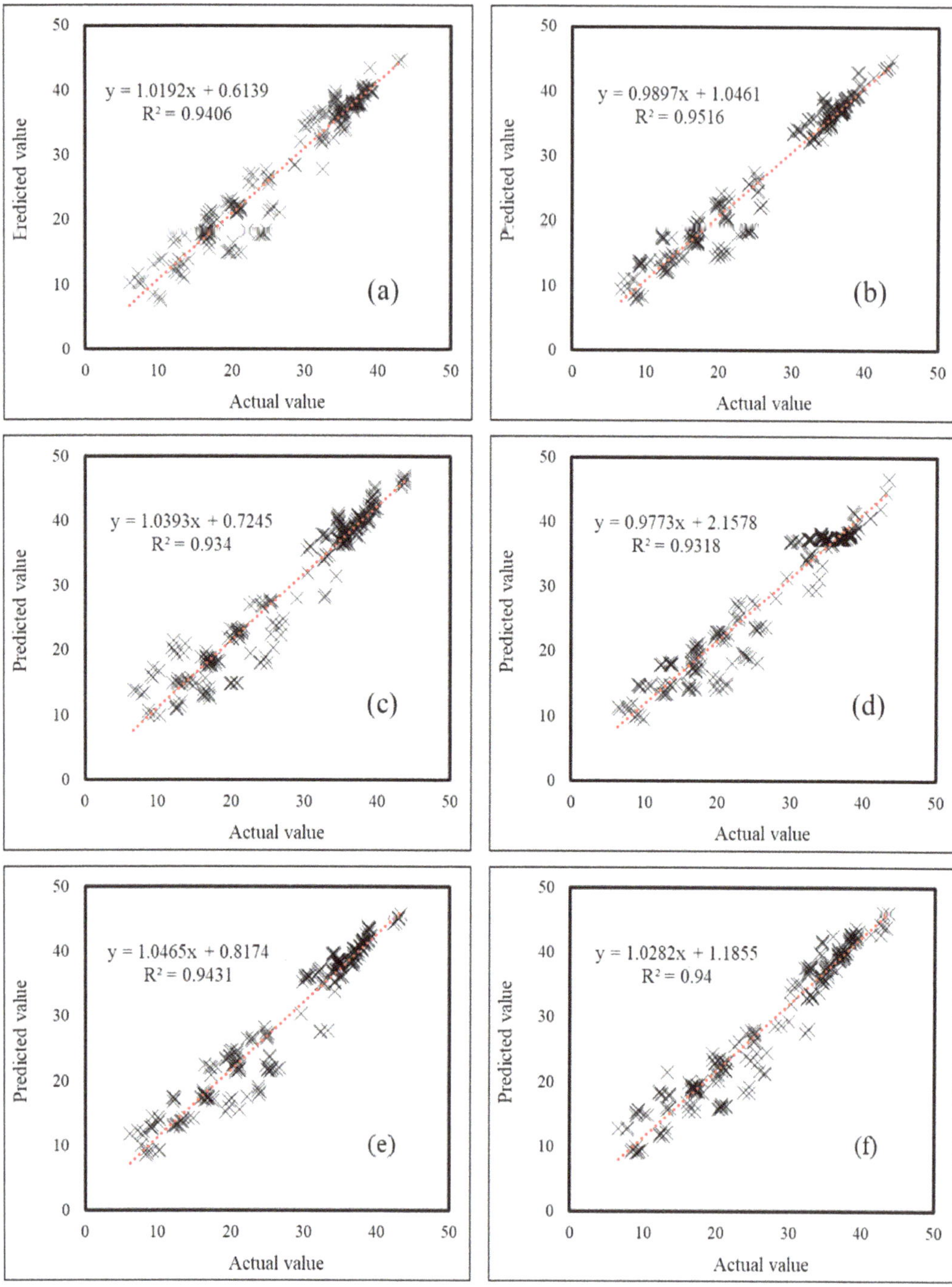


**Figure 6.** The $R^2$ results calculated in the testing phase of the (**a**) ALO-MLP, (**b**) BBO-MLP, (**c**) DA-MLP, (**d**) ES-MLP, (**e**) IWO-MLP, and (**f**) LCA-MLP models.

### 4.4. Score-Based Comparison and Time Efficiency

Table 1 summarizes the values of the *RMSE*, *MAE*, and $R^2$ obtained for the training and testing phases. In this section, the comparison between the performance of the used predictors is carried out to determine the most reliable one. For this purpose, by taking into consideration all three accuracy criteria, a ranking system is developed. In this way, a score is calculated for each criterion based on the relative performance of the proposed model. The summation of these scores gives an overall score (OS) to rank the models. Table 2 gives the scores assigned to each model.

**Table 1.** The results of accuracy assessment.

| Ensemble Models | Network Results | | | | | |
|---|---|---|---|---|---|---|
| | Training Phase | | | Testing Phase | | |
| | *RMSE* | *MAE* | $R^2$ | *RMSE* | *MAE* | $R^2$ |
| **ALO-MLP** | 2.6054 | 2.0992 | 0.9539 | 2.7162 | 2.1865 | 0.9406 |
| **BBO-MLP** | 2.5359 | 2.0846 | 0.9596 | 2.4807 | 1.8284 | 0.9516 |
| **DA-MLP** | 3.4314 | 2.9402 | 0.9222 | 3.3998 | 2.8713 | 0.9340 |
| **ES-MLP** | 2.7146 | 2.0848 | 0.9357 | 3.0958 | 2.5072 | 0.9318 |
| **IWO-MLP** | 3.2506 | 2.8709 | 0.9547 | 3.3524 | 2.9702 | 0.9431 |
| **LCA-MLP** | 3.8297 | 3.4091 | 0.9386 | 3.2954 | 2.7807 | 0.9400 |

**Table 2.** The executed ranking system.

| Models | Scores | | | | | | | | | |
|---|---|---|---|---|---|---|---|---|---|---|
| | Training | | | | | Testing | | | | |
| | *RMSE* | *MAE* | $R^2$ | Overall Score | Rank | *RMSE* | *MAE* | $R^2$ | Overall Score | Rank |
| **ALO-MLP** | 5 | 4 | 4 | 13 | 2 | 5 | 5 | 4 | 14 | 2 |
| **BBO-MLP** | 6 | 6 | 6 | 18 | 1 | 6 | 6 | 6 | 18 | 1 |
| **DA-MLP** | 2 | 2 | 1 | 5 | 5 | 1 | 2 | 2 | 5 | 6 |
| **ES-MLP** | 4 | 5 | 2 | 11 | 3 | 4 | 4 | 1 | 9 | 3 |
| **IWO-MLP** | 3 | 3 | 5 | 11 | 3 | 2 | 1 | 5 | 8 | 5 |
| **LCA-MLP** | 1 | 1 | 3 | 5 | 5 | 3 | 3 | 3 | 9 | 3 |

According to the results, the most significant OS (=18) is obtained for the BBO-MLP in both the training and testing phases. The ALO and ES-based ensembles emerged as the second and third most accurate ones, respectively. However, the IWO in the training phase and the LCA in the testing phase gained a similar rank to the ES. In addition, it can be seen that the results of the DA-MLP are less consistent than other colleagues.

Moreover, Figure 7 illustrates the time required for implementing the used models. This item is also measured for other well-known optimization techniques (including Harris hawks optimization (HHO) [131], GWO [132], whale optimization algorithm (WOA) [133], artificial bee colony (ABC) [134], ant colony optimization (ACO) [135], elephant herding optimization (EHO) [136], genetic algorithm (GA) [137], imperialist competitive algorithm (ICA) [138], particle swarm optimization (PSO) [139], and wind driven optimization (WDO) [140]) to be compared with ALO, BBO, DA, ES, IWO, and LCA. This figure indicates that the metaheuristic algorithms used in this study present a good time-efficiency in comparison with other models. Moreover, it was observed that the ABC, HHO, and DA take the greatest amount of time for almost all of the population sizes.

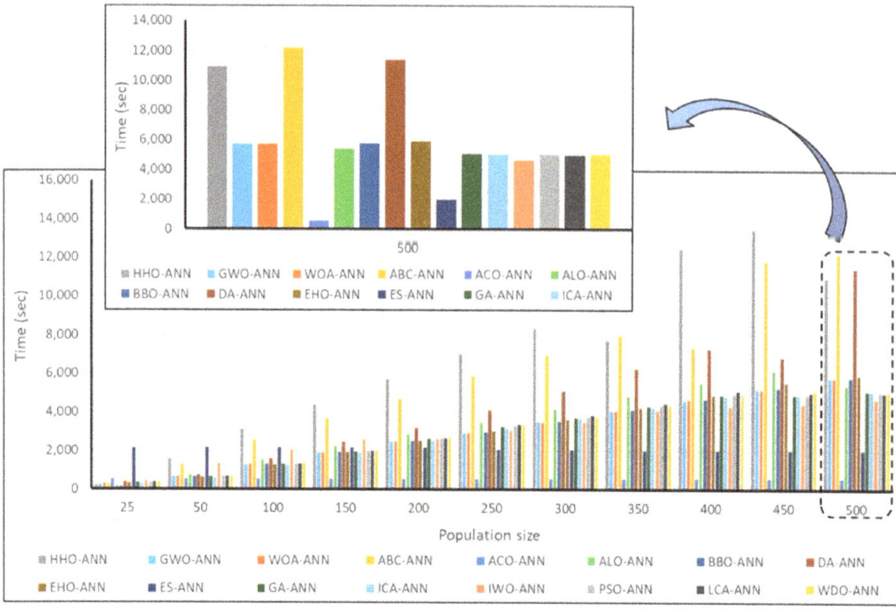


**Figure 7.** The computation time needed for various hybrid methods.

### 4.5. Presenting the HL Predictive Equation

In the previous section, it was concluded that the BBO constructs the most reliable neural network. This means that the biases and connecting weights optimized by this technique can analyze and predict the HL more accurately compared to other metaheuristic algorithms. Therefore, the governing relationships in the BBO-MLP ensemble are extracted and presented as the best HL predictive formula (Equation (9)). As is seen, there are five parameters (Z1, Z2, ... , Z5) in this equation, which need to be calculated by Equation (10). Basically, the response of the neurons in the hidden layer are represented by Z1, Z2, ... , Z5. Remarkably, the term *Tansig* is the network activation function, which is expressed by Equation (11).

$$HL_{BBO\text{-}MLP} = 0.9076 \times Z1 + 0.0050 \times Z2 - 0.3986 \times Z3 - 0.4754 \times Z4 - 0.2692 \times Z5 + 0.0283 \quad (9)$$

$$\begin{bmatrix} Z1 \\ Z2 \\ Z3 \\ Z4 \\ Z5 \end{bmatrix} = Tansig\left(\left(\begin{bmatrix} -0.8459 & 0.2944 & -0.7562 & 0.1225 & -0.2456 & 0.3266 & -1.0020 & 0.6090 \\ -0.2863 & 0.4134 & -0.1649 & -0.8857 & 0.8828 & -0.9327 & 0.1703 & 0.4336 \\ 0.7094 & -0.5079 & -0.6916 & 0.6346 & -0.3142 & -0.0794 & -0.4306 & 0.9990 \\ -1.1274 & -0.0470 & -0.1336 & 0.6061 & 0.0406 & 0.3088 & -0.8939 & -0.6135 \\ 0.1514 & 0.2735 & -0.8389 & 0.1982 & -0.6465 & -1.0777 & 0.2336 & 0.6753 \end{bmatrix}\begin{bmatrix} RC \\ SA \\ WA \\ RA \\ OH \\ Orientation \\ GA \\ GAD \end{bmatrix}\right) + \begin{bmatrix} 1.7120 \\ 0.8560 \\ 0.0000 \\ -0.8560 \\ 1.7120 \end{bmatrix}\right) \quad (10)$$

$$Tansig\,(x) = \frac{2}{1 + e^{-2x}} - 1 \quad (11)$$

### 4.6. Further Discussion and Future Works

Due to the fact that the dataset used in this study is a prepared dataset dedicated to residential buildings, the applicability of the used methods is derived for this type of building. However, there are many studies that have successfully employed machine learning tools for predicting the thermal loads of buildings with other usages, such as office, commercial, and industrial ones [141]. Hence, utilizing multi-usage datasets for future works can overcome this limitation.

Another idea may be evaluating the accuracy of the new generation of hybrid models which can be divided into (a) the combination of the existing metaheuristic tools with other intelligent models, e.g., ANFIS and SVM, or (b) utilizing more recent optimizers for the existing ANN models. Both ideas are helpful to possibly recognize more efficient predictive methods. Moreover, a practical use of the implemented models is also of interest. In order to evaluate the generalizability of the methods, they can be applied to the information taken from real-world buildings noting that the input parameters considered for predicting the HL should be the same as those used in this study; otherwise, it would be a new development.

## 5. Conclusions

The high competency of optimization techniques in various engineering fields motivated the authors to employ and compare the efficacy of six novel metaheuristic techniques, namely ant lion optimization, biogeography-based optimization, dragonfly algorithm, evolutionary strategy, invasive weed optimization, and league champion optimization, in hybridizing the neural network for accurate estimation of the heating load. The proper structure of all seven methods was determined by sensitivity analysis and it was shown that the most appropriate population size could be varied from one algorithm to another. The smallest and largest populations were 50 and 500 hired by the IWO and ES, respectively. The high rate of accuracy observed for all models indicated that metaheuristic techniques could successfully establish a non-linear ANN-based relationship that predicts the HL from the building characteristics. Comparison based on the used accuracy indices revealed that the BBO, ALO, and ES (with around 94% correlation of the results) are able to construct more reliable ANNs in comparison with IWO, LCA, and DA. In addition, the models enjoy a good time efficiency relative to some other existing algorithms. However, the authors believe that, due to recent advances in metaheuristic science, further comparative studies may be required for outlining the most efficient predictive method.

**Author Contributions:** H.M. methodology; software validation, writing—original draft preparation, A.M. writing—review and editing, visualization, supervision, project administration. All authors have read and agreed to the published version of the manuscript.

**Funding:** This research was not funded.

**Institutional Review Board Statement:** Not applicable.

**Informed Consent Statement:** Not applicable.

**Data Availability Statement:** Not applicable.

**Acknowledgments:** This research in part is supported by the Alexander von Humboldt Foundation.

**Conflicts of Interest:** The authors declare no conflict of interest.

## References

1. IEA. *Key World Energy Statistics*; International Energy Agency: Paris, France, 2015.
2. McQuiston, F.C.; Parker, J.D.; Spitler, J.D. *Heating, Ventilating, and Air Conditioning: Analysis and Design*; John Wiley & Sons: Hoboken, NJ, USA, 2004.
3. Turhan, C.; Kazanasmaz, T.; Uygun, I.E.; Ekmen, K.E.; Akkurt, G.G. Comparative study of a building energy performance software (KEP-IYTE-ESS) and ANN-based building heat load estimation. *Energy Build.* **2014**, *85*, 115–125. [CrossRef]
4. Catalina, T.; Iordache, V.; Caracaleanu, B. Multiple regression model for fast prediction of the heating energy demand. *Energy Build.* **2013**, *57*, 302–312. [CrossRef]
5. Geysen, D.; De Somer, O.; Johansson, C.; Brage, J.; Vanhoudt, D. Operational thermal load forecasting in district heating networks using machine learning and expert advice. *Energy Build.* **2018**, *162*, 144–153. [CrossRef]
6. Liu, S.; Chan, F.T.; Ran, W. Decision making for the selection of cloud vendor: An improved approach under group decision-making with integrated weights and objective/subjective attributes. *Expert Syst. Appl.* **2016**, *55*, 37–47. [CrossRef]

7. Tian, P.; Lu, H.; Feng, W.; Guan, Y.; Xue, Y. Large decrease in streamflow and sediment load of Qinghai–Tibetan Plateau driven by future climate change: A case study in Lhasa River Basin. *Catena* **2020**, *187*, 104340. [CrossRef]
8. Sun, Y.; Wang, J.; Wu, J.; Shi, W.; Ji, D.; Wang, X.; Zhao, X. Constraints Hindering the Development of High-Rise Modular Buildings. *Appl. Sci.* **2020**, *10*, 7159. [CrossRef]
9. Liu, S.; Yu, W.; Chan, F.T.S.; Niu, B. A variable weight-based hybrid approach for multi-attribute group decision making under interval-valued intuitionistic fuzzy sets. *Int. J. Intell. Syst.* **2021**, *36*, 1015–1052. [CrossRef]
10. Liu, Z.; Shao, J.; Xu, W.; Chen, H.; Zhang, Y. An extreme learning machine approach for slope stability evaluation and prediction. *Nat. Hazards* **2014**, *73*, 787–804. [CrossRef]
11. Piotrowski, A.P.; Osuch, M.; Napiorkowski, M.J.; Rowinski, P.M.; Napiórkowski, J.J. Comparing large number of metaheuristics for artificial neural networks training to predict water temperature in a natural river. *Comput. Geosci.* **2014**, *64*, 136–151. [CrossRef]
12. Liu, J.; Wu, C.; Wu, G.; Wang, X. A novel differential search algorithm and applications for structure design. *Appl. Math. Comput.* **2015**, *268*, 246–269. [CrossRef]
13. Feng, S.; Lu, H.; Tian, P.; Xue, Y.; Lu, J.; Tang, M.; Feng, W. Analysis of microplastics in a remote region of the Tibetan Plateau: Implications for natural environmental response to human activities. *Sci. Total. Environ.* **2020**, *739*, 140087. [CrossRef]
14. Fu, X.; Fortino, G.; Pace, P.; Aloi, G.; Li, W. Environment-fusion multipath routing protocol for wireless sensor networks. *Inf. Fusion* **2020**, *53*, 4–19. [CrossRef]
15. Han, X.; Zhang, D.; Yan, J.; Zhao, S.; Liu, J. Process development of flue gas desulphurization wastewater treatment in coal-fired power plants towards zero liquid discharge: Energetic, economic and environmental analyses. *J. Clean. Prod.* **2020**, *261*, 121144. [CrossRef]
16. He, L.; Chen, Y.; Zhao, H.; Tian, P.; Xue, Y.; Chen, L. Game-based analysis of energy-water nexus for identifying environmental impacts during Shale gas operations under stochastic input. *Sci. Total. Environ.* **2018**, *627*, 1585–1601. [CrossRef]
17. He, L.; Shen, J.; Zhang, Y. Ecological vulnerability assessment for ecological conservation and environmental management. *J. Environ. Manag.* **2018**, *206*, 1115–1125. [CrossRef]
18. Liu, J.; Liu, Y.; Wang, X. An environmental assessment model of construction and demolition waste based on system dynamics: A case study in Guangzhou. *Environ. Sci. Pollut. Res.* **2019**, *27*, 37237–37259. [CrossRef]
19. Wang, Y.; Yuan, Y.; Wang, Q.; Liu, C.; Zhi, Q.; Cao, J. Changes in air quality related to the control of coronavirus in China: Implications for traffic and industrial emissions. *Sci. Total. Environ.* **2020**, *731*, 139133. [CrossRef]
20. Liu, L.; Li, J.; Yue, F.; Yan, X.; Wang, F.; Bloszies, S.; Wang, Y. Effects of arbuscular mycorrhizal inoculation and biochar amendment on maize growth, cadmium uptake and soil cadmium speciation in Cd-contaminated soil. *Chemosphere* **2018**, *194*, 495–503. [CrossRef] [PubMed]
21. Yang, Y.; Liu, J.; Yao, J.; Kou, J.; Li, Z.; Wu, T.; Zhang, K.; Zhang, L.; Sun, H. Adsorption behaviors of shale oil in kerogen slit by molecular simulation. *Chem. Eng. J.* **2020**, *387*, 124054. [CrossRef]
22. Liu, J.; Yi, Y.; Wang, X. Exploring factors influencing construction waste reduction: A structural equation modeling approach. *J. Clean. Prod.* **2020**, *276*, 123185. [CrossRef]
23. Sun, L.; Li, C.; Zhang, C.; Liang, T.; Zhao, Z. The Strain Transfer Mechanism of Fiber Bragg Grating Sensor for Extra Large Strain Monitoring. *Sensors* **2019**, *19*, 1851. [CrossRef] [PubMed]
24. Su, Z.; Liu, E.; Xu, Y.; Xie, P.; Shang, C.; Zhu, Q. Flow field and noise characteristics of manifold in natural gas transportation station. *Oil Gas Sci. Technol. Rev. IFP* **2019**, *74*, 70. [CrossRef]
25. Liu, E.; Lv, L.; Yi, Y.; Xie, P. Research on the Steady Operation Optimization Model of Natural Gas Pipeline Considering the Combined Operation of Air Coolers and Compressors. *IEEE Access* **2019**, *7*, 83251–83265. [CrossRef]
26. Keshtegar, B.; Heddam, S.; Sebbar, A.; Zhu, S.-P.; Trung, N.-T. SVR-RSM: A hybrid heuristic method for modeling monthly pan evaporation. *Environ. Sci. Pollut. Res.* **2019**, *26*, 35807–35826. [CrossRef] [PubMed]
27. Ghaemi, A.; Rezaie-Balf, M.; Adamowski, J.; Kisi, O.; Quilty, J. On the applicability of maximum overlap discrete wavelet transform integrated with MARS and M5 model tree for monthly pan evaporation prediction. *Agric. For. Meteorol.* **2019**, *278*, 107647. [CrossRef]
28. Kisi, O.; Heddam, S. Evaporation modelling by heuristic regression approaches using only temperature data. *Hydrol. Sci. J.* **2019**, *64*, 653–672. [CrossRef]
29. Roy, D.K.; Barzegar, R.; Quilty, J.; Adamowski, J. Using ensembles of adaptive neuro-fuzzy inference system and optimization algorithms to predict reference evapotranspiration in subtropical climatic zones. *J. Hydrol.* **2020**, *591*, 125509. [CrossRef]
30. Zhang, B.; Xu, D.; Liu, Y.; Li, F.; Cai, J.; Du, L. Multi-scale evapotranspiration of summer maize and the controlling meteorological factors in north China. *Agric. For. Meteorol.* **2016**, *216*, 1–12. [CrossRef]
31. Chao, L.; Zhang, K.; Li, Z.; Zhu, Y.; Wang, J.; Yu, Z. Geographically weighted regression based methods for merging satellite and gauge precipitation. *J. Hydrol.* **2018**, *558*, 275–289. [CrossRef]
32. Chen, Y.; He, L.; Guan, Y.; Lu, H.; Li, J. Life cycle assessment of greenhouse gas emissions and water-energy optimization for shale gas supply chain planning based on multi-level approach: Case study in Barnett, Marcellus, Fayetteville, and Haynesville shales. *Energy Convers. Manag.* **2017**, *134*, 382–398. [CrossRef]
33. He, L.; Chen, Y.; Li, J. A three-level framework for balancing the tradeoffs among the energy, water, and air-emission implications within the life-cycle shale gas supply chains. *Resour. Conserv. Recycl.* **2018**, *133*, 206–228. [CrossRef]

34. Lu, H.; Tian, P.; He, L. Evaluating the global potential of aquifer thermal energy storage and determining the potential worldwide hotspots driven by socio-economic, geo-hydrologic and climatic conditions. *Renew. Sustain. Energy Rev.* **2019**, *112*, 788–796. [CrossRef]
35. Wu, C.; Wu, P.; Wang, J.; Jiang, R.; Chen, M.; Wang, X. Critical review of data-driven decision-making in bridge operation and maintenance. *Struct. Infrastruct. Eng.* **2020**, 1–24. [CrossRef]
36. Zhu, L.; Kong, L.; Zhang, C. Numerical Study on Hysteretic Behaviour of Horizontal-Connection and Energy-Dissipation Structures Developed for Prefabricated Shear Walls. *Appl. Sci.* **2020**, *10*, 1240. [CrossRef]
37. Zhao, X.; Ye, Y.; Ma, J.; Shi, P.; Chen, H. Construction of electric vehicle driving cycle for studying electric vehicle energy consumption and equivalent emissions. *Environ. Sci. Pollut. Res.* **2020**, *27*, 37395–37409. [CrossRef]
38. Deng, Y.; Zhang, T.; Sharma, B.K.; Nie, H. Optimization and mechanism studies on cell disruption and phosphorus recovery from microalgae with magnesium modified hydrochar in assisted hydrothermal system. *Sci. Total. Environ.* **2019**, *646*, 1140–1154. [CrossRef]
39. Zhang, T.; Wu, X.; Fan, X.; Tsang, D.C.; Li, G.; Shen, Y. Corn waste valorization to generate activated hydrochar to recover ammonium nitrogen from compost leachate by hydrothermal assisted pretreatment. *J. Environ. Manag.* **2019**, *236*, 108–117. [CrossRef]
40. Chen, Y.; Li, J.; Lu, H.; Yan, P. Coupling system dynamics analysis and risk aversion programming for optimizing the mixed noise-driven shale gas-water supply chains. *J. Clean. Prod.* **2021**, *278*, 123209. [CrossRef]
41. Cheng, X.; He, L.; Lu, H.; Chen, Y.; Ren, L. Optimal water resources management and system benefit for the Marcellus shale-gas reservoir in Pennsylvania and West Virginia. *J. Hydrol.* **2016**, *540*, 412–422. [CrossRef]
42. Li, L.-L.; Liu, Y.-W.; Tseng, M.-L.; Lin, G.-Q.; Ali, M.H. Reducing environmental pollution and fuel consumption using optimization algorithm to develop combined cooling heating and power system operation strategies. *J. Clean. Prod.* **2020**, *247*, 119082. [CrossRef]
43. Qian, J.; Feng, S.; Tao, T.; Hu, Y.; Li, Y.; Chen, Q.; Zuo, C. Deep-learning-enabled geometric constraints and phase unwrapping for single-shot absolute 3D shape measurement. *APL Photon* **2020**, *5*, 46105. [CrossRef]
44. Quan, Q.; Hao, Z.; Huang, X.; Lei, J. Research on water temperature prediction based on improved support vector regression. *Neural Comput. Appl.* **2020**, 1–10. [CrossRef]
45. Yang, M.; Sowmya, A. An Underwater Color Image Quality Evaluation Metric. *IEEE Trans. Image Process.* **2015**, *24*, 6062–6071. [CrossRef]
46. Cao, B.; Fan, S.; Zhao, J.; Yang, P.; Muhammad, K.; Tanveer, M. Quantum-enhanced multiobjective large-scale optimization via parallelism. *Swarm Evol. Comput.* **2020**, *57*, 100697. [CrossRef]
47. Qian, J.M.; Feng, S.; Li, Y.; Tao, T.; Han, J.; Chen, Q.; Zuo, C. Single-shot absolute 3D shape measurement with deep-learning-based color fringe projection profilometry. *Opt. Lett.* **2020**, *45*, 1842–1845. [CrossRef]
48. Lyu, Z.; Chai, J.; Xu, Z.; Qin, Y.; Cao, J. A Comprehensive Review on Reasons for Tailings Dam Failures Based on Case History. *Adv. Civ. Eng.* **2019**, *2019*, 1–18. [CrossRef]
49. Feng, W.; Lu, H.; Yao, T.; Yu, Q. Drought characteristics and its elevation dependence in the Qinghai–Tibet plateau during the last half-century. *Sci. Rep.* **2020**, *10*, 1–11. [CrossRef]
50. Yan, J.; Pu, W.; Zhou, S.; Liu, H.; Bao, Z. Collaborative detection and power allocation framework for target tracking in multiple radar system. *Inf. Fusion* **2020**, *55*, 173–183. [CrossRef]
51. Xiong, Z.; Tang, Z.; Chen, X.; Zhang, X.-M.; Zhang, K.; Ye, C. Research on Image Retrieval Algorithm Based on Combination of Color and Shape Features. *J. Signal Process. Syst.* **2019**, 1–8. [CrossRef]
52. Zhu, Q. Research on Road Traffic Situation Awareness System Based on Image Big Data. *IEEE Intell. Syst.* **2019**, *35*, 18–26. [CrossRef]
53. Xu, S.; Wang, J.; Shou, W.; Ngo, T.; Sadick, A.-M.; Wang, X. Computer Vision Techniques in Construction: A Critical Review. *Arch. Comput. Methods Eng.* **2020**, 1–15. [CrossRef]
54. Sun, L.; Yang, Z.; Jin, Q.; Yan, W. Effect of Axial Compression Ratio on Seismic Behavior of GFRP Reinforced Concrete Columns. *Int. J. Struct. Stab. Dyn.* **2020**, *20*, 2040004. [CrossRef]
55. Chao, M.; Kai, C.; Zhang, Z. Research on tobacco foreign body detection device based on machine vision. *Trans. Inst. Meas. Control.* **2020**, *42*, 2857–2871. [CrossRef]
56. Liu, D.; Wang, S.; Huang, D.; Deng, G.; Zeng, F.; Chen, H. Medical image classification using spatial adjacent histogram based on adaptive local binary patterns. *Comput. Biol. Med.* **2016**, *72*, 185–200. [CrossRef] [PubMed]
57. Abedini, M.; Mutalib, A.A.; Zhang, C.; Mehrmashhadi, J.; Raman, S.N.; Alipour, R.; Momeni, T.; Mussa, M.H. Large deflection behavior effect in reinforced concrete columns exposed to extreme dynamic loads. *Front. Struct. Civ. Eng.* **2020**, *14*, 532–553. [CrossRef]
58. Yang, C.; Gao, F.; Dong, M. Energy Efficiency Modeling of Integrated Energy System in Coastal Areas. *J. Coast. Res.* **2020**, *103*, 995–1001. [CrossRef]
59. Xiong, Z.; Xiao, N.; Xu, F.; Zhang, X.; Xu, Q.; Zhang, K.; Ye, C. An Equivalent Exchange Based Data Forwarding Incentive Scheme for Socially Aware Networks. *J. Signal Process. Syst.* **2020**, 1–15. [CrossRef]
60. Abedini, M.; Zhang, C. Performance Assessment of Concrete and Steel Material Models in LS-DYNA for Enhanced Numerical Simulation, A State of the Art Review. *Arch. Comput. Methods Eng.* **2020**, 1–22. [CrossRef]

61. Mou, B.; Li, X.; Bai, Y.; Wang, L. Shear Behavior of Panel Zones in Steel Beam-to-Column Connections with Unequal Depth of Outer Annular Stiffener. *J. Struct. Eng.* **2019**, *145*, 4018247. [CrossRef]
62. Mou, B.; Zhao, F.; Qiao, Q.; Wang, L.; Li, H.; He, B.; Hao, Z. Flexural behavior of beam to column joints with or without an overlying concrete slab. *Eng. Struct.* **2019**, *199*, 109616. [CrossRef]
63. Gholipour, G.; Zhang, C.; Mousavi, A.A. Numerical analysis of axially loaded RC columns subjected to the combination of impact and blast loads. *Eng. Struct.* **2020**, *219*, 110924. [CrossRef]
64. Zhang, H.; Qu, S.; Li, H.; Luo, J.; Xu, W. A Moving Shadow Elimination Method Based on Fusion of Multi-Feature. *IEEE Access* **2020**, *8*, 63971–63982. [CrossRef]
65. Yang, W.; Pudasainee, D.; Gupta, R.; Li, W.; Wang, B.; Sun, L. An overview of inorganic particulate matter emission from coal/biomass/MSW combustion: Sampling and measurement, formation, distribution, inorganic composition and influencing factors. *Fuel Process. Technol.* **2020**, *213*, 106657. [CrossRef]
66. Zhang, C.-W.; Ou, J.-P.; Zhang, J.-Q. Parameter optimization and analysis of a vehicle suspension system controlled by magnetorheological fluid dampers. *Struct. Control. Health Monit.* **2006**, *13*, 885–896. [CrossRef]
67. Yue, H.; Wang, H.; Chen, H.; Cai, K.; Jin, Y. Automatic detection of feather defects using Lie group and fuzzy Fisher criterion for shuttlecock production. *Mech. Syst. Signal Process.* **2020**, *141*, 106690. [CrossRef]
68. Zhu, G.; Wang, S.; Sun, L.; Ge, W.; Zhang, X. Output Feedback Adaptive Dynamic Surface Sliding-Mode Control for Quadrotor UAVs with Tracking Error Constraints. *Complexity* **2020**, *2020*, 8537198. [CrossRef]
69. Xiong, Q.; Zhang, X.; Wang, W.-F.; Gu, Y. A Parallel Algorithm Framework for Feature Extraction of EEG Signals on MPI. *Comput. Math. Methods Med.* **2020**, *2020*, 9812019. [CrossRef] [PubMed]
70. Zhang, J.; Liu, B. A review on the recent developments of sequence-based protein feature extraction methods. *Curr. Bioinform.* **2019**, *14*, 190–199. [CrossRef]
71. Zhang, X.; Fan, M.; Wang, D.; Zhou, P.; Tao, D. Top-k Feature Selection Framework Using Robust 0-1 Integer Programming. *IEEE Trans. Neural Netw. Learn. Syst.* **2020**, 1–15. [CrossRef]
72. Zhao, X.; Li, D.; Yang, B.; Chen, H.; Yang, X.; Yu, C.; Liu, S. A two-stage feature selection method with its application. *Comput. Electr. Eng.* **2015**, *47*, 114–125. [CrossRef]
73. Cao, B.; Zhao, J.; Lv, Z.; Gu, Y.; Yang, P.; Halgamuge, S.K. Multiobjective Evolution of Fuzzy Rough Neural Network via Distributed Parallelism for Stock Prediction. *IEEE Trans. Fuzzy Syst.* **2020**, *28*, 939–952. [CrossRef]
74. Shi, K.; Wang, J.; Tang, Y.; Zhong, S. Reliable asynchronous sampled-data filtering of T–S fuzzy uncertain delayed neural networks with stochastic switched topologies. *Fuzzy Sets Syst.* **2020**, *381*, 1–25. [CrossRef]
75. Shi, K.; Wang, J.; Zhong, S.; Tang, Y.; Cheng, J. Non-fragile memory filtering of T-S fuzzy delayed neural networks based on switched fuzzy sampled-data control. *Fuzzy Sets Syst.* **2020**, *394*, 40–64. [CrossRef]
76. Yang, S.; Deng, B.; Wang, J.; Li, H.; Lu, M.; Che, Y.; Wei, X.; Loparo, K.A. Scalable Digital Neuromorphic Architecture for Large-Scale Biophysically Meaningful Neural Network With Multi-Compartment Neurons. *IEEE Trans. Neural Netw. Learn. Syst.* **2020**, *31*, 148–162. [CrossRef]
77. Hornik, K.; Stinchcombe, M.; White, H. Multilayer feedforward networks are universal approximators. *Neural Netw.* **1989**, *2*, 359–366. [CrossRef]
78. Lv, Z.; Qiao, L. Deep belief network and linear perceptron based cognitive computing for collaborative robots. *Appl. Soft Comput.* **2020**, *92*, 106300. [CrossRef]
79. Adeli, H. Neural Networks in Civil Engineering: 1989–2000. *Comput. Civ. Infrastruct. Eng.* **2001**, *16*, 126–142. [CrossRef]
80. Xu, M.; Li, T.; Wang, Z.; Deng, X.; Yang, R.; Guan, Z. Reducing Complexity of HEVC: A Deep Learning Approach. *IEEE Trans. Image Process.* **2018**, *27*, 5044–5059. [CrossRef] [PubMed]
81. Li, T.; Xu, M.; Zhu, C.; Yang, R.; Wang, Z.; Guan, Z. A Deep Learning Approach for Multi-Frame In-Loop Filter of HEVC. *IEEE Trans. Image Process.* **2019**, *28*, 5663–5678. [CrossRef] [PubMed]
82. Qiu, T.; Shi, X.; Wang, J.; Li, Y.; Qu, S.; Cheng, Q.; Cui, T.; Sui, S. Deep Learning: A Rapid and Efficient Route to Automatic Metasurface Design. *Adv. Sci.* **2019**, *6*, 1900128. [CrossRef] [PubMed]
83. Chen, H.-L.; Wang, G.; Ma, C.; Cai, Z.-N.; Liu, W.-B.; Wang, S.-J. An efficient hybrid kernel extreme learning machine approach for early diagnosis of Parkinson's disease. *Neurocomputing* **2016**, *184*, 131–144. [CrossRef]
84. Hu, L.; Hong, G.; Ma, J.; Wang, X.; Chen, H. An efficient machine learning approach for diagnosis of paraquat-poisoned patients. *Comput. Biol. Med.* **2015**, *59*, 116–124. [CrossRef]
85. Wang, S.-J.; Chen, H.-L.; Yan, W.-J.; Chen, Y.-H.; Fu, X. Face Recognition and Micro-expression Recognition Based on Discriminant Tensor Subspace Analysis Plus Extreme Learning Machine. *Neural Process. Lett.* **2013**, *39*, 25–43. [CrossRef]
86. Xia, J.; Chen, H.; Li, Q.; Zhou, M.; Chen, L.; Cai, Z.; Fang, Y.; Zhou, H. Ultrasound-based differentiation of malignant and benign thyroid Nodules: An extreme learning machine approach. *Comput. Methods Progr. Biomed.* **2017**, *147*, 37–49. [CrossRef] [PubMed]
87. Li, Q.; Meng, Q.; Cai, J.; Yoshino, H.; Mochida, A. Applying support vector machine to predict hourly cooling load in the building. *Appl. Energy* **2009**, *86*, 2249–2256. [CrossRef]
88. Urdaneta, S.; Zapata, E.; Contreras, J. Fuzzy Model for Estimation of Energy Performance of Residential Buildings. *Int. J. Appl. Eng. Res.* **2017**, *12*, 2766–2771.
89. Fan, C.; Xiao, F.; Zhao, Y. A short-term building cooling load prediction method using deep learning algorithms. *Appl. Energy* **2017**, *195*, 222–233. [CrossRef]

90. Xie, L. The Heat load Prediction Model based on BP Neural Network-markov Model. *Procedia Comput. Sci.* **2017**, *107*, 296–300. [CrossRef]
91. Chou, J.-S.; Bui, D.-K. Modeling heating and cooling loads by artificial intelligence for energy-efficient building design. *Energy Build.* **2014**, *82*, 437–446. [CrossRef]
92. Roy, S.S.; Roy, R.; Balas, V.E. Estimating heating load in buildings using multivariate adaptive regression splines, extreme learning machine, a hybrid model of MARS and ELM. *Renew. Sustain. Energy Rev.* **2018**, *82*, 4256–4268. [CrossRef]
93. Sholahudin, S.; Han, H. Simplified dynamic neural network model to predict heating load of a building using Taguchi method. *Energy* **2016**, *115*, 1672–1678. [CrossRef]
94. Tsanas, A.; Xifara, A. Accurate quantitative estimation of energy performance of residential buildings using statistical machine learning tools. *Energy Build.* **2012**, *49*, 560–567. [CrossRef]
95. Gao, W.; Alsarraf, J.; Moayedi, H.; Shahsavar, A.; Nguyen, H. Comprehensive preference learning and feature validity for designing energy-efficient residential buildings using machine learning paradigms. *Appl. Soft Comput.* **2019**, *84*, 105748. [CrossRef]
96. Nilashi, M.; Dalvi-Esfahani, M.; Ibrahim, O.; Bagherifard, K.; Mardani, A.; Zakuan, N. A soft computing method for the prediction of energy performance of residential buildings. *Measurement* **2017**, *109*, 268–280. [CrossRef]
97. Moayedi, H.; Hayati, S. Artificial intelligence design charts for predicting friction capacity of driven pile in clay. *Neural Comput. Appl.* **2019**, *31*, 7429–7445. [CrossRef]
98. Fu, X.; Pace, P.; Aloi, G.; Yang, L.; Fortino, G. Topology optimization against cascading failures on wireless sensor networks using a memetic algorithm. *Comput. Netw.* **2020**, *177*, 107327. [CrossRef]
99. Moayedi, H.; Nguyen, H.; Foong, L.K. Nonlinear evolutionary swarm intelligence of grasshopper optimization algorithm and gray wolf optimization for weight adjustment of neural network. *Eng. Comput.* **2019**, 1–11. [CrossRef]
100. Amar, M.N.; Zeraibi, N. A combined support vector regression with firefly algorithm for prediction of bottom hole pressure. *SN Appl. Sci.* **2020**, *2*, 1–12.
101. Le, L.T.; Nguyen, H.; Dou, J.; Zhou, J. A Comparative Study of PSO-ANN, GA-ANN, ICA-ANN, and ABC-ANN in Estimating the Heating Load of Buildings' Energy Efficiency for Smart City Planning. *Appl. Sci.* **2019**, *9*, 2630. [CrossRef]
102. Cao, B.; Dong, W.; Lv, Z.; Gu, Y.; Singh, S.; Kumar, P. Hybrid Microgrid Many-Objective Sizing Optimization With Fuzzy Decision. *IEEE Trans. Fuzzy Syst.* **2020**, *28*, 2702–2710. [CrossRef]
103. Cao, B.; Wang, X.; Zhang, W.; Song, H.; Lv, Z. A Many-Objective Optimization Model of Industrial Internet of Things Based on Private Blockchain. *IEEE Netw.* **2020**, *34*, 78–83. [CrossRef]
104. Cao, B.; Zhao, J.; Yang, P.; Gu, Y.; Muhammad, K.; Rodrigues, J.J.P.C.; De Albuquerque, V.H.C. Multiobjective 3-D Topology Optimization of Next-Generation Wireless Data Center Network. *IEEE Trans. Ind. Informatics* **2020**, *16*, 3597–3605. [CrossRef]
105. Qu, S.; Han, Y.; Wu, Z.; Raza, H. Consensus Modeling with Asymmetric Cost Based on Data-Driven Robust Optimization. *Group Decis. Negot.* **2020**, 1–38. [CrossRef]
106. Roberts, A.; Marsh, A. *ECOTECT: Environmental Prediction in Architectural Education*; Cardiff University: Cardiff, UK, 2001.
107. McCulloch, W.; Pitts, W. A Logical Calculus of the Ideas Immanent in Nervous Activity (1943). *Ideas That Creat. Future* **2021**, *5*, 79–88. [CrossRef]
108. Seyedashraf, O.; Mehrabi, M.; Akhtari, A.A. Novel approach for dam break flow modeling using computational intelligence. *J. Hydrol.* **2018**, *559*, 1028–1038. [CrossRef]
109. Moayedi, H.; Mehrabi, M.; Mosallanezhad, M.; Rashid, A.S.A.; Pradhan, B. Modification of landslide susceptibility mapping using optimized PSO-ANN technique. *Eng. Comput.* **2019**, *35*, 967–984. [CrossRef]
110. Hornik, K. Approximation capabilities of multilayer feedforward networks. *Neural Netw.* **1991**, *4*, 251–257. [CrossRef]
111. Orr, M.J. *Introduction to Radial Basis Function Networks*; Technical Report; Center for Cognitive Science, University of Edinburgh: Edinburgh, UK, 1996.
112. Specht, D. A general regression neural network. *IEEE Trans. Neural Netw.* **1991**, *2*, 568–576. [CrossRef]
113. Marquardt, D.W. An Algorithm for Least-Squares Estimation of Nonlinear Parameters. *J. Soc. Ind. Appl. Math.* **1963**, *11*, 431–441. [CrossRef]
114. El-Bakry, M. Feed forward neural networks modeling for K–P interactions. *Chaos Solitons Fractals* **2003**, *18*, 995–1000. [CrossRef]
115. Cigizoglu, H.K.; Kişi, Ö. Flow prediction by three back propagation techniques using k-fold partitioning of neural network training data. *Hydrol. Res.* **2005**, *36*, 49–64. [CrossRef]
116. Mirjalili, S. The Ant Lion Optimizer. *Adv. Eng. Softw.* **2015**, *83*, 80–98. [CrossRef]
117. Simon, D. Biogeography-Based Optimization. *IEEE Trans. Evol. Comput.* **2008**, *12*, 702–713. [CrossRef]
118. Mirjalili, S.; Mirjalili, S.M.; Lewis, A. Let a biogeography-based optimizer train your multi-layer perceptron. *Inf. Sci.* **2014**, *269*, 188–209. [CrossRef]
119. Mirjalili, S. Dragonfly algorithm: A new meta-heuristic optimization technique for solving single-objective, discrete, and multi-objective problems. *Neural Comput. Appl.* **2016**, *27*, 1053–1073. [CrossRef]
120. Schwefel, H.-P.P. *Evolution and Optimum Seeking: The Sixth Generation*; John Wiley & Sons, Inc.: Hoboken, NJ, USA, 1993.
121. Mehrabian, A.; Lucas, C. A novel numerical optimization algorithm inspired from weed colonization. *Ecol. Inform.* **2006**, *1*, 355–366. [CrossRef]

122. Kashan, A.H. League Championship Algorithm: A New Algorithm for Numerical Function Optimization. In Proceedings of the 2009 International Conference of Soft Computing and Pattern Recognition, Malacca, Malaysia, 4–7 December 2009; IEEE: New York, NY, USA, 2009; pp. 43–48.
123. Spoljaric, T.; Pavic, I. Performance analysis of an ant lion optimizer in tuning generators' excitation controls in multi machine power system. In Proceedings of the 2018 41st International Convention on Information and Communication Technology, Electronics and Microelectronics (MIPRO), Opatija, Croatia, 21–25 May 2018; IEEE: New York, NY, USA, 2018; pp. 1040–1045.
124. Kose, U. An Ant-Lion Optimizer-Trained Artificial Neural Network System for Chaotic Electroencephalogram (EEG) Prediction. *Appl. Sci.* **2018**, *8*, 1613. [CrossRef]
125. Bhattacharya, A.; Chattopadhyay, P.K. Solving complex economic load dispatch problems using biogeography-based optimization. *Expert Syst. Appl.* **2010**, *37*, 3605–3615. [CrossRef]
126. Vanishree, J.; Ramesh, V. Optimization of size and cost of static var compensator using dragonfly algorithm for voltage profile improvement in power transmission systems. *Int. J. Renew. Energy Res.* **2018**, *8*, 56–66.
127. Yuan, C.; Moayedi, H. The performance of six neural-evolutionary classification techniques combined with multi-layer perception in two-layered cohesive slope stability analysis and failure recognition. *Eng. Comput.* **2020**, *36*, 1705–1714. [CrossRef]
128. Ghasemi, M.; Ghavidel, S.; Akbari, E.; Vahed, A.A. Solving non-linear, non-smooth and non-convex optimal power flow problems using chaotic invasive weed optimization algorithms based on chaos. *Energy* **2014**, *73*, 340–353. [CrossRef]
129. Kashan, A.H. League Championship Algorithm (LCA): An algorithm for global optimization inspired by sport championships. *Appl. Soft Comput.* **2014**, *16*, 171–200. [CrossRef]
130. Kashan, A.H. An efficient algorithm for constrained global optimization and application to mechanical engineering design: League championship algorithm (LCA). *Comput. Des.* **2011**, *43*, 1769–1792. [CrossRef]
131. Heidari, A.A.; Mirjalili, S.; Faris, H.; Aljarah, I.; Mafarja, M.; Chen, H. Harris hawks optimization: Algorithm and applications. *Future Gener. Comput. Syst.* **2019**, *97*, 849–872. [CrossRef]
132. Mirjalili, S.; Mirjalili, S.M.; Lewis, A. Grey wolf optimizer. *Adv. Eng. Softw.* **2014**, *69*, 46–61. [CrossRef]
133. Mirjalili, S.; Lewis, A. The Whale Optimization Algorithm. *Adv. Eng. Softw.* **2016**, *95*, 51–67. [CrossRef]
134. Karaboga, D. *An Idea Based on Honey Bee Swarm for Numerical Optimization*; Technical Report; Citeseer Technical Report-tr06; Erciyes University, Engineering Faculty, Computer Engineering Department Kayseri Turkey: Kayseri, Turkey, 2005.
135. Colorni, A.; Dorigo, M.; Maniezzo, V. Distributed optimization by ant colonies. In Proceedings of the First European Conference on Artificial Life, Paris, France, 11 December 1991; MIT Press: Cambridge, MA, USA, 1992; pp. 134–142.
136. Wang, G.-G.; Deb, S.; Coelho, L.D.S. Elephant herding optimization. In Proceedings of the 2015 3rd International Symposium on Computational and Business Intelligence (ISCBI), Bali, Indonesia, 7–9 December 2015; IEEE: New York, NY, USA, 2015; pp. 1–5.
137. Holland, J.H. *Adaptation in Natural and Artificial Systems: An Introductory Analysis with Applications to Biology, Control, and Artificial Intelligence*; University of Michigan Press: Ann Arbor, MI, USA, 1975.
138. Atashpaz-Gargari, E.; Lucas, C. Imperialist competitive algorithm: An algorithm for optimization inspired by imperialistic competition. In Proceedings of the 2007 IEEE Congress on Evolutionary Computation, Singapore, 25–28 September 2007. [CrossRef]
139. Kennedy, J.; Eberhart, R. Particle swarm optimization. In Proceedings of the ICNN'95—International Conference on Neural Networks, 27 November–1 December 1995; IEEE: New York, NY, USA, 1995; Volume 4, pp. 1942–1948.
140. Bayraktar, Z.; Komurcu, M.; Werner, D.H. Wind Driven Optimization (WDO): A novel nature-inspired optimization algorithm and its application to electromagnetics. In Proceedings of the 2010 IEEE Antennas and Propagation Society International Symposium, Toronto, ON, Canada, 11–17 July 2010; IEEE: New York, NY, USA, 2010; pp. 1–4.
141. Deng, H.; Fannon, D.; Eckelman, M.J. Predictive modeling for US commercial building energy use: A comparison of existing statistical and machine learning algorithms using CBECS microdata. *Energy Build.* **2018**, *163*, 34–43. [CrossRef]

*sustainability*

MDPI

*Article*

# Machine Learning-Based Node Characterization for Smart Grid Demand Response Flexibility Assessment

**Rostislav Krč [1], Martina Kratochvílová [1], Jan Podroužek [1,2,*], Tomáš Apeltauer [1], Václav Stupka [2] and Tomáš Pitner [2]**

[1] Institute of Computer Aided Engineering and Computer Science, Faculty of Civil Engineering, Brno University of Technology, 602 00 Brno, Czech Republic; rostislav.krc@vut.cz (R.K.); Martina.Kratochvilova1@vut.cz (M.K.); tomas.apeltauer@vut.cz (T.A.)

[2] Centre for Education, Research and Innovation in Information and Communication Technologies-ExecUnit, Faculty of Informatics, Masaryk University, 602 00 Brno, Czech Republic; stupka@fi.muni.cz (V.S.); tomp@fi.muni.cz (T.P.)

* Correspondence: podrouzek.j@fce.vutbr.cz

**Abstract:** As energy distribution systems evolve from a traditional hierarchical load structure towards distributed smart grids, flexibility is increasingly investigated as both a key measure and core challenge of grid balancing. This paper contributes to the theoretical framework for quantifying network flexibility potential by introducing a machine learning based node characterization. In particular, artificial neural networks are considered for classification of historic demand data from several network substations. Performance of the resulting classifiers is evaluated with respect to clustering analysis and parameter space of the models considered, while the bootstrapping based statistical evaluation is reported in terms of mean confusion matrices. The resulting meta-models of individual nodes can be further utilized on a network level to mitigate the difficulties associated with identifying, implementing and actuating many small sources of energy flexibility, compared to the few large ones traditionally acknowledged.

**Keywords:** smart grid; electricity network; flexibility assessment; renewable energy sources; machine learning; network simulation; artificial neural networks; convolutional neural networks

**Citation:** Krč, R.; Kratochvílová, M.; Podroužek, J.; Apeltauer, T.; Stupka, V.; Pitner, T. Machine Learning-Based Node Characterization for Smart Grid Demand Response Flexibility Assessment. *Sustainability* **2021**, *13*, 2954. https://doi.org/10.3390/su13052954

Academic Editor: Mehdi Seyedmahmoudian, Alex Stojcevski, Ben Horan, Saad Mekhilef and Eklas Hossain

Received: 26 January 2021
Accepted: 1 March 2021
Published: 9 March 2021



## 1. Introduction

Safe and reliable operation of energy systems depends on maintaining a balance between consumption and production in real-time, while an increasingly large part of the production portfolio depends on the inherently variable non-stationary climate, such as renewable wind or solar sources [1].

The variability of the consumption, on the other hand, can be relatively easily controlled, when compared to the climate-dependent part of the production portfolio, by purposefully influencing the consumption behavior, i.e., targeting the reducible (epistemic) uncertainty components on the consumer side, rather than focusing only on mostly aleatory (irreducible) uncertainties related to long-term weather forecasting.

The ability to purposefully influence both the production and consumption behavior of selected elements of the electricity system is therefore gradually gaining in importance (e.g., [1–4], as it can effectively reduce fluctuations in the overall load diagram and thus reduce the demands on available power and dynamics of Support Services, as well as the associated costs.

The electricity market is currently at a crossroads. The current market model assumes that the market will ensure both short-term optimization, such as effective allocation of the necessary production among existing capacities, and long-term investment signals for the construction of new capacities. However, the significant degree of market distortions in the sector practically paralyzed this function of the market model. Such development leads to a situation where investors are only looking for the construction of sources with guaranteed (subsidized) prices. Investments in resources and networks are thus driven by state incentives, instead of the market. Under such conditions, market development without adjustments by the state leads to an unbalanced resource mix with a number of strategic and systemic risks for the future [1].

An important part of any state's critical infrastructure is its electricity network. This is traditionally based on centralized generation in large power plants, however, as the share of renewable energy production increases, the grid will need to adapt to a large number of smaller sources. Decentralized production growth is enabled by the spread of new technologies and typically benefits the local economy.

However, the transformation of energy must meet the basic conditions. These are secure supplies during normal operation as well as in the event of sudden changes in external conditions, and competitive prices. At the same time, energy must be sustainable in the sense that it does not harm the environment, is able to provide raw materials for its operation and the whole sector is economically stable.

The aim of this work is to contribute to the theoretical framework for quantifying network flexibility potential by introducing a machine learning-based node characterization. It is unique in the successful utilization of state-of-the-art convolutional neural network models for the classification of historic demand data from the Ausgrid distribution zone substation data. After introducing the related concepts of smart grid, grid flexibility and network modeling, demand interval data used for this study are introduced together with the clustering analysis performed. Next, machine learning-based time-series classification and surrogate resampling concepts are discussed, together with various architectures of convolutional neural network models. Finally, the statistics of resulting classifiers are discussed.

### 1.1. Smart Grid

Decentralizing the energy system and thus at least partially replacing large-scale energy production (e.g., fossil, nuclear or hydroelectric) is an increasingly common effort. These facilities are usually far from the end consumer and therefore require an extensive and reliable high-voltage transmission network. The global tendency to achieve a sustainable economy and improve the environment leads to higher use of energy from renewable sources and thus, for example, to reduce the global temperature disruption [5]. Carbon dioxide emissions during energy production account for about two-thirds of all greenhouse gases [6]. Power plants using renewable sources are usually smaller in format and closer to the end-user. Thus, energy is not transmitted over such long distances and the transmission network has a decentralized structure. This results in fewer losses during transmission and the network is less vulnerable because it does not depend on a small number of remote large power plants. The whole system is therefore composed of smaller subsystems, which do not have to be interdependent, but still communicate with each other and can help each other.

An ideal (smart) grid is a modernized electrical self-monitoring grid that can combine conventional central sources with alternative sources of electricity [7]. This includes an intelligent control system that monitors and adjusts the operation of the network in real-time, including the self-healing capabilities and supported by intelligent elements, without the need for human intervention. Smart grids communicate with the customer in real-time and help to optimize the consumption with regard to the current price of electricity and the burden on the environment, allowing better integration of renewable electricity sources and improving the efficiency, reliability, economics, and sustainability of the production and distribution of electricity.

As the production of energy using renewable sources is difficult to predict in the long run (climate-dependent production), it is necessary to be able to target consumers and ensure communication between individual entities. In this way, the demands on the peak loads and the operational cost and costs of providing support services can be reduced [8].

This can be done through a combination of technical and economic tools. Various smart metrics for measuring, communication, synchronization, forecasting and control are being developed [9]. Among the economic instruments, it is possible to name, for example, the real-time pricing [10] or the Adaptive Billing Mechanism [11], which can work, with negative energy prices and thus flatten the oscillation of the overall load diagram. A whole new market with new entrants can be expected.

Information security is discussed in [12], where risk propagation model based on the Susceptible–Exposed–Infected–Recovered (SEIR) infectious disease model is proposed for a smart grid. The high volatility and uncertainty of load profiles and the tremendous communication pressure are discussed in a two-stage household electricity demand estimation study by [13]. Investigation of voltage control at consumers connection points based on smart approach has recently been carried out by [14], proposing a voltage control system for use in the Russian distribution grid.

### 1.2. Grid Flexibility

Flexibility is considered a key enabler for the smart grid according to O'Connell et al. [15], and is required to facilitate Demand-Side Management (DSM) programs, manage electrical consumption to reduce peaks, balance renewable generation and provide ancillary services to the grid. The ISO 50002:2014 [16] specifies the process requirements for carrying out an energy audit in relation to energy performance. It is applicable to all types of establishments and organizations, and all forms of energy and energy use. This standard can be used to assess flexibility and formulate optimization requirements [2,15]. According to a given scale, flexibility analysis can help to identify and quantify the available electrical load at a network or node level, i.e., substation, site or building.

U.S. Energy Information Administration (EIA) [17] defines DSM programs as those including planning, implementing, and monitoring activities of electric utilities which are designed to encourage consumers to modify their level and pattern of electricity usage. In its international energy outlook or other EIA annual reports, projected and actual energy production can be compared with the global changes in manufacturing and services share, an important component in any flexibility analysis for smart grid DSM.

The primary objective of most DSM programs in the past was to provide cost-effective energy and capacity resources in order to help defer the need for new sources of power, including generating facilities, power purchases, and transmission and distribution capacity additions. However, due to changes that are occurring within the industry, electric utilities are also using DSM as a way to enhance customer service. According to EIA, DSM refers to only energy and load-shape modifying activities that are undertaken in response to utility-administered programs. It does not refer to energy and load-shape changes arising from the normal operation of the marketplace or from government-mandated energy-efficiency standards.

Moreover, the European Commission's (EC) 2020 targets [18] to generate 20% of Europe's energy from renewable energy and reduce greenhouse gasses emissions by 20% have already resulted in increased climate-dependent production. In order to further increase this production to 25%, all aspects of grid flexibility have to be carefully addressed to ensure grid resilience and stability. This includes, among others, the ability to balance non-dispatchable sources and managing the power locally.

In the International Energy Agency's annex 67 [19], energy flexibility is also presented as a key asset in the smart building future, where buildings can manage itself, interact with their users and take part in demand response.

### 1.3. Network Modeling

In the current state of discussions, energy flexibility is typically associated with "smartness" and evaluated either in a qualitative framework according to the number and type of services provided by its components, or, as presented in this paper, by quantitative and physical indicators, utilizing measured (historic) data and network-level simulations.

To better understand the functioning of Smart grids and investigate the possibilities of optimization of their functions, it is appropriate to create a mathematical meta-model of individual network nodes and simulate the operation of the whole network, where the nodes are connected according to real network topology and edge capacity. Because it is a decentralized system, decision intelligence is divided between the individual nodes of the network. Agent models are commonly used for simulations, where each agent has its own decision-making power and none of them depends on any central authority [20]. Relationships and connections between agents are usually modeled using network theory [21]. The agent decision-making process and behavior prediction can be modeled, for example, using machine learning (ML) [22].

Among recent contributions to the integrated simulation of power and communication networks for smart grid, applications can be found [4], where the smart grid discrete-event simulator is implemented in C++ using the open-source OMNeT++ simulation environment. In [23], a comprehensive real-time simulation of the smart grid is presented, including a microgrid model of a small community. A recent overview of simulation and modeling application to residential demand response can be found in [24].

In order to simulate the behavior of the entire network, and to evaluate the impact of various control strategies on the power grid, it is necessary to validate the behavior of the individual nodes first. In this paper, a historic 15 min interval demand data from Ausgrid substations [3,25] have been classified using machine learning methods.

## 2. Interval Demand Data

Publicly available distribution zone substation 15-min interval demand data from the Australian network operator Ausgrid [25] have been used for the machine learning-based node characterization in order to support computational reproducibility of this research. In particular, historic data from the year 2019 (between May 2018 and April 2019) and from the year 2020 (between May 2019 and April 2020) from 185 substations from distribution networks around Sydney and Newcastle have been used. These substations form the boundary between the sub-transmission network and the distribution (11 kV) network. The time is in Australian Eastern Standard Time (AEST) format during the winter period and Australian Eastern Daylight Time (AEDT) during the summer period. Figure 1 shows the irregular topology of the investigated distribution network, where real node positions (red dots) correspond to population density clusters, resulting in large variability of edge length (black lines).

Data from individual substations were sampled at 15 min and divided into time series by days, resulting in 96 data points. The proposed classification cannot intentionally distinguish between workdays, weekends or holidays, as there is no information on the actual date attached to the individual time series, although the dataset exhibit typical daily, weekly and seasonal fluctuations in electricity demand. Nevertheless, the achieved accuracies are far from prohibitive, as discussed in the next chapter.

The daily fluctuations include morning and afternoon peaks throughout all four seasons, including workdays and weekends, see Figures 2–5 for an example from Harbord substation, with group averages in dark color. In summer, a lower overall demand can be observed, including less distinct morning and evening peaks.

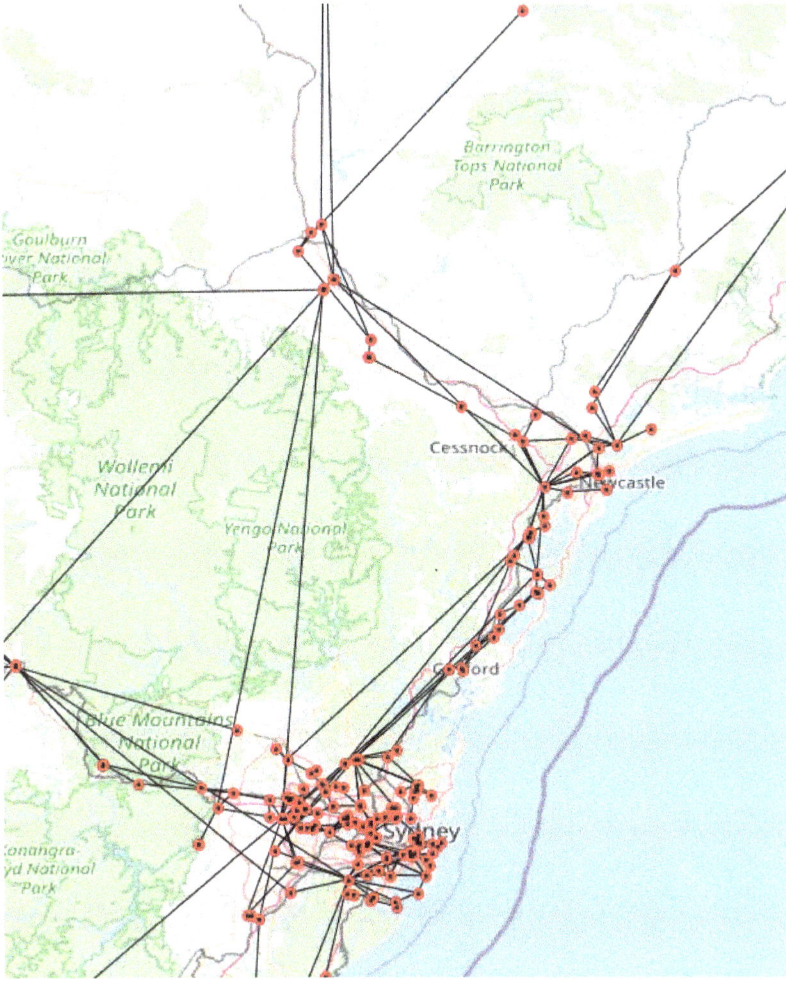


**Figure 1.** Irregular topology of the investigated distribution network (map source [26]).

*Clustering Analysis*

In general, the goal of clustering is to identify structure in an unlabeled data set by objectively organizing data into homogeneous groups where the within-group-object similarity is minimized and the between-group-object dissimilarity is maximized [27]. The time-series demand data presented in this section have been clustered by the basic k-means algorithm [28] in order to split substations into several groups. K-means clustering is a renowned heuristic method for crisp partitions (i.e., each object belongs to exactly one cluster, as opposed to fuzzy if one object is allowed to be in more than one cluster to a different degree), where each cluster is represented by the mean value of the objects in the cluster.

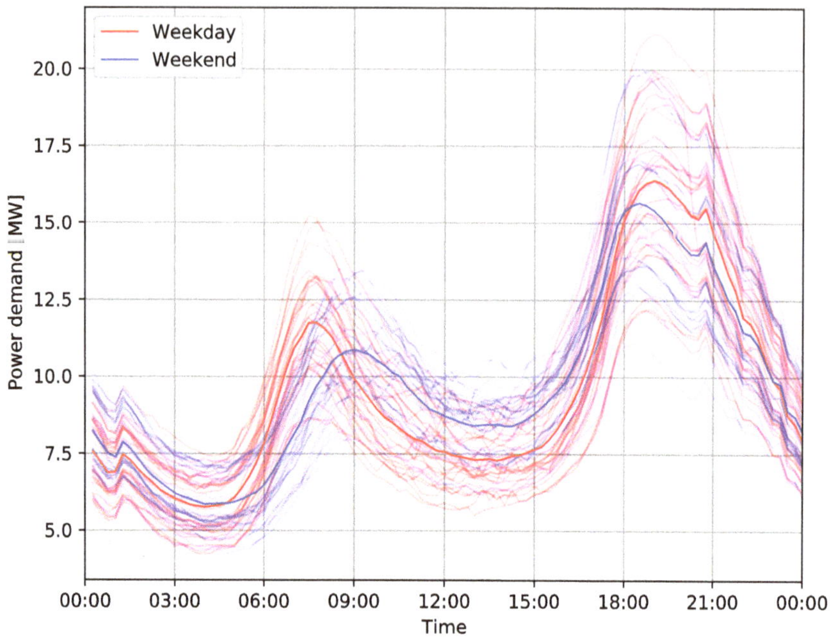


**Figure 2.** Daily fluctuations and seasonal scattering at Harbord substation, May–July 2018.

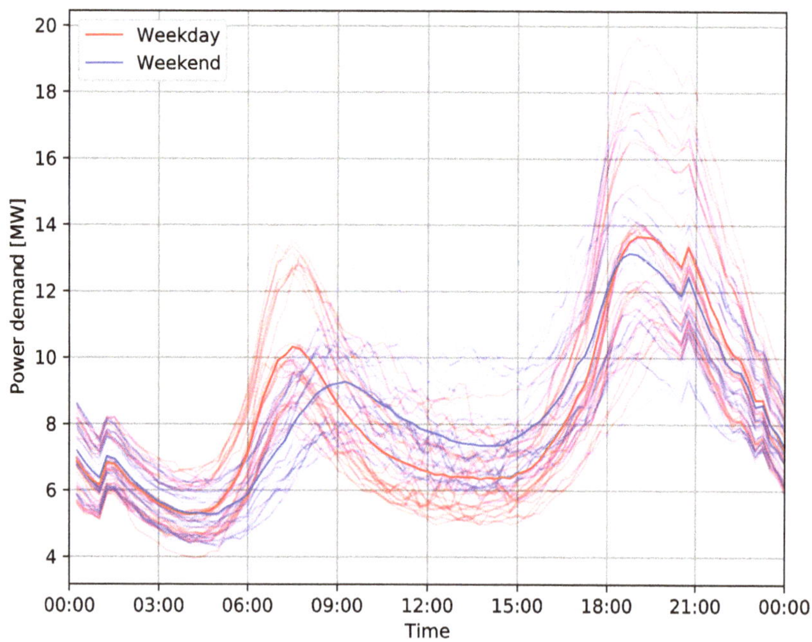


**Figure 3.** Daily fluctuations and seasonal scattering at Harbord substation, August–October 2018.

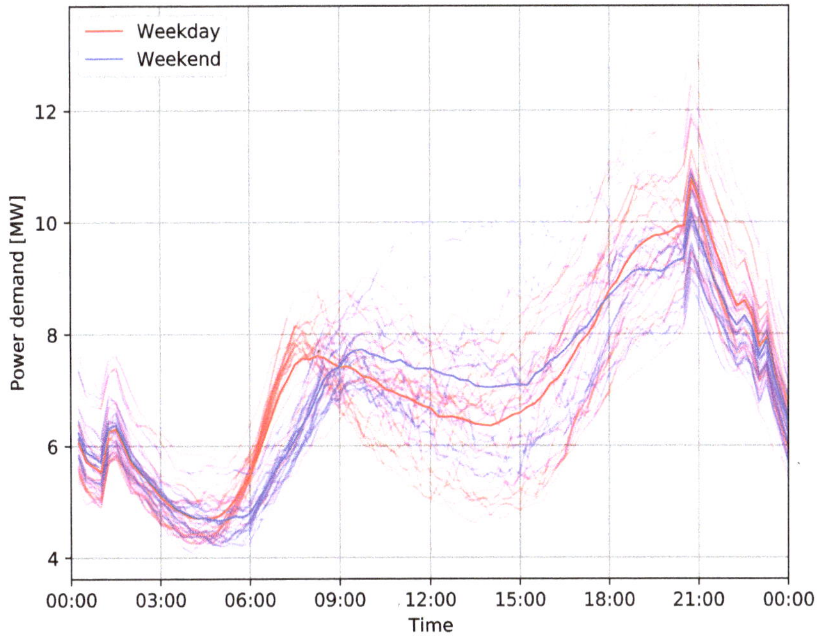


**Figure 4.** Daily fluctuations and seasonal scattering at Harbord substation, November 2018–January 2019.

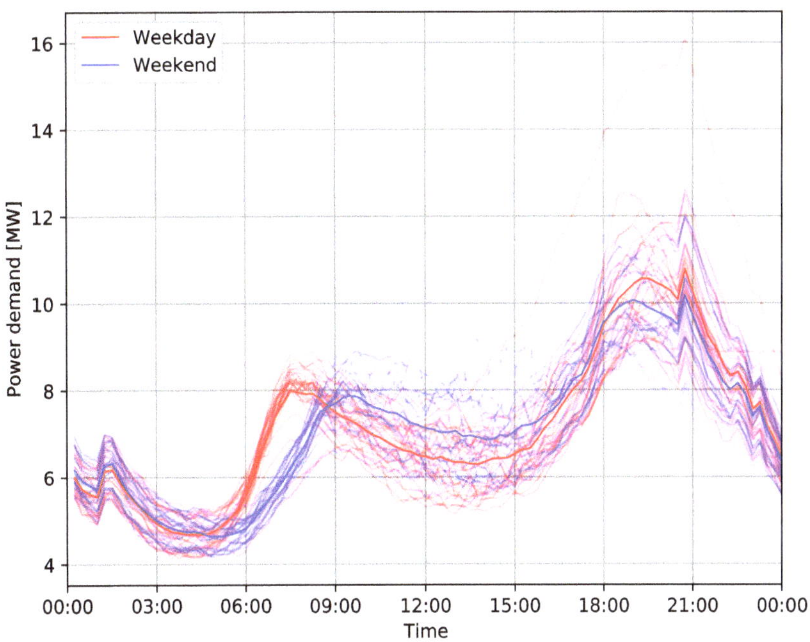


**Figure 5.** Daily fluctuations and seasonal scattering at Harbord substation, February–April 2019.

For effective optimization of the distribution grid, and for any network-level simulations in general, the behavior of nodes (i.e., stations and substations) must be understood. Daily fluctuations of power demand may depend on the type of supplied neighborhood. Three basic types of substation neighborhoods—residential, business and combined—were considered as only 96 equally spaced data points were available per signal. K-means clustering analysis with k = 3 identified the following clusters (Figures 6–8):

- C1: 48 substations, residential, morning and evening demand peak,
- C2: 66 substations, combined, morning and evening peaks less distinct than in the clusters C1,
- C3: 71 substations, business, high and flat distribution of energy demand during the day.

Note that the qualitative evaluation of characteristics of the three identified clusters is based only on the assumption of daily peak distribution and has not been verified by any other on-site investigation as it was not the goal of this research. Note the normalization of amplitudes of daily signals to the maximum value in order to ensure classification based on demand patterns instead of absolute values of demand.

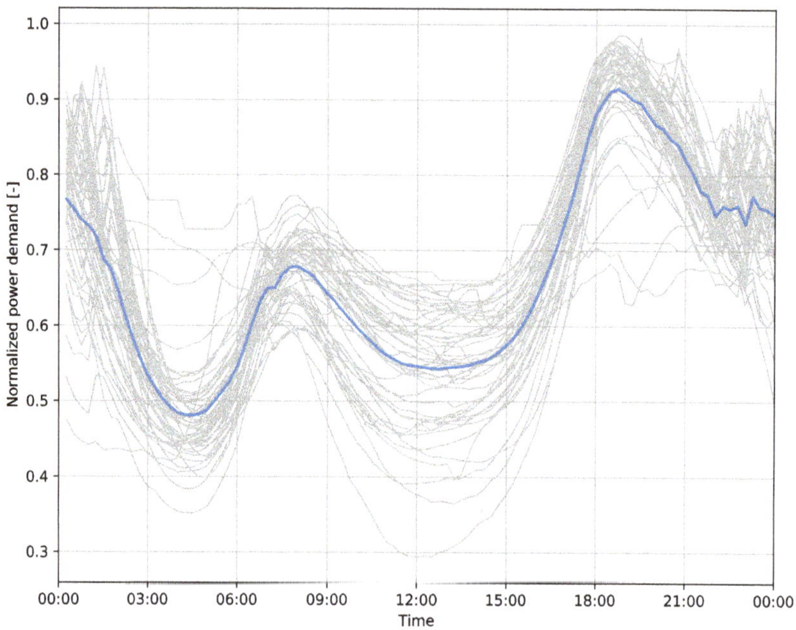


**Figure 6.** Cluster C1 with average daily power demand for each substation (grey) and overall average demand (blue).

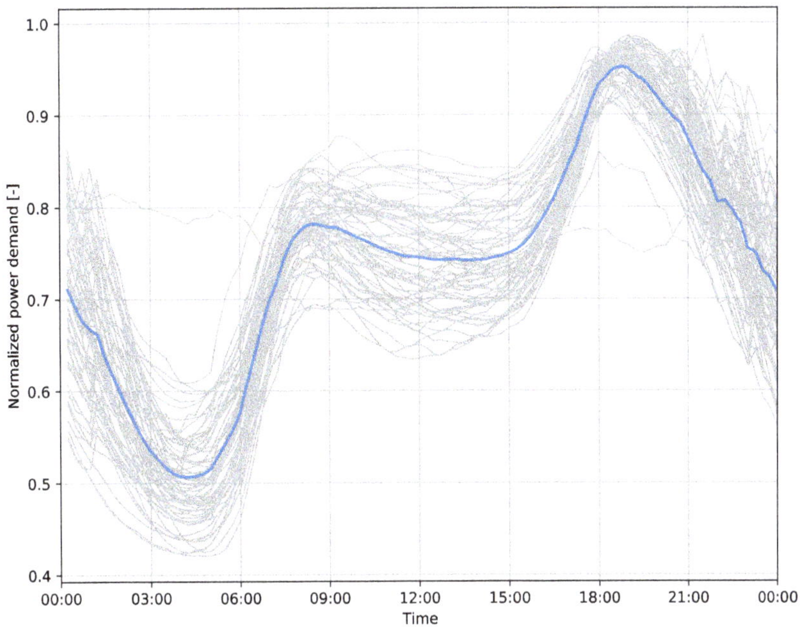


**Figure 7.** Cluster C2 with average power demand for each substation (grey) and overall average demand (blue).

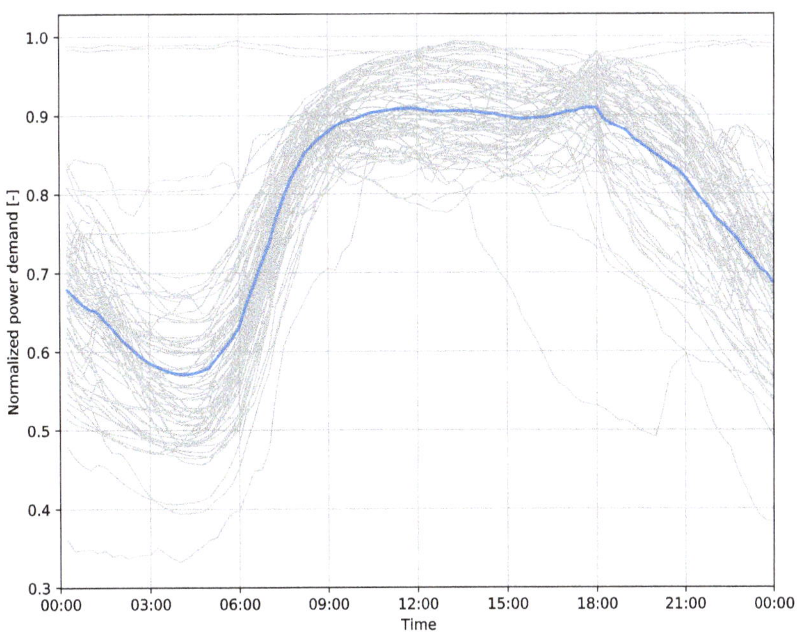


**Figure 8.** Cluster C3 with average power demand for each substation (grey) and overall average demand (blue).

## 3. Machine Learning Based Classification

### 3.1. Surrogates Resampling

In each identified cluster, 14 substations have been randomly selected (see Table 1) for the machine learning-based classification. The goal is to identify a substation based on its daily demand, all selected substations being type 33/11 kV. Since data for all substations were not complete, a criterion for the minimal number of measurement days per substation has been set to 350 days of a year.

**Table 1.** Selected substations for each cluster.

| C1 | C2 | C3 |
|---|---|---|
| Avondale | Baerami | Auburn |
| Cessnock South | Blakehurst | Blackwattle Bay |
| Edgeworth | Campsie | Brookvale |
| Harbord | Dulwich Hill | Camperdown |
| Jannali | Enfield | City East |
| Killarney | Gateshead | Darlinghurst |
| Medowie | Lindfield | Graving Dock |
| Mt Hutton | Maitland | Kotara |
| Newport | Nelson Bay | Lucas Heights |
| Paxton | Peats Ridge | New Lambton |
| Sans Souci | Riverwood | Newcastle CBD |
| Swansea | St Ives | Paddington |
| Tanilba Bay | Thornton | Surry Hills |
| Tomaree | Williamtown | Tomago |

Three scenarios were evaluated considering data from year the 2019, 2020 and both years respectively. In the first two scenarios, data were divided between training and testing set using a typical split ratio 80%/20%. Data were selected randomly from the original dataset in such a way that the same split ratio is ensured for each substation. In order to avoid over-fitting and evaluate the reliability of the models, 10 splits of the dataset were randomly generated. This technique is often referred to as bootstrapping [29] and has been preferred over cross-validation due to the limited number of time series available, as cross-validation resamples without replacement and thus produces surrogate data sets that are smaller than the original. Bootstrapping used here resamples without replacement, produces surrogate datasets with the same number of time series as the original dataset, therefore statistical evaluation of the performance of the model becomes available, as represented here by the average confusion matrices and reported variability of accuracies. Each row in a confusion matrix represents the instances in a real class, while each column represents the instances in a predicted class, so whether the system is confusing two classes can easily be visible.

In the third scenario, denoted as 2019–2020, data from the year 2019 were used for training and data from the year 2020 for testing. Five runs for each cluster and model were done and their results averaged in order to consider variance due to random initialization of network parameters and data shuffling. This scenario aims to evaluate generalization capability of the used model for future years. Setup of all three scenarios is listed in Table 2.

**Table 2.** Setup of the evaluated years and clusters.

| Year | Cluster | No. of Classes | Dataset Size | Training Size | Testing Size | No. of Repeats |
|---|---|---|---|---|---|---|
| 2019 | C1 | 14 | 5085 | 4070 | 1015 | 10 |
| | C2 | 14 | 5081 | 4068 | 1013 | 10 |
| | C3 | 14 | 5085 | 4069 | 1016 | 10 |
| 2020 | C1 | 14 | 5109 | 4091 | 1018 | 10 |
| | C2 | 14 | 5098 | 4082 | 1016 | 10 |
| | C3 | 14 | 5105 | 4087 | 1018 | 10 |
| 2019–2020 | C1 | 14 | 10,194 | 5085 | 5109 | 5 |
| | C2 | 14 | 10,179 | 5081 | 5098 | 5 |
| | C3 | 14 | 10,190 | 5085 | 5105 | 5 |

### *3.2. CNN Models*

Commonly used machine learning methods for the classification of time-series data are Support Vector Machines (SVM) and Artificial Neural Networks (ANN), as recently reported in [30–32].

SVM is a simple algorithm that looks for a hyperplane that divides the n-dimensional input space into two or more categories and assigns an output value accordingly. However, there is an infinite number of such planes, and the goal of the algorithm is to find a plane that has the maximum distance from the points of both (all) classes. Multidimensional problems are usually transformed using the so-called kernel transformation, so the non-linear problem is converted to linear. That means, from the original space to the Euclidean space. Thus, it is clear that the correct function of the SVM depends on the correct choice of the kernel function. This method is computationally expensive if large amounts of data are to be considered.

The best known and probably also the most universal ML algorithm is the Artificial Neural Network (ANN) method. This algorithm is inspired by the decision-making processes of the human brain. It is composed of several million neurons, which evaluate and pass information to each other. Likewise, an artificial neural network is composed of layers, which are composed of neurons. Each layer has given rules, based on which it evaluates the input information from the previous layer and passes the output to the next layer. Input and output can be of different formats. Within ANN, however, the information is transmitted using an internal weighing system. The number of layers and neurons in them is arbitrary, as well as the type of layers. However, all these parameters affect the reliability of ANN.

Both SVM and ANN are sensitive to the subjective choice of parameters, in the case of SVM, these are the describing (scalar) features of the time-series, such as the number of peaks, total energy or various Fourier transform properties. In the case of ANN, the subjective choice of its architecture can significantly influence both its performance as well as computational requirements. Given the relatively small size of the time series (96 data points), compared to applications that differ by two orders of magnitude, where high-frequency components have to be maintained, such as the dynamic response of railway track due to a passing train, which, if resampled to a lower resolution, looses its most important characteristics (see e.g., [30]), finding proper characterization for SVM input vector would make a little sense, since the entire time series vector can be directly processed by a more general ANN model.

Some advanced time series classification techniques can be used such as Least-Squares Wavelet (LSWAVE) [33]. The spectral and wavelet analyses are very useful for estimating trends and seasonal components of any time series and identifying their patterns in the time-frequency domain [34]. Herein, we directly classify the time series data and shall leave the use of wavelets to future research.

A specific type of ANN, the convolutional neural networks (CNN) were selected for data classification as they provide state-of-the-art performance for computer vision or time-series classification and are also widely adopted for end-to-end learning [35]. End-to-end approach utilizes raw time-series data without any preprocessing and manual feature extraction which often introduces unnecessary bias as extracted features are often domain-specific. Convolutional layers in CNN also enhance pattern recognition capabilities of the network.

In the following, three CNN models with different architectures have been considered:

- CNN1 has the same architecture as the best performing model CNN in [30]. It contains one convolutional layer with 64 filters followed by max-pooling layer, fully connected hidden layer and an output layer;
- CNN2 contains three convolutional layers with 128, 64 and 32 filters, respectively. Batch normalization and dropout with 25% probability is applied to the output of the last convolutional layer before passed to a max-pooling layer. The result of the pooling layer is flattened to one fully connected output layer. It is the deepest architecture with the largest number of layers with trainable parameters;
- CNN3 contains one convolutional layer with 64 filters followed by an average pooling layer with output size 20 and fully connected output layer. This architecture contains the lowest number of trainable parameters.

Rectified linear unit (ReLU) has been used between layers as an activation function for all presented architectures. Overview of parameters for evaluated architectures is presented in Table 3.

**Table 3.** Number of layers and number of trainable parameters for the evaluated convolutional neural networks.

| **Model** | **CNN1** | **CNN2** | **CNN3** |
|---|---|---|---|
| Number of layers with trainable parameters | 4 | 6 | 3 |
| Number of trainable parameters | 123,498 | 122,830 | 18,446 |
| Average training time (s) [1,2] | 96.4 | 99.5 | 93.6 |

[1] C2 dataset in year 2019 (4069 training samples), [2] NVIDIA GeForce GTX 1050 Max-Q with 640 cores and 3 GB of memory.

Training has been executed in 20 epochs. Data were forwarded through the networks in batches of size 32. The learning rate has been set to 0.001, while Adam optimizer has been used to minimize cross-entropy loss function (which is a composition of negative log-likelihood and logarithmic softmax function).

A graphics processing unit (GPU) has been used for training, approximately 0.6–0.8 GB of memory has been required, and models with a lower number of parameters (CNN3) had only slightly lower training time.

## 4. Results

The most accurate model with mean accuracy over 88.8% and deviation less than 1.5% for all clusters and all scenarios is the CNN2 due to the incorporation of multiple convolutional layers. Slightly lower accuracy can be seen in the scenario 2019–2020, compared to the scenarios 2019 and 2020.

It was also shown that CNN2 model has the best ability to generalize future years (trained 2019, tested 2020), C1 cluster shows higher accuracy compared to C2 and C3.

The highest mean accuracy can be observed in scenario 2019–2020 for CNN2 model and cluster C1 at Jannali substation (99.3%), while the lowest mean accuracy occurred in cluster C2 at Maitland substation (67.3%), which was often confused with Thornton.

The mean model accuracies are presented in Table 4 or in graphical representation, Figure 9, where the accuracy of independent and random selection results (base) are also included for reference.

Results are presented by means of mean confusion matrices for scenario three (training on 2019 data and testing on 2020 data) and the best performing CNN2 classifier model for individual clusters C1 to C3, see Figures 10–12. These matrices show how the model is able to cope with the classification of new data in future years and which classes (substations) are often confused.

**Table 4.** Mean model accuracies for each year and cluster.

| Year | Cluster | Base (%) | Mean (%) | CNN1 (%) | CNN2 (%) | CNN3 (%) |
|---|---|---|---|---|---|---|
| 2019 | C1 | 7.1 | 83.7 ± 10.0 | 80.7 ± 2.5 | 96.8 ± 1.4 | 73.6 ± 2.8 |
| | C2 | 7.1 | 80.9 ± 10.0 | 75.3 ± 2.4 | 94.6 ± 1.4 | 72.8 ± 2.3 |
| | C3 | 7.1 | 81.6 ± 8.8 | 78.7 ± 2.4 | 93.3 ± 1.4 | 72.9 ± 1.7 |
| 2020 | C1 | 7.1 | 83.3 ± 10.7 | 79.7 ± 4.7 | 97.2 ± 0.9 | 72.9 ± 1.8 |
| | C2 | 7.1 | 75.0 ± 13.8 | 64.5 ± 6.9 | 93.5 ± 1.4 | 66.9 ± 2.7 |
| | C3 | 7.1 | 79.0 ± 9.9 | 74.7 ± 2.9 | 92.4 ± 1.3 | 70.0 ± 2.7 |
| 2019–2020 | C1 | 7.1 | 79.1 ± 12.0 | 76.8 ± 0.9 | 94.8 ± 0.7 | 65.7 ± 1.4 |
| | C2 | 7.1 | 74.2 ± 10.4 | 68.0 ± 1.5 | 88.8 ± 1.4 | 65.9 ± 1.0 |
| | C3 | 7.1 | 79.9 ± 7.7 | 77.4 ± 3.6 | 89.9 ± 1.5 | 72.4 ± 1.3 |

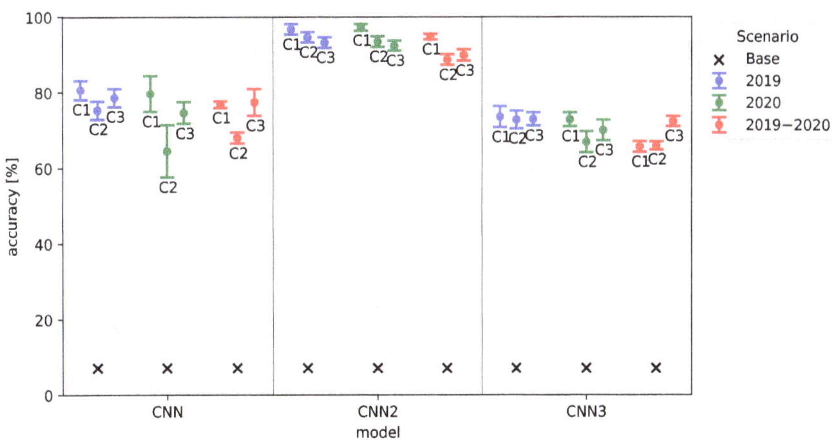


**Figure 9.** Accuracy of evaluated neural network models for different clusters in different years (mean +− standard deviation).

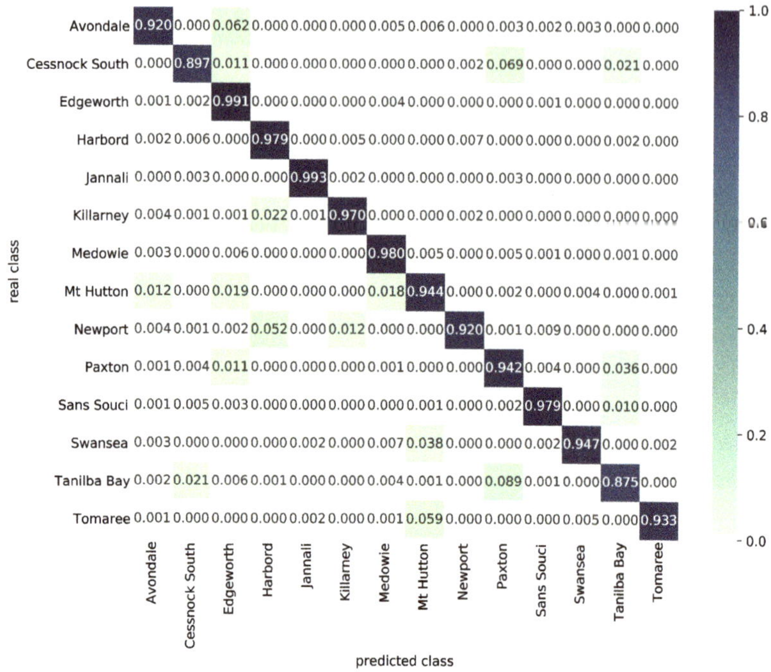


**Figure 10.** Mean confusion matrix for best performing model convolutional neural networks CNN2 and cluster C1.

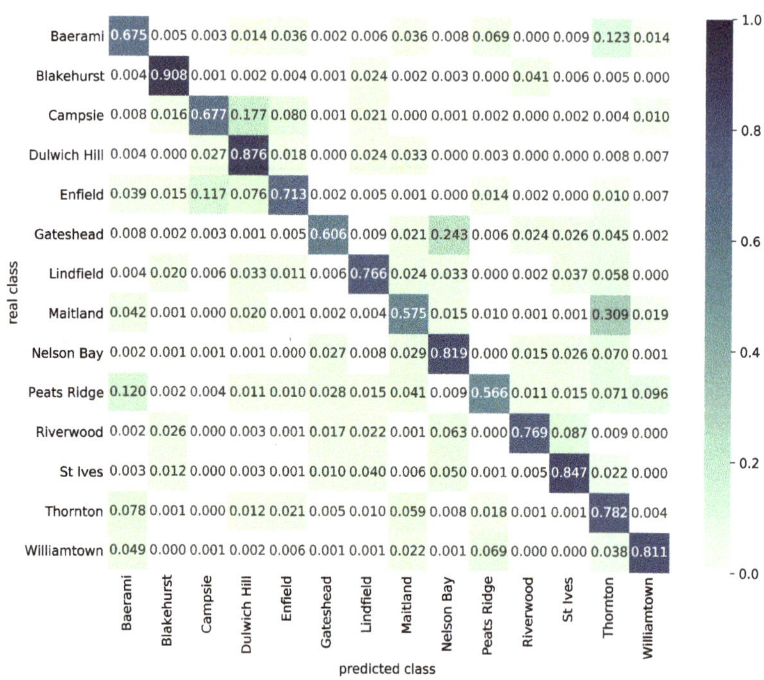


**Figure 11.** Mean confusion matrix for best performing model CNN2 and cluster C2.

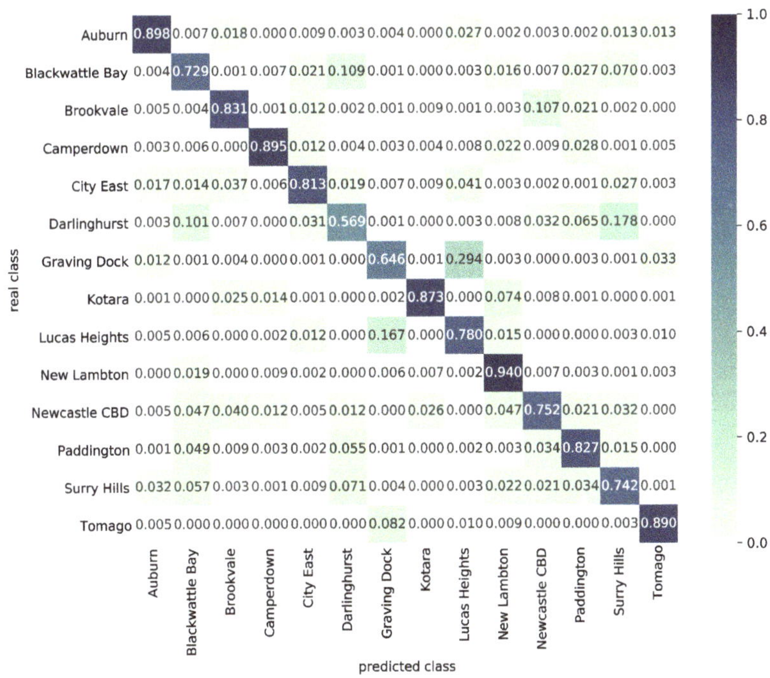


**Figure 12.** Mean confusion matrix for best performing model CNN2 and cluster C3.

## 5. Discussion

The machine learning-based substation node energy demand characterization represents a first logical step in a network-level flexibility assessment based on simulations or optimizations of DSM programs, ensuring resilient and stable operation of smart grids.

The resulting meta-models of individual substations can be further utilized to mitigate the difficulties associated with identifying, implementing and actuating various sources of energy flexibility, such as those related to the technology of indoor environmental comfort, compared to the few large ones traditionally acknowledged.

As explained and demonstrated in the paper, the ML-based classification of substations can be further utilized to study different scales of smart grid applications, and verify new control strategies. In particular, this study shows that:

(1) Clustering analysis can effectively help to understand the type of supplied neighbourhood, such as residential, mixed or business, and is used in this paper for benchmarking categorization of the three convolutional neural network models.

(2) Despite inherent daily (accounted timestamp), weekly (unaccounted directly) and seasonal (unaccounted directly) fluctuations in historic node demand data, the proposed CNN2 model yields relatively reliable results even when validated on future data, with mean model accuracy ranging from $88.8 + -1.4$ (combined cluster C2) to $94.8 + -0.7$ (residential cluster C1) in case of scenario 2019–2020.

(3) Given the relatively high ratio of trainable parameters (e.g., 122,830 for CNN2) to the input size (96 data points), over-fitting and over-determinism can clearly represent a problem and has to be carefully acknowledged in general, however, due to the proposed state-of-the-art ANN architecture, including multiple convolutional layers accompanied with regularization techniques such as batch normalization and dropout and conservatively set learning rate, the presented sparse mean confusion matrices based on bootstrap (10× resampling) demonstrate a rather robust fit, if relatively small accuracy standard deviation of 1.4% (CNN2) and dominant diagonal are considered.

(4) Certain classes may be mutually confused due to similarities in substations and their neighbourhoods. Some types of neighbourhood such as tourist or holiday locations may also increase demand variability in different parts of a year. The mutual confusions are similar for all investigated CNN models, and since these models have a different number of convolutional layers, the confusion is more likely to stem from substations variability rather than from the different architecture of CNN. For example, Swansea has fishing and tourism, Campsie has business and commercial areas, while Darlinghurst is a vibrant central district dependent on high season.

Limitations of this study include a limited range of considered years due to computational intensity. In order to optimize the distribution grid as a whole, network topology including its inner dependencies and boundary conditions must be considered (e.g., using hierarchical neural networks). In this study, only classification of network nodes was presented.

## 6. Conclusions

After introducing the importance of quantifying network flexibility potential for the safe integration of renewable energy sources and sustainable economy, the background on the smart grid and network modeling has been presented together with state-of-the-art machine learning techniques and their application to the classification of historical demand data. The resulting characterization of individual substations is important for future work on network-level simulations, the aim of which is to mitigate the difficulties associated with identifying, implementing and actuating many small sources of energy flexibility, compared to the few large ones traditionally acknowledged.

The proposed CNN models do not require any pre-processing of the 15 min interval demand data, the only subjective choice associated with the classification is the architecture of the neural network. Three scenarios were evaluated considering data from year 2019, 2020 and both years respectively. In the first two scenarios, data were divided between training and testing set using a typical split ratio 80/20, and in the third scenario, denoted as 2019–2020, data from the year 2019 were used for training and data from the year 2020 for testing. This enabled the verification of both statistical significance of the classifier, based on bootstrapping, as well as generalization of the resulting meta-models.

The sparse mean confusion matrices indicate a robust modeling approach, considering very similar structures across the investigated architectures, and relatively small standard deviation of accuracies. A more detailed (qualitative) assessment of substations and their neighbourhoods was beyond the scope of this paper, as well as the effects of boundary conditions on real network topology. Future work will continue with hierarchical neural networks modeling of the entire network.

**Author Contributions:** Conceptualization, methodology and writing J.P. and R.K.; programming, validation R.K. and J.P.; formal analysis, M.K.; project administration, funding acquisition and review, T.P., V.S. and T.A.; All authors have read and agreed to the published version of the manuscript.

**Funding:** This research was funded by The SecureFlex project TK01030078 and by the TJ04000232 project, both supported by the Technology Agency of the Czech Republic.

**Data Availability Statement:** Publicly available datasets were analyzed in this study. This data can be found here: https://www.ausgrid.com.au/Industry/Our-Research/Data-to-share/Distribution-zone-substation-data (accessed on 11 January 2021).

**Conflicts of Interest:** The authors declare no conflict of interest.

## References

1. Střelec, M.; Mamula, O.; Pitner, T.; Kasl, F.; Hrycej, D. Technical aspects of usability aggregated flexibility for business and technical services. *Energetika* **2020**, *70*, 1–20.
2. O'Connell, S.; Reynders, G.; Seri, F.; Sterling, R.; Keane, M.M. A standardised flexibility assessment methodology for demand response. *International Journal of Building Pathology and Adaptation* **2019**, *38*, 20–37. [CrossRef]

3. Xenophon, A.; Hill, D. Open grid model of Australia's National Electricity Market allowing backtesting against historic data. *Sci. Data* **2018**, *5*, 180203. [CrossRef]
4. Mets, K.; Verschueren, T.; Develder, C.; Vandoorn, T.L.; Vandevelde, L. Integrated simulation of power and communication networks for smart grid applications. In Proceedings of the 2011 IEEE 16th International Workshop on Computer Aided Modeling and Design of Communication Links and Networks (CAMAD), Kyoto, Japan, 10–11 June 2011; pp. 61–65.
5. Gielen, D. Perspectives for the Energy Transition Investment Needs for a Low-Carbon Energy System. 2017. Available online: https://iki-alliance.mx/de/download/Mexirec%20decarbonization(4).pdf (accessed on 11 January 2021).
6. Field, C.B. *Climate Change 2014–Impacts, Adaptation and Vulnerability: Regional Aspects*; Cambridge University Press: Cambridge, UK, 2014.
7. EUROPA–Smart Grids | SETIS–European Commission. Available online: https://setis.ec.europa.eu/publications/setis-magazine/smart-grids (accessed on 5 January 2021).
8. Shayeghi, H.; Shahryari, E.; Moradzadeh, M.; Siano, P. A Survey on Microgrid Energy Management Considering Flexible Energy Sources. *Energies* **2019**, *12*, 2156. [CrossRef]
9. Avancini, D.B.; Rodrigues, J.J.P.C.; Martins, S.G.B.; Rabêlo, R.A.L.; Al-Muhtadi, J.; Solic, P. Energy meters evolution in smart grids: A review. *J. Clean. Prod.* **2019**, *217*, 702–715. [CrossRef]
10. Tabors, R.D.; Parker, G.; Caramanis, M.C. Development of the smart grid: Missing elements in the policy process. In Proceedings of the 2010 43rd Hawaii International Conference on System Sciences, Kauai, HI, USA, 5–8 January 2010; pp. 1–7.
11. Atănăsoae, P.; Pentiuc, R.D.; Hopulele, E. Considerations Regarding the Negative Prices on the Electricity Market. *Proceedings* **2020**, *63*, 26. [CrossRef]
12. Zhu, B.; Deng, S.; Xu, Y.; Yuan, X.; Zhang, Z. Information security risk propagation model based on the SEIR infectious disease model for smart grid. *Information* **2019**, *10*, 323. [CrossRef]
13. Liu, Y.; Sun, Y.; Li, B. A two-stage household electricity demand estimation approach based on edge deep sparse coding. *Information* **2019**, *10*, 224. [CrossRef]
14. Vanin, A.; Aleshin, S.; Nasirov, R.; Novikov, D.; Tulsky, V. Investigation of voltage control at consumers connection points based on smart approach. *Information* **2016**, *7*, 42. [CrossRef]
15. O'Connell, S.; Riverso, S. Flexibility analysis for smart grid demand response. *arXiv* **2017**, arXiv:1704.01308.
16. ISO, I. *50002: 2014 Energy Audits—Requirements with Guidance for Use*; ISO: Geneva, Switzerland, 2014.
17. Capuano, L. *International energy outlook 2018 (IEO2018)*; US Energy Information Administration (EIA): Washington, DC, USA, 2018; Volume 2018, p. 21.
18. Vinois, J. DG ENER Working Paper: The Future Role and Challenges of Energy Storage. 2012. Available online: https://ec.europa.eu/energy/sites/ener/files/energy_storage.pdf (accessed on 9 January 2021).
19. Jensen, S.Ø.; Madsen, H.; Lopes, R.; Junker, R.; Aelenei, D.; Li, R. *Annex67: Energy Flexible Buildings-Energy Flexibility as a Key Asset in a Smart Building Future*; IEA Energy in Buildings and Communities Program (EBC): Birmingham, UK, 2017.
20. Monroe, J.G.; Hansen, P.; Sorell, M.; Berglund, E.Z. Agent-Based Model of a Blockchain Enabled Peer-to-Peer Energy Market: Application for a Neighborhood Trial in Perth, Australia. *Smart Cities* **2020**, *3*, 1072–1099. [CrossRef]
21. Nakayama, K.; Shinomiya, N. Distributed control based on tie-set graph theory for smart grid networks. In *International Congress on Ultra Modern Telecommunications and Control Systems*; IEEE: Moscow, Russia, 2010; pp. 957–964. [CrossRef]
22. Naz, A.; Javed, M.U.; Javaid, N.; Saba, T.; Alhussein, M.; Aurangzeb, K. Short-Term Electric Load and Price Forecasting Using Enhanced Extreme Learning Machine Optimization in Smart Grids. *Energies* **2019**, *12*, 866. [CrossRef]
23. Guo, F.; Herrera, L.; Murawski, R.; Inoa, E.; Wang, C.L.; Beauchamp, P.; Ekici, E.; Wang, J. Comprehensive real-time simulation of the smart grid. *IEEE Trans. Ind. Appl.* **2013**. *49*, 899–908. [CrossRef]
24. Wang, Z.; Paranjape, R.; Sadanand, A.; Chen, Z. Residential demand response: An overview of recent simulation and modeling applications. In Proceedings of the 2013 26th IEEE Canadian Conference on Electrical and Computer Engineering (CCECE), Regina, SK, Canada, 5–8 May 2013; pp. 1–6.
25. Distribution Zone Substation Data—Ausgrid. Available online: https://www.ausgrid.com.au:443/Industry/Our-Research/Data-to-share/Distribution-zone-substation-data (accessed on 30 December 2020).
26. OpenStreetMap. Available online: https://www.openstreetmap.org/copyright/en (accessed on 8 January 2021).
27. Liao, T.W. Clustering of time series data—A survey. *Pattern Recognit.* **2005**, *38*, 1857–1874. [CrossRef]
28. MacQueen, J.; others. Some methods for classification and analysis of multivariate observations. In Proceedings of the Fifth Berkeley Symposium on Mathematical Statistics and Probability, Oakland, CA, USA, 21 June–18 July 1967; Volume 1, pp. 281–297.
29. Franke, J.; Neumann, M.H. Bootstrapping neural networks. *Neural Comput.* **2000**, *12*, 1929–1949. [CrossRef] [PubMed]
30. Krč, R.; Podroužek, J.; Kratochvílová, M.; Vukušič, I.; Plášek, O. Neural Network-Based Train Identification in Railway Switches and Crossings Using Accelerometer Data. *J. Adv. Transp.* **2020**, *2020*. [CrossRef]
31. Kratochvílová, M.; Podroužek, J.; Apeltauer, J.; Vukušič, I.; Plášek, O. Train Type Identification at S&C. *J. Adv. Transp.* **2020**, *2020*. [CrossRef]
32. Yoon, H.; Hyun, Y.; Ha, K.; Lee, K.K.; Kim, G.B. A method to improve the stability and accuracy of ANN-and SVM-based time series models for long-term groundwater level predictions. *Comput. Geosci.* **2016**, *90*, 144–155. [CrossRef]
33. Ghaderpour, E.; Pagiatakis, S.D. LSWAVE: a MATLAB software for the least-squares wavelet and cross-wavelet analyses. *GPS Solut.* **2019**, *23*, 50–58. [CrossRef]

34. Podrouzek, J.; Bucher, C.; Deodatis, G. Identification of critical samples of stochastic processes towards feasible structural reliability applications. *Struct. Saf.* **2014**, *47*, 39–47. [CrossRef]
35. Ismail Fawaz, H.; Forestier, G.; Weber, J.; Idoumghar, L.; Muller, P.A. Deep learning for time series classification: a review. *Data Min. Knowl. Disc.* **2019**, *33*, 917–963. [CrossRef]

sustainability

MDPI

Article

# Electrical Power Prediction through a Combination of Multilayer Perceptron with Water Cycle Ant Lion and Satin Bowerbird Searching Optimizers

**Hossein Moayedi [1,2] and Amir Mosavi [3,4,5,6,*]**

1 Informetrics Research Group, Ton Duc Thang University, Ho Chi Minh City, Vietnam; hossein.moayedi@tdtu.edu.vn
2 Faculty of Civil Engineering, Ton Duc Thang University, Ho Chi Minh City, Vietnam
3 School of Economics and Business, Norwegian University of Life Sciences, 1430 Ås, Norway
4 School of the Built Environment, Oxford Brookes University, Oxford OX3 0BP, UK
5 John von Neumann Faculty of Informatics, Obuda University, 1034 Budapest, Hungary
6 Department of Informatics, Selye Janos University, 94501 Komarom, Slovakia
* Correspondence: amir.mosavi@nmbu.no

**Abstract:** Predicting the electrical power ($P_E$) output is a significant step toward the sustainable development of combined cycle power plants. Due to the effect of several parameters on the simulation of $P_E$, utilizing a robust method is of high importance. Hence, in this study, a potent metaheuristic strategy, namely, the water cycle algorithm (WCA), is employed to solve this issue. First, a nonlinear neural network framework is formed to link the $P_E$ with influential parameters. Then, the network is optimized by the WCA algorithm. A publicly available dataset is used to feed the hybrid model. Since the WCA is a population-based technique, its sensitivity to the population size is assessed by a trial-and-error effort to attain the most suitable configuration. The results in the training phase showed that the proposed WCA can find an optimal solution for capturing the relationship between the $P_E$ and influential factors with less than 1% error. Likewise, examining the test results revealed that this model can forecast the $P_E$ with high accuracy. Moreover, a comparison with two powerful benchmark techniques, namely, ant lion optimization and a satin bowerbird optimizer, pointed to the WCA as a more accurate technique for the sustainable design of the intended system. Lastly, two potential predictive formulas, based on the most efficient WCAs, are extracted and presented.

**Keywords:** power plant; electrical power modeling; metaheuristic optimization; water cycle algorithm; machine learning; deep learning; big data; energy; deep learning

**Citation:** Moayedi, H.; Mosavi, A. Electrical Power Prediction through a Combination of Multilayer Perceptron with Water Cycle Ant Lion and Satin Bowerbird Searching Optimizers. *Sustainability* **2021**, *13*, 2336. https://doi.org/10.3390/su13042336

Academic Editor: Mehdi Seyedmahmoudian

Received: 25 January 2021
Accepted: 10 February 2021
Published: 21 February 2021



## 1. Introduction

The accurate forecast of power generation capacity is a significant task for power plants [1]. This task concerns the efficiency of plants toward an economically beneficial performance [2]. Due to the nonlinear effect of several factors on thermodynamic systems [3,4] and related parameters like electrical power ($P_E$), many scholars have updated earlier solutions by using machine learning. As a matter of fact, there are diverse types of machine learning methods (e.g., regression [5], neural systems [6,7], fuzzy-based approaches [8],) that have presented reliable solutions to various problems. Liao [9] could successfully predict the output power of a plant using a regression model. The model attained 99% accuracy and was introduced as a promising approach for this purpose. Wood [10] employed a transparent open box algorithm for the $P_E$ output approximation of a combined cycle power plant (CCPP). The evaluations revealed the suitability of this algorithm as it provided an efficient and optimal prediction. Besides, as discussed by many scholars, intelligence techniques have a high capability to undertake nonlinear and complicated calculations [11–16]. A large number artificial intelligence-based practices are studied, for example, in the subjects of environmental concerns [17–21], pan evaporation and soil precipitation prediction [22,23], sustainability [24], water and groundwater supply chains [25–32], natural gas consumption [33], optimizing energy systems [34–45], air quality [46], image classification and processing [47–49], face or particular pattern recognition [50–52], structural health monitoring [53], target tracking and computer vision [54–56], building and structural design analysis [57–59], soil-pile analysis and landslide assessment [60–64], quantifying climatic contributions [65], structural material (e.g., steel and concrete) behaviors [66–71], or even some complex concerns such as signal processing [72,73] as well as feature selection and extraction problems [74–78]. Similar to deep learning-based applications [79–84], many decision-making applications are related to complicated engineering problems as well [85–91]. In another sense, the technique of the artificial neural network (ANN) is a sophisticated nonlinear processor that has attracted massive attention for sensitive engineering modeling [92]. In this sense, the multi-layer perceptron (MLP) [93,94] is composed of a minimum of three layers, each of which contains some neurons for handling the computations—noting that a more complicated ANN-based solution is known as deep learning [95]. For instance, Chen, et al. [96], Hu, et al. [97], Wang, et al. [98], and Xia, et al. [99] employed the use of extreme machine learning techniques in the field of medical sciences. As new advanced prediction techniques, hybrid searching algorithms have been widely developed to have more accurate prediction outputs; namely, harris hawks optimization [100–102], fruit fly optimization [103], multi-swarm whale optimizer [104,105], ant colony optimization [57,106], grasshopper optimizer [107], bacterial foraging optimization [108], many-objective optimization [109,110], and chaos enhanced grey wolf optimization [111,112].

In machine learning, ANNs have been widely used for analyzing diverse energy-related parameters in power plants [113–115]. Akdemir [116], for example, suggested the use of ANNs for predicting the hourly power of combined gas and steam turbine power plants. Regarding the coefficient of determination ($R^2$) of nearly 0.97, the products of the ANN were found to be in great agreement with real data. The successful use of two machine learning models, namely, recurrent ANN and a neuro-fuzzy system, was reported by Bandić et al. [117], who applied three popular machine learning approaches, namely, random forest, random tree, and an adaptive neuro-fuzzy inference system (ANFIS), to the same problem. Their findings indicated that the random forest outperforms other models. They also took a feature selection measure. It was shown that the original and changed data led to root mean square errors (RMSEs) of 3.0271 and 3.0527 MW, respectively. Mohammed et al. [118] used an ANFIS to find the thermal efficiency and optimal power output of combined cycle gas turbines which were 61% and 1540 MW, respectively.

Metaheuristic techniques have effectively assisted engineers and scholars in optimizing diverse problems [23,119–128], especially energy-related parameters such as solar energy [129], building thermal load [130], wind turbine interconnections [131], and green computing awareness [132]. Seyedmahmoudian et al. [133] used a differential evolution and particle swarm optimization (DEPSO) method to analyze the output power for a building-integrated photovoltaic system. These algorithms have also gained a lot of attention for optimally supervising conventional predictors like ANNs. Hu et al. [134] proposed a sophisticated hybrid composed of an ANN with a genetic algorithm (GA) and the PSO for predicting short-term electric load. With a relative error of 0.77%, this model performed better than the GA-ANN and PSO-ANN. Another application of the GA was studied by Lorencin et al. [135]. They tuned an ANN to estimate the $P_E$ output of a CCPP. Since the proposed model achieved a noticeably smaller error than a typical ANN, it was concluded that the GA is a nice optimizer for this system. Ghosh et al. [136] used a metaheuristic algorithm called beetle antennae search (BAS) to exploit a cascade feed-forward neural network applied to simulate the $P_E$ output of a CCPP. Due to the suitable performance of the developed model, they introduced it as an effective method for $P_E$ analysis. Chatterjee et al. [137] combined the ANN with cuckoo search (CS) and the PSO for electrical energy modeling at a combined cycle gas turbine. Their findings showed the superiority of the CS-trained ANN (with an average RMSE of approximately 2.6%) over the conventional ANN and PSO-trained version.

Due to the crucial role of power generation forecast in the sustainability of systems like gas turbines [138], selecting an appropriate predictive model is of great importance. On the other hand, the above literature reflects the high potential of metaheuristic algorithms for supervising the ANN. However, a significant gap in the knowledge emerges when the literature of $P_E$ analysis relies mostly on the first generation of these techniques (e.g., PSO and GA). Hence, this study is concerned with the application of a novel metaheuristic technique, namely, the water cycle algorithm (WCA) for the accurate prediction of the $P_E$ of a base load operated CCPP. Moreover, the performance of this algorithm is comparatively validated by ant lion optimization (ALO) and satin bowerbird optimizer (SBO) as benchmarks. These techniques are applied to this problem through a neural network framework. Some previous studies have shown the competency of the WCA [139], ALO [140], and SBO [141] in optimizing intelligent models like ANNs and ANFIS. The main contribution of these algorithms to the $P_E$ estimation lies in finding the optimal relationship between this parameter and influential factors.

## 2. Materials and Methods

### 2.1. Data Provision

When it comes to intelligent learning, the models acquire knowledge by mining the data. In ANN-based models, this knowledge draws on a group of tunable weights, as well as biases. The data should represent records of one (or a number of) input parameter(s) and their corresponding target(s).

In this work, the data are downloaded from a publicly available repository at: http://archive.ics.uci.edu/ml/datasets/Combined+Cycle+Power+Plant, based on studies by Tüfekci [138] and Kaya et al. [142]. The 6 years of records (2006–2011) of a CCPP working with full load (nominal generating capacity of 480 MW, made up of 2 × 160 MW ABB 13E2 gas turbines, 2× dual pressure heat recovery steam generators, and 1 × 160 MW ABB steam turbine) form this dataset [138]. It gives full load electrical power output as the target parameter, along with four input parameters, namely, ambient temperature (AT), exhaust steam pressure (vacuum, V), atmospheric pressure (AP), and relative humidity (RH). Figure 1 shows the relationship between the $P_E$ and input parameters. According to the drawn trendlines, a meaningful correlation can be seen in the figures of $P_E$-AT and $P_E$-V ($R^2$ of 0.8989 and 0.7565, respectively), while the values of AP and RH do not indicate an explicit correlation. Both AT and V are adversely proportional to the $P_E$.

Table 1 describes the dataset statistically. The values of AT, V, AP, and RH range in [1.8, 37.1] °C, [25.4, 81.6] cm Hg, [992.9, 1033.3] mbar, and [25.6, 100.2] % with average values of 19.7 °C, 54.3 cm Hg, 1013.3 mbar, and 73.3%, respectively. Additionally, the minimum and maximum recorded $P_E$s are 420.3 and 495.8 MW. The dataset comprises a total of 9568 samples, out of which 7654 samples are selected as training data and the remaining 1914 samples form the testing data. To do this, a random selection with an 80:20 ratio is applied.

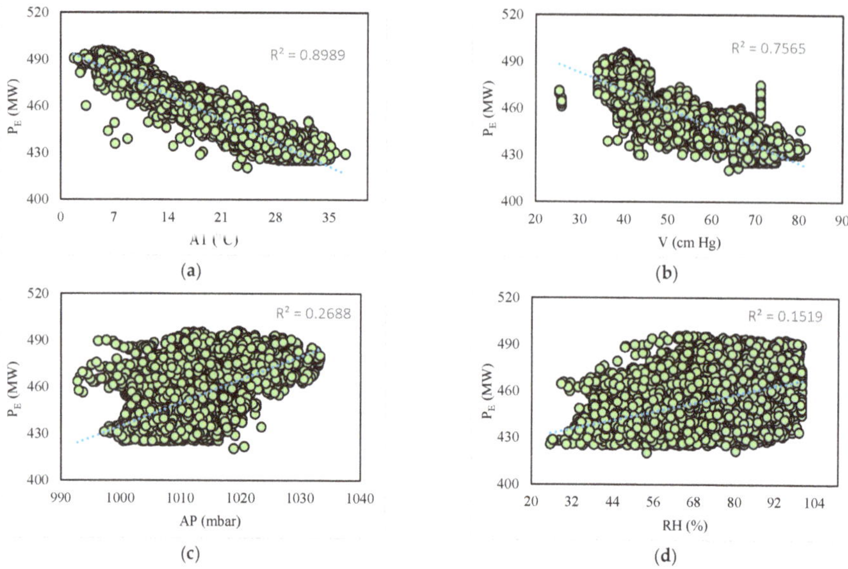

**Figure 1.** The graphical situation of $P_E$ versus (**a**) AT, (**b**) V, (**c**) AP, and (**d**) RH.

**Table 1.** Descriptive statistics of the $P_E$ and input parameters.

| Factor | Unit | Descriptive Indicator | | | | | |
|---|---|---|---|---|---|---|---|
| | | Mean | Std. Error | Std. Deviation | Sample Variance | Minimum | Maximum |
| **AT** | °C | 19.7 | 0.1 | 7.5 | 55.5 | 1.8 | 37.1 |
| **V** | cm Hg | 54.3 | 0.1 | 12.7 | 161.5 | 25.4 | 81.6 |
| **AP** | mbar | 1013.3 | 0.1 | 5.9 | 35.3 | 992.9 | 1033.3 |
| **RH** | % | 73.3 | 0.1 | 14.6 | 213.2 | 25.6 | 100.2 |
| $\mathbf{P_E}$ | MW | 454.4 | 0.2 | 17.1 | 291.3 | 420.3 | 495.8 |

### 2.2. *Methodology*

The overall methodology used in this study is shown in Figure 2.

#### 2.2.1. The WCA

Simulating the water cycle process was the main idea of the WCA algorithm, which was designed by Eskandar et al. [143]. In studies like [144], scholars have used this algorithm for sustainable energy issues. When the algorithm gets started, a population with the size $N_{pop}$ is generated from raindrops. Among the individuals, the best one is designated as the sea whose solution is shown by $X_{sea}$. Additionally, individuals with promising solutions ($X_r$s) are considered as rivers. The number of rivers is determined based on the parameter $N_{sr}$ that gives the number of rivers plus the unique sea. The residual individuals form the stream group ($X_s$s). The number of streams is the difference between $N_{pop}$ and $N_{sr}$.

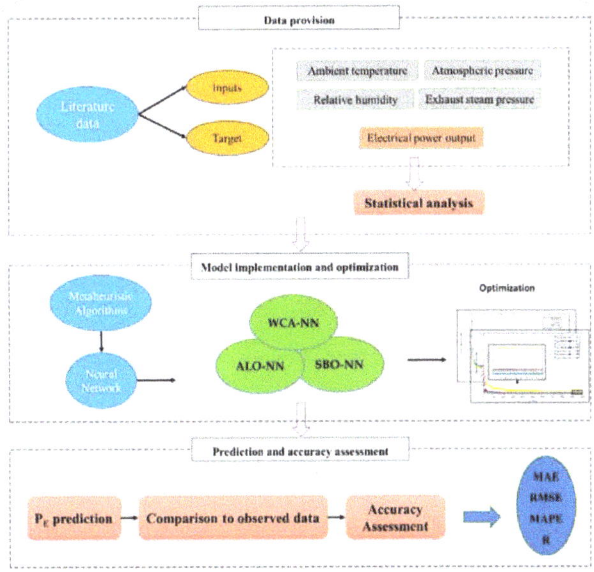


**Figure 2.** The general path of the study.

The description of the used algorithms is presented below.

The population can be expressed as follows:

$$\begin{bmatrix} x_1^1 & x_1^2 & \cdots & x_1^D \\ x_2^1 & x_2^2 & \cdots & x_2^D \\ \vdots & \vdots & \vdots & \vdots \\ x_{N_{pop}}^1 & x_{N_{pop}}^2 & \cdots & x_{N_{pop}}^D \end{bmatrix} = \begin{bmatrix} X_{sea} \\ X_r^1 \\ X_r^2 \\ \vdots \\ X_s^1 \\ \vdots \\ X_s^{N_{pop}-N_{sr}} \end{bmatrix}, \tag{1}$$

Concerning the function value of $X_r$ and $X_{sea}$ in the beginning, a number of $X_s$ are designated to each $X_r$ and $X_{sea}$ based on the following relationship:

$$C_n = f(n) - f\left(X_s^1\right), \tag{2}$$

$$NS(n) = round\left\{\left|C_n / \sum_{j=1}^{N_{sr}} C_j\right| \times \left(N_{pop} - N_{sr}\right)\right\}, \tag{3}$$

in which $f$ stands for the function value and $n = X_{sea}, X_r^1, \ldots, X_r^{N_{sr}-1}$.

Despite the typical procedure in nature (stream → river → sea), some streams may flow straight to the sea. The new values of $X_r$ and $X_s$ are obtained from the below equations:

$$X_r^{t+1} = X_r^t + rand \times cons \times \left(X_{sea}^t - X_r^t\right), \tag{4}$$

$$X_s^{t+1} = X_s^t + rand \times cons \times \left(X_r^t - X_s^t\right), \tag{5}$$

$$X_s^{t+1} = X_s^t + rand \times cons \times \left(X_{sea}^t - X_s^t\right), \tag{6}$$

where *rand* is a random number (in [0, 1]), *cons* gives a positive constant value (in [1, 2]), $t$ signifies the iteration number. $X_r$ and $X_s$ are evaluated and compared. If the quality of $X_s$ is better than that of $X_r$, they exchange their positions. A similar process happens between $X_r$ and $X_{sea}$ [145,146]. By performing the evaporation part of the water cycle, the algorithm is again implemented to improve the solution iteratively.

#### 2.2.2. The Benchmarks

The first benchmark algorithm is the ALO. Mirjalili [147] designed this algorithm as a robust nature-inspired strategy. Additionally, it has attracted the attention of experts for tasks like load shifting in analyzing sustainable renewable resources [148]. The pivotal idea of this algorithm is simulating the idealized hunting actions of the antlion. They build a cone-shaped fosse and wait for prey (often ants) to fall into the trap. The prey makes some movements to escape from antlions. The fitness of the solution is evaluated by a roulette wheel selection function. In this sense, the more powerful the hunter is, the better the prey is [149]. The details of the ALO and its application for optimizing intelligent models like ANNs can be found in earlier literature [150].

The SBO is considered as the second benchmark for the WCA. Inspired by the lifestyle of satin bowerbirds, Moosavi and Bardsiri [141] developed the SBO. Scholars like Zhang et al. [151] and Chintam and Daniel [152] have confirmed the successful performance of this algorithm in dealing with structural and energy-related optimization issues. In this strategy, there is a bower-making competition between male birds to attract a mate. The population is randomly created and the fitness of each bower is calculated. By making an elitism decision, the most promising individual is considered as the best solution. After determining the changes in the positions, a mutation operation is applied, followed by a step to combine the solutions of the old and new (updated) population [153]. A mathematical description of the SBO can be found in studies like [154].

## 3. Results and Discussion

### *3.1. Accuracy Assessment Measures*

Two essential error criteria, namely, the RMSE and mean absolute error (MAE), are defined to return different forms of the prediction error. Another error indicator called mean absolute percentage error (MAPE) is also defined to report the relative (percentage) error. Given $P_{E\ i_{expected}}$ and $P_{E\ i_{predicted}}$ as the expected and predicted electrical power outputs, Equations (7) to (9) denote the calculation of these indicators.

$$RMSE = \sqrt{\frac{1}{N}\sum_{i=1}^{N}\left[\left(P_{E i_{expected}} - P_{E i_{predicted}}\right)\right]^2}, \tag{7}$$

$$MAE = \frac{1}{N}\sum_{i=1}^{N}\left|P_{E i_{expected}} - P_{E i_{predicted}}\right|, \tag{8}$$

$$MAPE = \frac{1}{N}\sum_{i=1}^{N}\left|\frac{P_{E i_{expected}} - P_{E i_{predicted}}}{P_{E i_{expected}}}\right| \times 100, \tag{9}$$

where the number of samples (i.e., 7654 and 1914 in the training and testing groups, respectively) is signified by $N$.

Moreover, a correlation indicator called the Pearson correlation coefficient ($R$) is used. According to Equation (10), it reports the consistency between $P_{E\ expected}$ and $P_{E\ predicted}$. Note that the ideal value for this indicator is 1.

$$R = \frac{\sum_{i=1}^{N}\left(P_{E i_{predicted}} - \overline{P}_{E predicted}\right)\left(P_{E i_{expected}} - \overline{P}_{E expected}\right)}{\sqrt{\sum_{i=1}^{N}\left(P_{E i_{predicted}} - \overline{P}_{E predicted}\right)^2}\sqrt{\sum_{i=1}^{N}\left(P_{E i_{expected}} - \overline{P}_{E expected}\right)^2}}, \tag{10}$$

### 3.2. Hybridizing and Training

It was earlier stated that this study pursues a novel forecasting method for the problem of $P_E$ modeling. To this end, the water cycle algorithm explores the relationship between this parameter and four inputs through an MLP neural network. This skeleton is used to establish nonlinear equations between the mentioned parameters. A three-layer MLP is considered wherein the number of neurons lying in the first, second, and third layer (also known as input, hidden, and output layers) equals four (the number of inputs), nine (obtained by trial and error practice), and one (the number of outputs only), respectively. Figure 3 shows this structure:

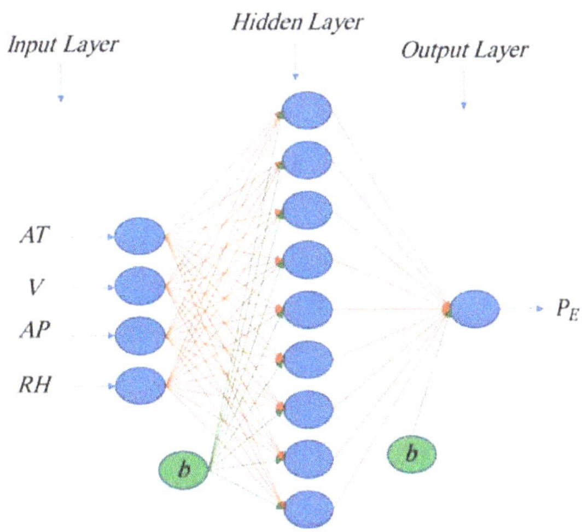


**Figure 3.** The used artificial neural network (ANN).

There are two kinds of tunable computational parameters in an MLP: (a) weights (*W*) that are designated to each input factor and (b) bias terms. Equation (11) shows the calculation of a neuron with a given input (*I*).

$$Response = Tansig(W \times I + b)\,, \tag{11}$$

where Tansig signifies an activation function which is defined as follows:

$$\text{Tansig}\ (x) = \frac{2}{1 + e^{-2x}} - 1, \tag{12}$$

Each neuron of the ANN applies an activation functions to a linear combination of inputs and network parameters (i.e., *W* and *b*) to give its specific response. There are a number of functions (e.g., Logsig, Purelin, etc.) that can be used for this purpose. However, many studies have stated the superiority of Tansig for hidden neurons [155–157].

The WCA finds the optimal values of the parameters in Equation (11) in an iterative procedure. In this way, the suitability of each response (in each iteration) is reported by an objective function (OF). This study uses the RMSE of training data for this purpose. So, the lower the OF is, the better the optimization is. Figure 4a shows the optimization curves of the WCA for the given problem. The reduction of the OF in this figure shows that the RMSE error is being reduced consecutively.

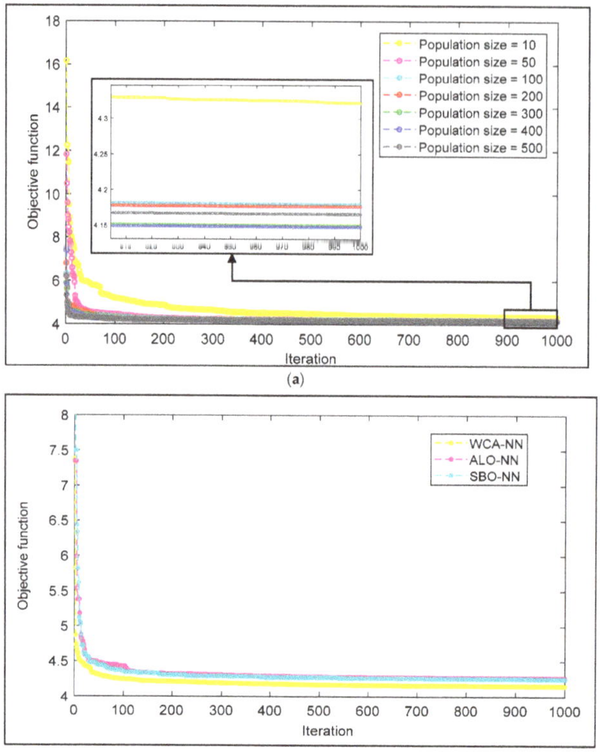


**Figure 4.** (**a**) Convergence curves belonging to the tested PS of WCA-ANN and (**b**) comparison between the convergence behaviors of the chosen networks.

Famously, the size of the population can greatly impact the quality of optimization. The convergence curves are plotted for seven different WCA-NN networks distinguished by different population sizes (PS of 10, 50, 100, 200, 300, 400, 500). As is seen, the curve of PS = 400 is finally below the others. Therefore, this network is the representative of the WCA-NN for further evaluations. Note that a total of 1000 iterations were considered for all tested PSs.

The same strategy (i.e., the same PSs and number of iterations) was executed for the benchmark models. It was shown that ALO-NN and SBO-NN with PSs of 400 and 300 are superior. Figure 4b depicts and compares the convergence behavior of the selected networks. According to this figure, all three algorithms have a similar performance in dealing with error minimization. The OF is chiefly reduced over the initial iterations.

Figure 4b also says that the OF of the WCA-NN is below both benchmarks. In this sense, the RMSEs of 4.1468, 4.2656, and 4.2484 are calculated for the WCA-NN, ALO-NN, and SBO-NN, respectively. Additionally, the corresponding MAEs (3.2112, 3.3389, and 3.3075) can support this claim.

Subtracting $P_{E\ predicted}$ from $P_{E\ expected}$ returns an error value for each sample. Figure 5 shows these errors. It can be seen that close-to-zero values are obtained for the majority of training samples. Concerning peak values, the errors lie in the ranges [−18.4548, 42.4231], [−18.9855, 43.2264], and [−19.1242, 42.8160]. With respect to the range of $P_E$ (Table 1), these values indicate a very good prediction for all models. Moreover, the calculated MAPEs report less than 1% relative errors (0.7076%, 0.7359%, and 0.7289%).

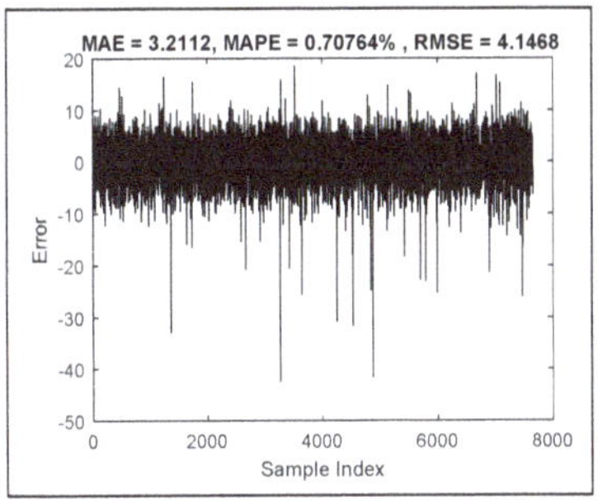


(**a**)

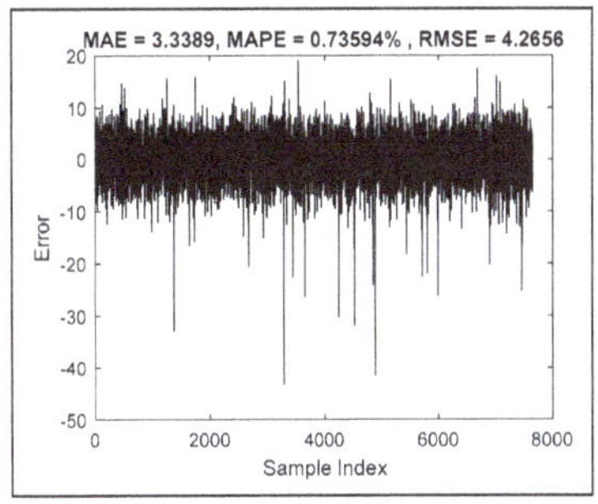


(**b**)

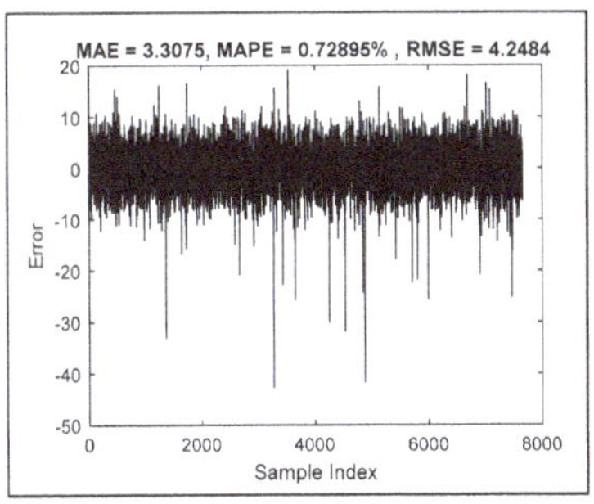


(**c**)

**Figure 5.** The magnitude of error over the training dataset obtained by (**a**) WCA-NN, (**b**) ALO-NN, and (**c**) SBO-NN.

Moreover, the R values of 0.96985, 0.96807, and 0.96834 profess an excellent correlation between the products of the used models and the observed $P_E$. This favorable performance means that the WCA, ALO, and SBO have nicely understood the dependence of the $P_E$ on AT, V, AP, and RH and, accordingly, they have optimally tuned the parameters of the MLP system.

### 3.3. Testing Performance

The testing ability of a forecasting model illustrates the generalizability of the captured knowledge for unfamiliar conditions. The weights and bias terms tuned by the WCA, ALO, and SBO created three separate methods that predicted the $P_E$ for testing samples. The quality of the results is assessed in this section.

Figure 6 presents two charts for each model. First, the correlation between the $P_{E\ expected}$ and $P_{E\ predicted}$ is graphically shown. Along with it, the frequency of errors ($P_{E\ expected} - P_{E\ predicted}$) is shown in the form of histogram charts. At a glance, the results of all three models demonstrate promising generalizability, due to the aggregation of points around the ideal line (i.e., $x = y$) in Figure 6a,c,e. Additionally, as a general trend in Figure 6b,d,f, small errors (zero and close-to-zero ranges) have a higher frequency compared to large values. Remarkably, testing errors range within [−16.6585, 44.7929], [−15.8225, 45.7482], and [−16.3683, 45.8428].

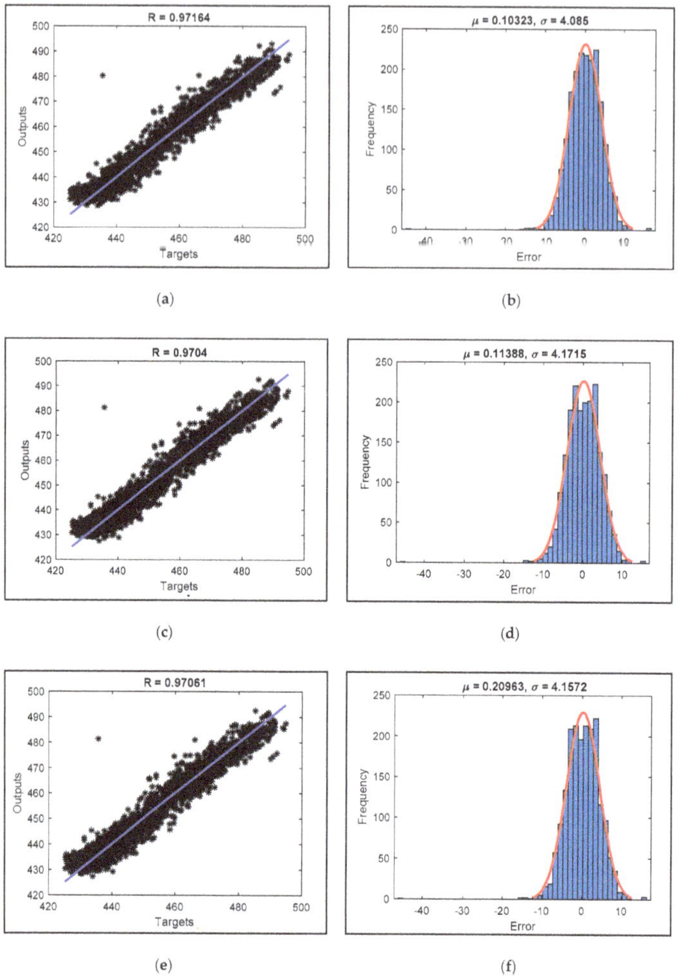


**Figure 6.** The testing results in terms of (**a**,**c**,**e**) correlation and (**b**,**d**,**f**) histogram of errors for the WCA-NN, ALO-NN, and SBO-NN, respectively.

The RMSE and MAE of the WCA-NN, ALO-NN, and SBO-NN were 4.0852 and 3.1996, 4.1719 and 3.3028, and 4.1614 and 3.2802, respectively. These values are close to those of the training phase. Hence, all three models enjoy a high accuracy in dealing with out-of-data situations. Furthermore, a desirable level of relative error can be represented by the MAPEs of 0.7045%, 0.7272%, and 0.7221%.

According to the obtained R values (0.97164, 0.97040, and 0.97061), all three hybrids are able to predict the $P_E$ of a CCPP with highly reliable accuracy. In all regression charts, there is an outlying value, $P_E$ = 435.58 (obtained for AT = 7.14 °C, V = 41.22 cm Hg, AP = 1016.6 mbar, and RH = 97.09%) that is predicted to be 480.3728513, 481.3282482, and 481.4228308.

### 3.4. WCA vs. ALO and SBO

The quality of the results showed that the WCA, ALO, and SBO metaheuristic algorithms benefit from potential search strategies for exploring and mapping the $P_E$ pattern. However, comparative evaluation using the RMSE, MAE, MAPE, and R pointed out noticeable distinctions in the performance of these algorithms.

Figure 7 depicts and compares the accuracies in the form of radar charts. The shape of the produced triangles indicates the superiority of the WCA-based model over the benchmark algorithms in both training and testing phases. In terms of all four indicators, this model could predict the $P_E$ with the best quality. It means that the ANN supervised by the WCA is constructed of more promising parameters. Following the proposed algorithm, the SBO won the competition with ALO. It is noteworthy that the accuracy of these two algorithms in the testing phase was closer compared to the training results.

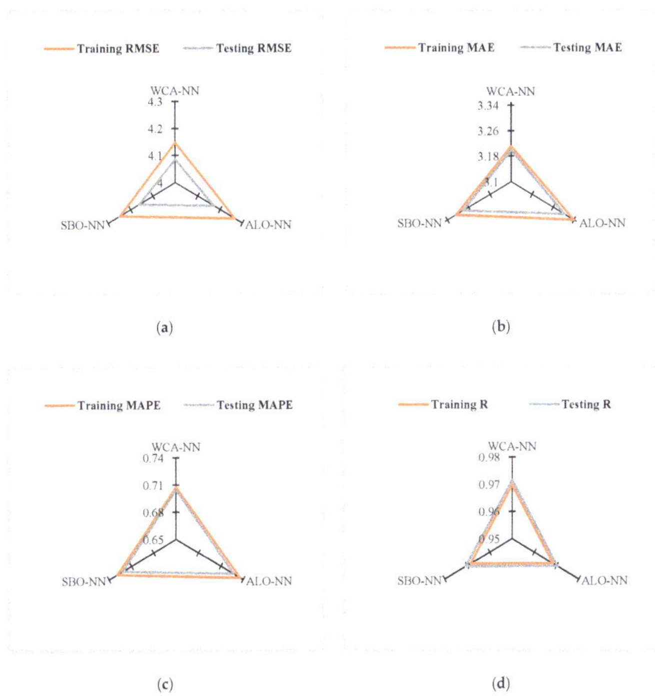


**Figure 7.** Radar charts for comparing the calculated (**a**) RMSE, (**b**) MPE, (**c**) MAPE, and (**d**) R.

From the time-efficiency point of view, computations of the ALO were shorter than the two other methods. The elapsed times for tuning the ANN parameters were nearly 14,261.1, 12,928.1, and 14,871.3 s by the WCA, ALO, and SBO, respectively. It should be also noted that the WCA and ALO used PS = 400, while this value was 300 for the SBO.

According to the above results, the WCA provides both an accurate and efficient solution to the problem of $P_E$ approximation, and thus, sustainable development of the CCPPs. It is true that the ALO could optimize the neural network in a shorter time, but smaller PSs of the WCA (i.e., 300, 200, ...) were far faster. On the other hand, back to Figure 4, the PS of 300 produced a solution almost as good as that of 400. It is interesting to know that the prediction of PS = 300 was slightly better than PS = 400 (testing RMSEs 4.0760 vs. 4.0852). The computation time of this configuration was around 3186.9 seconds which is considerably smaller than the two other algorithms. Thus, for time-sensitive projects, less complex configurations of the WCA are efficiently applicable.

### 3.5. Predictive Formulas

Due to the comparisons in the previous section, the solutions found by WCAs with the PSs of 300 and 400 are presented here in the form of two separate (different) formulas for forecasting the electrical power. Equations (13) and (14) give the $P_E$ through a linear relationship.

$$P_{E_{PS\,=\,300}} = 0.814 \times Y_1 - 0.543 \times Y_2 + 0.825 \times Y_3 - 0.584 \times Y_4 - 0.509 \times Y_5 - 0.220 \times Y_6 + 0.296 \times Y_7 + 0.039 \times Y_8 + 0.542 \times Y_9. - 0.076, \quad (13)$$

$$P_{E_{PS=400}} = -0.782 \times Z_1 + 0.627 \times Z_2 - 0.569 \times Z_3 - 0.594 \times Z_4 - 0.891 \times Z_5 - 0.548 \times Z_6 + 0.661 \times Z_7 + 0.416 \times Z_8 + 0.383 \times Z_9 - 0.696, \quad (14)$$

where $Y_i$ and $Z_i$ ($i$ = 1, 2, ..., 9) symbolize the output of the hidden neurons. These parameters are calculated using a generic equation as follows:

$$Y_i \text{ and } Z_i = Tansig\ (W_{i1} \times AT + W_{i2} \times V + W_{i3} \times AP + W_{i4} \times RH + b_i), \quad (15)$$

and with the help of Table 2.

**Table 2.** The optimized parameters of the WCA configurations.

| i | For $Z_i$ (PS = 400) | | | | | For $Y_i$(PS = 300) | | | | |
|---|---|---|---|---|---|---|---|---|---|---|
| | $W_{i1}$ | $W_{i2}$ | $W_{i3}$ | $W_{i4}$ | $b_i$ | $W_{i1}$ | $W_{i2}$ | $W_{i3}$ | $W_{i4}$ | $b_i$ |
| 1 | −1.238 | 0.344 | 1.240 | −1.640 | 2.425 | 0.887 | −1.670 | 1.517 | 0.068 | −2.425 |
| 2 | 1.482 | −1.851 | 0.311 | 0.399 | −1.819 | −0.042 | 2.181 | −0.983 | −0.395 | 1.819 |
| 3 | −0.870 | 1.152 | −1.755 | −0.847 | 1.212 | 1.035 | 1.770 | 0.848 | 0.979 | −1.212 |
| 4 | −0.830 | 0.172 | 1.716 | 1.489 | 0.606 | 0.639 | 1.690 | 1.572 | −0.378 | −0.606 |
| 5 | 0.864 | −1.691 | −1.343 | 0.685 | 0.000 | −1.587 | −1.512 | −1.016 | −0.213 | 0.000 |
| 6 | −1.394 | −1.677 | −1.052 | −0.136 | −0.606 | 1.256 | 1.282 | −1.204 | 1.100 | 0.606 |
| 7 | −2.004 | −1.261 | 0.276 | −0.446 | −1.212 | −0.313 | 0.385 | −1.739 | −1.615 | −1.212 |
| 8 | 1.609 | 0.883 | 1.532 | 0.402 | 1.819 | 1.277 | 0.190 | −1.739 | −1.090 | 1.819 |
| 9 | −1.876 | −0.740 | 0.819 | −1.069 | −2.425 | −0.514 | −1.679 | 1.003 | −1.339 | −2.425 |

According to the above formulas, calculating the $P_E$ consists of two steps: First, recalling the MLP structure (Figure 3) and also Equation (11) from Section 3.2, Equation (15) is applied to produce the response of nine hidden neurons (e.g., $Y_1$, $Y_2$, ..., $Y_9$ for the formula corresponding to PS = 300). For instance, $W_{32}$ represents the weight of the 3rd neuron applied to the 2nd input (i.e., V). Thus, it equals 1.152 in Table 2 used for calculating $Y_3$. Next, these parameters are used by the output neuron (in Equation (13)) to yield the $P_E$. The same goes for the formula corresponding to PS = 400 ($Z_1$, $Z_2$, ..., $Z_9$ and Equation (14)).

## 4. Conclusions

This paper investigated the efficiency of three capable metaheuristic approaches for the accurate analysis of electrical power output. The water cycle algorithm was used to supervise the learning process of an ANN. This algorithm was compared with two other techniques, namely antlion optimization and a satin bowerbird optimizer. The results showed the superiority of the WCA in all cases and terms of all accuracy indicators. For example, the RMSEs of 4.1468 vs. 4.2656 and 4.2484 in the training phase and 4.0852 vs. 4.1719 and 4.1614 in the prediction phase. However, all three hybrids could understand and reproduce the $P_E$ pattern with less than 1% error. All in all, a significant sustainability issue was efficiently managed and solved by metaheuristic science. Thus, the presented hybrid models can be practically employed to forecast the electrical power output of combined cycle power plants by having the records of AT, V, AP, and RH. They can also be appropriate substitutes for time-consuming and costly methods. However, further efforts are recommended for future projects to compare the applicability of different metaheuristic techniques and also to present innovative measures that may improve the efficiency of the existing models in terms of both time and accuracy.

**Author Contributions:** H.M. and A.M., methodology; software validation, writing—original draft preparation. H.M. and A.M., writing—review and editing, visualization, supervision, project administration. All authors have read and agreed to the published version of the manuscript.

**Funding:** This research received no funding.

**Institutional Review Board Statement:** Not applicable.

**Informed Consent Statement:** Not applicable.

**Data Availability Statement:** Not applicable.

**Conflicts of Interest:** The authors declare no conflict of interest.

## References

1. Lee, J.H.; Kim, T.S.; Kim, E.-H. Prediction of power generation capacity of a gas turbine combined cycle cogeneration plant. *Energy* **2017**, *124*, 187–197. [CrossRef]
2. Sun, W.; Zhang, J.; Wang, R. Predicting electrical power output by using Granular Computing based Neuro-Fuzzy modeling method. In Proceedings of the The 27th Chinese Control and Decision Conference (2015 CCDC), Qingdao, China, 23–25 May 2015; pp. 2865–2870.
3. Han, X.; Chen, N.; Yan, J.; Liu, J.; Liu, M.; Karellas, S. Thermodynamic analysis and life cycle assessment of supercritical pulverized coal-fired power plant integrated with No. 0 feedwater pre-heater under partial loads. *J. Clean. Prod.* **2019**, *233*, 1106–1122. [CrossRef]
4. Han, X.; Zhang, D.; Yan, J.; Zhao, S.; Liu, J. Process development of flue gas desulphurization wastewater treatment in coal-fired power plants towards Zero Liquid Discharge: Energetic, economic and environmental analyses. *J. Clean. Prod.* **2020**, *261*, 121144. [CrossRef]
5. Xu, X.; Chen, L. Projection of long-term care costs in China, 2020–2050: Based on the Bayesian quantile regression method. *Sustainability* **2019**, *11*, 3530. [CrossRef]
6. Shi, K.; Wang, J.; Tang, Y.; Zhong, S. Reliable asynchronous sampled-data filtering of T–S fuzzy uncertain delayed neural networks with stochastic switched topologies. *Fuzzy Sets Syst.* **2020**, *381*, 1–25. [CrossRef]
7. Wang, M.; Chen, H.; Yang, B.; Zhao, X.; Hu, L.; Cai, Z.; Huang, H.; Tong, C.J.N. Toward an optimal kernel extreme learning machine using a chaotic moth-flame optimization strategy with applications in medical diagnoses. *Neurocomputing* **2017**, *267*, 69–84. [CrossRef]
8. Chen, H.; Qiao, H.; Xu, L.; Feng, Q.; Cai, K. A Fuzzy Optimization Strategy for the Implementation of RBF LSSVR Model in Vis–NIR Analysis of Pomelo Maturity. *IEEE Trans. Ind. Inform.* **2019**, *15*, 5971–5979. [CrossRef]
9. Liao, Y. Linear Regression and Gradient Descent Method for Electricity Output Power Prediction. *J. Comput. Commun.* **2019**, *7*, 31–36. [CrossRef]
10. Wood, D.A. Combined cycle gas turbine power output prediction and data mining with optimized data matching algorithm. *Sn Appl. Sci.* **2020**, *2*, 1–21. [CrossRef]
11. Liu, Z.; Shao, J.; Xu, W.; Chen, H.; Zhang, Y. An extreme learning machine approach for slope stability evaluation and prediction. *Nat. Hazards* **2014**, *73*, 787–804. [CrossRef]
12. Chen, Y.; He, L.; Li, J.; Zhang, S. Multi-criteria design of shale-gas-water supply chains and production systems towards optimal life cycle economics and greenhouse gas emissions under uncertainty. *Comput. Chem. Eng.* **2018**, *109*, 216–235. [CrossRef]
13. Zhu, J.; Shi, Q.; Wu, P.; Sheng, Z.; Wang, X. Complexity analysis of prefabrication contractors' dynamic price competition in mega projects with different competition strategies. *Complexity* **2018**, *2018*, 5928235. [CrossRef]
14. Hu, X.; Chong, H.-Y.; Wang, X. Sustainability perceptions of off-site manufacturing stakeholders in Australia. *J. Clean. Prod.* **2019**, *227*, 346–354. [CrossRef]
15. He, L.; Shao, F.; Ren, L. Sustainability appraisal of desired contaminated groundwater remediation strategies: An information-entropy-based stochastic multi-criteria preference model. *Environ. Dev. Sustain.* **2020**, 1–21. [CrossRef]
16. Li, C.; Hou, L.; Sharma, B.Y.; Li, H.; Chen, C.; Li, Y.; Zhao, X.; Huang, H.; Cai, Z.; Chen, H. Developing a new intelligent system for the diagnosis of tuberculous pleural effusion. *Comput. Methods Programs Biomed.* **2018**, *153*, 211–225. [CrossRef] [PubMed]
17. Liu, J.; Liu, Y.; Wang, X. An environmental assessment model of construction and demolition waste based on system dynamics: A case study in Guangzhou. *Environ. Sci. Pollut. Res.* **2020**, *27*, 37237–37259. [CrossRef]
18. Liu, L.; Li, J.; Yue, F.; Yan, X.; Wang, F.; Bloszies, S.; Wang, Y. Effects of arbuscular mycorrhizal inoculation and biochar amendment on maize growth, cadmium uptake and soil cadmium speciation in Cd-contaminated soil. *Chemosphere* **2018**, *194*, 495–503. [CrossRef]
19. Yang, Y.; Liu, J.; Yao, J.; Kou, J.; Li, Z.; Wu, T.; Zhang, K.; Zhang, L.; Sun, H. Adsorption behaviors of shale oil in kerogen slit by molecular simulation. *Chem. Eng. J.* **2020**, *387*, 124054. [CrossRef]
20. Feng, S.; Lu, H.; Tian, P.; Xue, Y.; Lu, J.; Tang, M.; Feng, W. Analysis of microplastics in a remote region of the Tibetan Plateau: Implications for natural environmental response to human activities. *Sci. Total Environ.* **2020**, *739*, 140087. [CrossRef]
21. Liu, J.; Yi, Y.; Wang, X. Exploring factors influencing construction waste reduction: A structural equation modeling approach. *J. Clean. Prod.* **2020**, *276*, 123185. [CrossRef]
22. Zhang, B.; Xu, D.; Liu, Y.; Li, F.; Cai, J.; Du, L. Multi-scale evapotranspiration of summer maize and the controlling meteorological factors in north China. *Agric. For. Meteorol.* **2016**, *216*, 1–12. [CrossRef]
23. Chao, L.; Zhang, K.; Li, Z.; Zhu, Y.; Wang, J.; Yu, Z. Geographically weighted regression based methods for merging satellite and gauge precipitation. *J. Hydrol.* **2018**, *558*, 275–289. [CrossRef]
24. Keshtegar, B.; Heddam, S.; Sebbar, A.; Zhu, S.-P.; Trung, N.-T. SVR-RSM: A hybrid heuristic method for modeling monthly pan evaporation. *Environ. Sci. Pollut. Res.* **2019**, *26*, 35807–35826. [CrossRef] [PubMed]

25. He, L.; Chen, Y.; Zhao, H.; Tian, P.; Xue, Y.; Chen, L. Game-based analysis of energy-water nexus for identifying environmental impacts during Shale gas operations under stochastic input. *Sci. Total Environ.* **2018**, *627*, 1585–1601. [CrossRef] [PubMed]
26. Chen, Y.; Li, J.; Lu, H.; Yan, P. Coupling system dynamics analysis and risk aversion programming for optimizing the mixed noise-driven shale gas-water supply chains. *J. Clean. Prod.* **2021**, *278*, 123209. [CrossRef]
27. Cheng, X.; He, L.; Lu, H.; Chen, Y.; Ren, L. Optimal water resources management and system benefit for the Marcellus shale-gas reservoir in Pennsylvania and West Virginia. *J. Hydrol.* **2016**, *540*, 412–422. [CrossRef]
28. Li, X.; Zhang, R.; Zhang, X.; Zhu, P.; Yao, T. Silver-Catalyzed Decarboxylative Allylation of Difluoroarylacetic Acids with Allyl Sulfones in Water. *Chem. Asian J.* **2020**, *15*, 1175–1179. [CrossRef] [PubMed]
29. Yang, M.; Sowmya, A. An Underwater Color Image Quality Evaluation Metric. *IEEE Trans. Image Process.* **2015**, *24*, 6062–6071. [CrossRef]
30. Qian, J.; Feng, S.; Li, Y.; Tao, T.; Han, J.; Chen, Q.; Zuo, C. Single-shot absolute 3D shape measurement with deep-learning-based color fringe projection profilometry. *Opt. Lett.* **2020**, *45*, 1842–1845. [CrossRef]
31. Lyu, Z.; Chai, J.; Xu, Z.; Qin, Y.; Cao, J. A Comprehensive Review on Reasons for Tailings Dam Failures Based on Case History. *Adv. Civ. Eng.* **2019**, *2019*, 4159306. [CrossRef]
32. Feng, W.; Lu, H.; Yao, T.; Yu, Q. Drought characteristics and its elevation dependence in the Qinghai–Tibet plateau during the last half-century. *Sci. Rep.* **2020**, *10*, 14323. [CrossRef] [PubMed]
33. Su, Z.; Liu, E.; Xu, Y.; Xie, P.; Shang, C.; Zhu, Q. Flow field and noise characteristics of manifold in natural gas transportation station. *Oil Gas Sci. Technol. Rev. D'ifp Energ. Nouv.* **2019**, *74*, 70. [CrossRef]
34. Chen, Y.; He, L.; Guan, Y.; Lu, H.; Li, J. Life cycle assessment of greenhouse gas emissions and water-energy optimization for shale gas supply chain planning based on multi-level approach: Case study in Barnett, Marcellus, Fayetteville, and Haynesville shales. *Energy Convers. Manag.* **2017**, *134*, 382–398. [CrossRef]
35. He, L.; Shen, J.; Zhang, Y. Ecological vulnerability assessment for ecological conservation and environmental management. *J. Environ. Manag.* **2018**, *206*, 1115–1125. [CrossRef] [PubMed]
36. Lu, H.; Tian, P.; He, L. Evaluating the global potential of aquifer thermal energy storage and determining the potential worldwide hotspots driven by socio-economic, geo-hydrologic and climatic conditions. *Renew. Sustain. Energy Rev.* **2019**, *112*, 788–796. [CrossRef]
37. Wang, Y.; Yao, M.; Ma, R.; Yuan, Q.; Yang, D.; Cui, B.; Ma, C.; Liu, M.; Hu, D. Design strategy of barium titanate/polyvinylidene fluoride-based nanocomposite films for high energy storage. *J. Mater. Chem. A* **2020**, *8*, 884–917. [CrossRef]
38. Zhao, X.; Ye, Y.; Ma, J.; Shi, P.; Chen, H. Construction of electric vehicle driving cycle for studying electric vehicle energy consumption and equivalent emissions. *Environ. Sci. Pollut. Res.* **2020**, *27*, 37395–37409. [CrossRef]
39. Zhu, L.; Kong, L.; Zhang, C. Numerical Study on Hysteretic Behaviour of Horizontal-Connection and Energy-Dissipation Structures Developed for Prefabricated Shear Walls. *Appl. Sci.* **2020**, *10*, 1240. [CrossRef]
40. Deng, Y.; Zhang, T.; Sharma, B.K.; Nie, H. Optimization and mechanism studies on cell disruption and phosphorus recovery from microalgae with magnesium modified hydrochar in assisted hydrothermal system. *Sci. Total Environ.* **2019**, *646*, 1140–1154. [CrossRef]
41. Zhang, T.; Wu, X.; Fan, X.; Tsang, D.C.W.; Li, G.; Shen, Y. Corn waste valorization to generate activated hydrochar to recover ammonium nitrogen from compost leachate by hydrothermal assisted pretreatment. *J. Environ. Manag.* **2019**, *236*, 108–117. [CrossRef]
42. Peng, S.; Zhang, Z.; Liu, E.; Liu, W.; Qiao, W. A new hybrid algorithm model for prediction of internal corrosion rate of multiphase pipeline. *J. Nat. Gas Sci. Eng.* **2021**, *85*, 103716. [CrossRef]
43. Peng, S.; Chen, Q.; Zheng, C.; Liu, E. Analysis of particle deposition in a new-type rectifying plate system during shale gas extraction. *Energy Sci. Eng.* **2020**, *8*, 702–717. [CrossRef]
44. Liu, E.; Wang, X.; Zhao, W.; Su, Z.; Chen, Q. Analysis and Research on Pipeline Vibration of a Natural Gas Compressor Station and Vibration Reduction Measures. *Energy Fuels* **2020**. [CrossRef]
45. Liu, E.; Guo, B.; Lv, L.; Qiao, W.; Azimi, M. Numerical simulation and simplified calculation method for heat exchange performance of dry air cooler in natural gas pipeline compressor station. *Energy Sci. Eng.* **2020**, *8*, 2256–2270. [CrossRef]
46. Wang, Y.; Yuan, Y.; Wang, Q.; Liu, C.; Zhi, Q.; Cao, J. Changes in air quality related to the control of coronavirus in China: Implications for traffic and industrial emissions. *Sci. Total Environ.* **2020**, *731*, 139133. [CrossRef]
47. Xu, M.; Li, C.; Zhang, S.; Callet, P.L. State-of-the-Art in 360° Video/Image Processing: Perception, Assessment and Compression. *IEEE J. Sel. Top. Signal Process.* **2020**, *14*, 5–26. [CrossRef]
48. Zhang, X.; Wang, T.; Luo, W.; Huang, P. Multi-level Fusion and Attention-guided CNN for Image Dehazing. *IEEE Trans. Circuits Syst. Video Technol.* **2020**. [CrossRef]
49. Zhang, X.; Wang, T.; Wang, J.; Tang, G.; Zhao, L. Pyramid Channel-based Feature Attention Network for image dehazing. *Comput. Vision Image Underst.* **2020**, *197*, 103003. [CrossRef]
50. Shi, K.; Wang, J.; Zhong, S.; Tang, Y.; Cheng, J. Non-fragile memory filtering of T-S fuzzy delayed neural networks based on switched fuzzy sampled-data control. *Fuzzy Sets Syst.* **2020**, *394*, 40–64. [CrossRef]
51. Mi, C.; Cao, L.; Zhang, Z.; Feng, Y.; Yao, L.; Wu, Y. A port container code recognition algorithm under natural conditions. *J. Coast. Res.* **2020**, *103*, 822–829. [CrossRef]

52. Salari, N.; Shohaimi, S.; Najafi, F.; Nallappan, M.; Karishnarajah, I. Application of pattern recognition tools for classifying acute coronary syndrome: An integrated medical modeling. *Theor. Biol. Med. Model.* **2013**, *10*, 57. [CrossRef]
53. Zhang, C.-W.; Ou, J.-P.; Zhang, J.-Q. Parameter optimization and analysis of a vehicle suspension system controlled by magnetorheological fluid dampers. *Struct. Control Health Monit.* **2006**, *13*, 885–896. [CrossRef]
54. Xu, S.; Wang, J.; Shou, W.; Ngo, T.; Sadick, A.-M.; Wang, X. Computer Vision Techniques in Construction: A Critical Review. *Arch. Comput. Methods Eng.* **2020**. [CrossRef]
55. Yan, J.; Pu, W.; Zhou, S.; Liu, H.; Bao, Z. Collaborative detection and power allocation framework for target tracking in multiple radar system. *Inf. Fusion* **2020**, *55*, 173–183. [CrossRef]
56. Liu, D.; Wang, S.; Huang, D.; Deng, G.; Zeng, F.; Chen, H. Medical image classification using spatial adjacent histogram based on adaptive local binary patterns. *Comput. Biol. Med.* **2016**, *72*, 185–200. [CrossRef] [PubMed]
57. Wang, B.; Zhang, B.F.; Liu, X.W.; Zou, F.C. Novel infrared image enhancement optimization algorithm combined with DFOCS. *Optik* **2020**, *224*, 165476. [CrossRef]
58. Abedini, M.; Mutalib, A.A.; Zhang, C.; Mehrmashhadi, J.; Raman, S.N.; Alipour, R.; Momeni, T.; Mussa, M.H. Large deflection behavior effect in reinforced concrete columns exposed to extreme dynamic loads. *Front. Struct. Civ. Eng.* **2020**, *14*, 532–553. [CrossRef]
59. Mou, B.; Li, X.; Bai, Y.; Wang, L. Shear behavior of panel zones in steel beam-to-column connections with unequal depth of outer annular stiffener. *J. Struct. Eng.* **2019**, *145*, 04018247. [CrossRef]
60. Wang, S.; Zhang, K.; van Beek, L.P.H.; Tian, X.; Bogaard, T.A. Physically-based landslide prediction over a large region: Scaling low-resolution hydrological model results for high-resolution slope stability assessment. *Environ. Model. Softw.* **2020**, *124*, 104607. [CrossRef]
61. Zhang, K.; Wang, Q.; Chao, L.; Ye, J.; Li, Z.; Yu, Z.; Yang, T.; Ju, Q. Ground observation-based analysis of soil moisture spatiotemporal variability across a humid to semi-humid transitional zone in China. *J. Hydrol.* **2019**, *574*, 903–914. [CrossRef]
62. Zhang, S.; Zhang, J.; Ma, Y.; Pak, R.Y.S. Vertical dynamic interactions of poroelastic soils and embedded piles considering the effects of pile-soil radial deformations. *Soils Found.* **2020**, *61*, 16–34. [CrossRef]
63. Pourya, K.; Abdolreza, O.; Brent, V.; Arash, H.; Hamid, R. Feasibility Study of Collapse Remediation of Illinois Loess Using Electrokinetics Technique by Nanosilica and Salt. In *Geo-Congress 2020*; American Society of Civil Engineers: Reston, VA, USA, 2020; pp. 667–675.
64. Baziar, M.H.; Rostami, H. Earthquake Demand Energy Attenuation Model for Liquefaction Potential Assessment. *Earthq. Spectra* **2017**, *33*, 757–780. [CrossRef]
65. Chao, M.; Kai, C.; Zhiwei, Z. Research on tobacco foreign body detection device based on machine vision. *Trans. Inst. Meas. Control* **2020**, *42*, 2857–2871. [CrossRef]
66. Abedini, M.; Zhang, C. Performance Assessment of Concrete and Steel Material Models in LS-DYNA for Enhanced Numerical Simulation, A State of the Art Review. *Arch. Comput. Methods Eng.* **2020**. [CrossRef]
67. Gholipour, G.; Zhang, C.; Mousavi, A.A. Numerical analysis of axially loaded RC columns subjected to the combination of impact and blast loads. *Eng. Struct.* **2020**, *219*, 110924. [CrossRef]
68. Mou, B.; Zhao, F.; Qiao, Q.; Wang, L.; Li, H.; He, B.; Hao, Z. Flexural behavior of beam to column joints with or without an overlying concrete slab. *Eng. Struct.* **2019**, *199*, 109616. [CrossRef]
69. Zhang, C.; Abedini, M.; Mehrmashhadi, J. Development of pressure-impulse models and residual capacity assessment of RC columns using high fidelity Arbitrary Lagrangian-Eulerian simulation. *Eng. Struct.* **2020**, *224*, 111219. [CrossRef]
70. Sun, Y.; Wang, J.; Wu, J.; Shi, W.; Ji, D.; Wang, X.; Zhao, X. Constraints hindering the development of high-rise modular buildings. *Appl. Sci.* **2020**, *10*, 7159. [CrossRef]
71. Liu, C.; Huang, X.; Wu, Y.-Y.; Deng, X.; Liu, J.; Zheng, Z.; Hui, D. Review on the research progress of cement-based and geopolymer materials modified by graphene and graphene oxide. *Nanotechnol. Rev.* **2020**, *9*, 155–169. [CrossRef]
72. Xiong, Z.; Xiao, N.; Xu, F.; Zhang, X.; Xu, Q.; Zhang, K.; Ye, C. An Equivalent Exchange Based Data Forwarding Incentive Scheme for Socially Aware Networks. *J. Signal Process. Syst.* **2020**. [CrossRef]
73. Zenggang, X.; Zhiwen, T.; Xiaowen, C.; Xue-min, Z.; Kaibin, Z.; Conghuan, Y. Research on Image Retrieval Algorithm Based on Combination of Color and Shape Features. *J. Signal Process. Syst.* **2019**, 1–8. [CrossRef]
74. Yue, H.; Wang, H.; Chen, H.; Cai, K.; Jin, Y. Automatic detection of feather defects using Lie group and fuzzy Fisher criterion for shuttlecock production. *Mech. Syst. Signal Process.* **2020**, *141*, 106690. [CrossRef]
75. Zhu, G.; Wang, S.; Sun, L.; Ge, W.; Zhang, X. Output Feedback Adaptive Dynamic Surface Sliding-Mode Control for Quadrotor UAVs with Tracking Error Constraints. *Complexity* **2020**, *2020*, 8537198. [CrossRef]
76. Xiong, Q.; Zhang, X.; Wang, W.-F.; Gu, Y. A Parallel Algorithm Framework for Feature Extraction of EEG Signals on MPI. *Comput. Math. Methods Med.* **2020**, *2020*, 9812019. [CrossRef] [PubMed]
77. Zhang, J.; Liu, B. A review on the recent developments of sequence-based protein feature extraction methods. *Curr. Bioinform.* **2019**, *14*, 190–199. [CrossRef]
78. Zhao, X.; Li, D.; Yang, B.; Chen, H.; Yang, X.; Yu, C.; Liu, S. A two-stage feature selection method with its application. *Comput. Electr. Eng.* **2015**, *47*, 114–125. [CrossRef]
79. Chen, H.; Chen, A.; Xu, L.; Xie, H.; Qiao, H.; Lin, Q.; Cai, K. A deep learning CNN architecture applied in smart near-infrared analysis of water pollution for agricultural irrigation resources. *Agric. Water Manag.* **2020**, *240*, 106303. [CrossRef]

80. Li, T.; Xu, M.; Zhu, C.; Yang, R.; Wang, Z.; Guan, Z. A Deep Learning Approach for Multi-Frame In-Loop Filter of HEVC. *IEEE Trans. Image Process.* **2019**, *28*, 5663–5678. [CrossRef]
81. Qian, J.; Feng, S.; Tao, T.; Hu, Y.; Li, Y.; Chen, Q.; Zuo, C. Deep-learning-enabled geometric constraints and phase unwrapping for single-shot absolute 3D shape measurement. *APL Photonics* **2020**, *5*, 046105. [CrossRef]
82. Qiu, T.; Shi, X.; Wang, J.; Li, Y.; Qu, S.; Cheng, Q.; Cui, T.; Sui, S. Deep Learning: A Rapid and Efficient Route to Automatic Metasurface Design. *Adv. Sci.* **2019**, *6*, 1900128. [CrossRef]
83. Xu, M.; Li, T.; Wang, Z.; Deng, X.; Yang, R.; Guan, Z. Reducing Complexity of HEVC: A Deep Learning Approach. *IEEE Trans. Image Process.* **2018**, *27*, 5044–5059. [CrossRef] [PubMed]
84. Zhu, Q. Research on Road Traffic Situation Awareness System Based on Image Big Data. *IEEE Intell. Syst.* **2020**, *35*, 18–26. [CrossRef]
85. Liu, S.; Chan, F.T.S.; Ran, W. Decision making for the selection of cloud vendor: An improved approach under group decision-making with integrated weights and objective/subjective attributes. *Expert Syst. Appl.* **2016**, *55*, 37–47. [CrossRef]
86. Tian, P.; Lu, H.; Feng, W.; Guan, Y.; Xue, Y. Large decrease in streamflow and sediment load of Qinghai–Tibetan Plateau driven by future climate change: A case study in Lhasa River Basin. *CATENA* **2020**, *187*, 104340. [CrossRef]
87. Yang, W.; Pudasainee, D.; Gupta, R.; Li, W.; Wang, B.; Sun, L. An overview of inorganic particulate matter emission from coal/biomass/MSW combustion: Sampling and measurement, formation, distribution, inorganic composition and influencing factors. *Fuel Process. Technol.* **2020**, 106657. [CrossRef]
88. Cao, B.; Dong, W.; Lv, Z.; Gu, Y.; Singh, S.; Kumar, P. Hybrid Microgrid Many-Objective Sizing Optimization with Fuzzy Decision. *IEEE Trans. Fuzzy Syst.* **2020**, *28*, 2702–2710. [CrossRef]
89. Cao, B.; Zhao, J.; Gu, Y.; Ling, Y.; Ma, X. Applying graph-based differential grouping for multiobjective large-scale optimization. *Swarm Evol. Comput.* **2020**, *53*, 100626. [CrossRef]
90. Qu, S.; Han, Y.; Wu, Z.; Raza, H. Consensus Modeling with Asymmetric Cost Based on Data-Driven Robust Optimization. *Group Decis. Negot.* **2020**, 1–38. [CrossRef]
91. Wu, C.; Wu, P.; Wang, J.; Jiang, R.; Chen, M.; Wang, X. Critical review of data-driven decision-making in bridge operation and maintenance. *Struct. Infrastruct. Eng.* **2020**, 1–24. [CrossRef]
92. Adeli, H. Neural networks in civil engineering: 1989–2000. *Comput. Aided Civ. Infrastruct. Eng.* **2001**, *16*, 126–142. [CrossRef]
93. Hornik, K.; Stinchcombe, M.; White, H. Multilayer feedforward networks are universal approximators. *Neural Netw.* **1989**, *2*, 359–366. [CrossRef]
94. Lv, Z.; Qiao, L. Deep belief network and linear perceptron based cognitive computing for collaborative robots. *Appl. Soft Comput.* **2020**, *92*, 106300. [CrossRef]
95. Yang, J.; Li, S.; Wang, Z.; Dong, H.; Wang, J.; Tang, S. Using Deep Learning to Detect Defects in Manufacturing: A Comprehensive Survey and Current Challenges. *Materials* **2020**, *13*, 5755. [CrossRef] [PubMed]
96. Chen, H.-L.; Wang, G.; Ma, C.; Cai, Z.-N.; Liu, W.-B.; Wang, S.-J. An efficient hybrid kernel extreme learning machine approach for early diagnosis of Parkinson's disease. *Neurocomputing* **2016**, *184*, 131–144. [CrossRef]
97. Hu, L.; Hong, G.; Ma, J.; Wang, X.; Chen, H. An efficient machine learning approach for diagnosis of paraquat-poisoned patients. *Comput. Biol. Med.* **2015**, *59*, 116–124. [CrossRef]
98. Wang, S.-J.; Chen, H.-L.; Yan, W.-J.; Chen, Y.-H.; Fu, X. Face recognition and micro-expression recognition based on discriminant tensor subspace analysis plus extreme learning machine. *Neural Process. Lett.* **2014**, *39*, 25–43. [CrossRef]
99. Xia, J.; Chen, H.; Li, Q.; Zhou, M.; Chen, L.; Cai, Z.; Fang, Y.; Zhou, H. Ultrasound-based differentiation of malignant and benign thyroid Nodules: An extreme learning machine approach. *Comput. Methods Programs Biomed.* **2017**, *147*, 37–49. [CrossRef]
100. Chen, H.; Heidari, A.A.; Chen, H.; Wang, M.; Pan, Z.; Gandomi, A.H. Multi-population differential evolution-assisted Harris hawks optimization: Framework and case studies. *Future Gener. Comput. Syst.* **2020**, *111*, 175–198. [CrossRef]
101. Zhang, Y.; Liu, R.; Wang, X.; Chen, H.; Li, C. Boosted binary Harris hawks optimizer and feature selection. *Eng. Comput.* **2020**, *25*, 26. [CrossRef]
102. Shen, L.; Chen, H.; Yu, Z.; Kang, W.; Zhang, B.; Li, H.; Yang, B.; Liu, D. Evolving support vector machines using fruit fly optimization for medical data classification. *Knowl. Based Syst.* **2016**, *96*, 61–75. [CrossRef]
103. Wang, M.; Chen, H. Chaotic multi-swarm whale optimizer boosted support vector machine for medical diagnosis. *Appl. Soft Comput. J.* **2020**, *88*, 105946. [CrossRef]
104. Tu, J.; Chen, H.; Liu, J.; Heidari, A.A.; Zhang, X.; Wang, M.; Ruby, R.; Pham, Q.-V.J.K.-B.S. Evolutionary biogeography-based whale optimization methods with communication structure: Towards measuring the balance. *Knowl. Based Syst.* **2021**, *212*, 106642. [CrossRef]
105. Zhao, X.; Li, D.; Yang, B.; Ma, C.; Zhu, Y.; Chen, H. Feature selection based on improved ant colony optimization for online detection of foreign fiber in cotton. *Appl. Soft Comput.* **2014**, *24*, 585–596. [CrossRef]
106. Zhao, D.; Liu, L.; Yu, F.; Heidari, A.A.; Wang, M.; Liang, G.; Muhammad, K.; Chen, H.J.K.-B.S. Chaotic random spare ant colony optimization for multi-threshold image segmentation of 2D Kapur entropy. *Knowl. Based Syst.* **2020**, 106510. [CrossRef]
107. Yu, C.; Chen, M.; Cheng, K.; Zhao, X.; Ma, C.; Kuang, F.; Chen, H. SGOA: Annealing-behaved grasshopper optimizer for global tasks. *Eng. Comput.* **2021**, 1–28. [CrossRef]
108. Xu, X.; Chen, H.-L. Adaptive computational chemotaxis based on field in bacterial foraging optimization. *Soft Comput.* **2014**, *18*, 797–807. [CrossRef]

109. Cao, B.; Wang, X.; Zhang, W.; Song, H.; Lv, Z. A Many-Objective Optimization Model of Industrial Internet of Things Based on Private Blockchain. *IEEE Netw.* **2020**, *34*, 78–83. [CrossRef]
110. Cao, B.; Fan, S.; Zhao, J.; Yang, P.; Muhammad, K.; Tanveer, M. Quantum-enhanced multiobjective large-scale optimization via parallelism. *Swarm Evol. Comput.* **2020**, *57*, 100697. [CrossRef]
111. Hu, J.; Chen, H.; Heidari, A.A.; Wang, M.; Zhang, X.; Chen, Y.; Pan, Z.J.K.-B.S. Orthogonal learning covariance matrix for defects of grey wolf optimizer: Insights, balance, diversity, and feature selection. *Knowl. Based Syst.* **2020**, *213*, 106684. [CrossRef]
112. Zhao, X.; Zhang, X.; Cai, Z.; Tian, X.; Wang, X.; Huang, Y.; Chen, H.; Hu, L. Chaos enhanced grey wolf optimization wrapped ELM for diagnosis of paraquat-poisoned patients. *Comput. Biol. Chem.* **2019**, *78*, 481–490. [CrossRef] [PubMed]
113. Zhang, X.; Wang, D.; Zhou, Z.; Ma, Y. Robust low-rank tensor recovery with rectification and alignment. *IEEE Trans. Pattern Anal. Mach. Intell.* **2019**. [CrossRef] [PubMed]
114. Zhang, X.; Jiang, R.; Wang, T.; Wang, J. Recursive neural network for video deblurring. *IEEE Trans. Circuits Syst. Video Technol.* **2020**. [CrossRef]
115. Zhang, Y.; Liu, R.; Heidari, A.A.; Wang, X.; Chen, Y.; Wang, M.; Chen, H.J.N. Towards augmented kernel extreme learning models for bankruptcy prediction: Algorithmic behavior and comprehensive analysis. *Neurocomputing* **2020**. [CrossRef]
116. Akdemir, B. Prediction of Hourly Generated Electric Power Using Artificial Neural Network for Combined Cycle Power Plant. *Int. J. Electr. Energy* **2016**, *4*, 91–95. [CrossRef]
117. Bandić, L.; Hasičić, M.; Kevrić, J. Prediction of Power Output for Combined Cycle Power Plant Using Random Decision Tree Algorithms and ANFIS. In *International Symposium on Innovative and Interdisciplinary Applications of Advanced Technologies*; Springer: Berlin/Heidelberg, Germany, 2019; pp. 406–416.
118. Mohammed, M.K.; Awad, O.I.; Rahman, M.; Najafi, G.; Basrawi, F.; Abd Alla, A.N.; Mamat, R. The optimum performance of the combined cycle power plant: A comprehensive review. *Renew. Sustain. Energy Rev.* **2017**, *79*, 459–474.
119. Moayedi, H.; Mehrabi, M.; Kalantar, B.; Abdullahi Mu'azu, M.; Rashid, A.S.A.; Foong, L.K.; Nguyen, H. Novel hybrids of adaptive neuro-fuzzy inference system (ANFIS) with several metaheuristic algorithms for spatial susceptibility assessment of seismic-induced landslide. *Geomat. Nat. Hazards Risk* **2019**, *10*, 1879–1911. [CrossRef]
120. Moayedi, H.; Mehrabi, M.; Mosallanezhad, M.; Rashid, A.S.A.; Pradhan, B. Modification of landslide susceptibility mapping using optimized PSO-ANN technique. *Eng. Comput.* **2019**, *35*, 967–984. [CrossRef]
121. Liu, J.; Wu, C.; Wu, G.; Wang, X. A novel differential search algorithm and applications for structure design. *Appl. Math. Comput.* **2015**, *268*, 246–269. [CrossRef]
122. Sun, G.; Yang, B.; Yang, Z.; Xu, G. An adaptive differential evolution with combined strategy for global numerical optimization. *Soft Comput.* **2019**, 1–20. [CrossRef]
123. Fu, X.; Pace, P.; Aloi, G.; Yang, L.; Fortino, G. Topology optimization against cascading failures on wireless sensor networks using a memetic algorithm. *Comput. Netw.* **2020**, *177*, 107327. [CrossRef]
124. Shan, W.; Qiao, Z.; Heidari, A.A.; Chen, H.; Turabieh, H.; Teng, Y.J.K.-B.S. Double adaptive weights for stabilization of moth flame optimizer: Balance analysis, engineering cases, and medical diagnosis. *Knowl. Based Syst.* **2020**, *214*, 106728. [CrossRef]
125. Yu, H.; Li, W.; Chen, C.; Liang, J.; Gui, W.; Wang, M.; Chen, H. Dynamic Gaussian bare-bones fruit fly optimizers with abandonment mechanism: Method and analysis. *Eng. Comput.* **2020**, 1–29. [CrossRef]
126. Xu, Y.; Chen, H.; Luo, J.; Zhang, Q.; Jiao, S.; Zhang, X.J.I.S. Enhanced Moth-flame optimizer with mutation strategy for global optimization. *Inf. Sci.* **2019**, *492*, 181–203. [CrossRef]
127. Zhang, X.; Wang, J.; Wang, T.; Jiang, R.; Xu, J.; Zhao, L.J.I.S. Robust Feature Learning for Adversarial Defense via Hierarchical Feature Alignment. *Inf. Sci.* **2020**. [CrossRef]
128. Zhang, X.; Fan, M.; Wang, D.; Zhou, P.; Tao, D. Top-k feature selection framework using robust 0-1 integer programming. *IEEE Trans. Neural Netw. Learn. Syst.* **2020**. [CrossRef] [PubMed]
129. Abedinia, O.; Amjady, N.; Ghadimi, N. Solar energy forecasting based on hybrid neural network and improved metaheuristic algorithm. *Comput. Intell.* **2018**, *34*, 241–260. [CrossRef]
130. Zhou, G.; Moayedi, H.; Bahiraei, M.; Lyu, Z. Employing artificial bee colony and particle swarm techniques for optimizing a neural network in prediction of heating and cooling loads of residential buildings. *J. Clean. Prod.* **2020**, *254*, 120082. [CrossRef]
131. El Mokhi, C.; Addaim, A. Optimization of Wind Turbine Interconnections in an Offshore Wind Farm Using Metaheuristic Algorithms. *Sustainability* **2020**, *12*, 5761. [CrossRef]
132. Okewu, E.; Misra, S.; Maskeliūnas, R.; Damaševičius, R.; Fernandez-Sanz, L. Optimizing green computing awareness for environmental sustainability and economic security as a stochastic optimization problem. *Sustainability* **2017**, *9*, 1857. [CrossRef]
133. Seyedmahmoudian, M.; Jamei, E.; Thirunavukkarasu, G.S.; Soon, T.K.; Mortimer, M.; Horan, B.; Stojcevski, A.; Mekhilef, S. Short-term forecasting of the output power of a building-integrated photovoltaic system using a metaheuristic approach. *Energies* **2018**, *11*, 1260. [CrossRef]
134. Hu, Y.; Li, J.; Hong, M.; Ren, J.; Lin, R.; Liu, Y.; Liu, M.; Man, Y. Short term electric load forecasting model and its verification for process industrial enterprises based on hybrid GA-PSO-BPNN algorithm—A case study of papermaking process. *Energy* **2019**, *170*, 1215–1227. [CrossRef]
135. Lorencin, I.; Anđelić, N.; Mrzljak, V.; Car, Z. Genetic algorithm approach to design of multi-layer perceptron for combined cycle power plant electrical power output estimation. *Energies* **2019**, *12*, 4352. [CrossRef]

136. Ghosh, T.; Martinsen, K.; Dan, P.K. Data-Driven Beetle Antennae Search Algorithm for Electrical Power Modeling of a Combined Cycle Power Plant. In *World Congress on Global Optimization*; Springer: Berlin/Heidelberg, Germany, 2019; pp. 906–915.
137. Chatterjee, S.; Dey, N.; Ashour, A.S.; Drugarin, C.V.A. Electrical energy output prediction using cuckoo search based artificial neural network. In *Smart Trends in Systems, Security and Sustainability*; Springer: Berlin/Heidelberg, Germany, 2018; pp. 277–285.
138. Tüfekci, P. Prediction of full load electrical power output of a base load operated combined cycle power plant using machine learning methods. *Int. J. Electr. Power Energy Syst.* **2014**, *60*, 126–140. [CrossRef]
139. Foong, L.K.; Moayedi, H.; Lyu, Z. Computational modification of neural systems using a novel stochastic search scheme, namely evaporation rate-based water cycle algorithm: An application in geotechnical issues. *Eng. Comput.* **2020**, 1–12. [CrossRef]
140. Moayedi, H.; Tien Bui, D.; Anastasios, D.; Kalantar, B. Spotted hyena optimizer and ant lion optimization in predicting the shear strength of soil. *Appl. Sci.* **2019**, *9*, 4738. [CrossRef]
141. Moosavi, S.H.S.; Bardsiri, V.K. Satin bowerbird optimizer: A new optimization algorithm to optimize ANFIS for software development effort estimation. *Eng. Appl. Artif. Intell.* **2017**, *60*, 1–15. [CrossRef]
142. Kaya, H.; Tüfekci, P.; Gürgen, F.S. Local and global learning methods for predicting power of a combined gas & steam turbine. In Proceedings of the International Conference on Emerging Trends in Computer and Electronics Engineering ICETCEE, Dubai, UAE, 24 March 2012; pp. 13–18.
143. Eskandar, H.; Sadollah, A.; Bahreininejad, A.; Hamdi, M. Water cycle algorithm—A novel metaheuristic optimization method for solving constrained engineering optimization problems. *Comput. Struct.* **2012**, *110*, 151–166. [CrossRef]
144. Mohamed, A.-A.A.; Ali, S.; Alkhalaf, S.; Senjyu, T.; Hemeida, A.M. Optimal Allocation of Hybrid Renewable Energy System by Multi-Objective Water Cycle Algorithm. *Sustainability* **2019**, *11*, 6550. [CrossRef]
145. Chen, C.; Wang, P.; Dong, H.; Wang, X. Hierarchical Learning Water Cycle Algorithm. *Appl. Soft Comput.* **2020**, *86*, 105935. [CrossRef]
146. M'zoughi, F.; Bouallègue, S.; Garrido, A.J.; Garrido, I.; Ayadi, M. Water cycle algorithm–based airflow control for oscillating water column–based wave energy converters. *Proc. Inst. Mech. Eng. Part I J. Syst. Control Eng.* **2020**, *234*, 118–133. [CrossRef]
147. Mirjalili, S. The ant lion optimizer. *Adv. Eng. Softw.* **2015**, *83*, 80–98. [CrossRef]
148. Gajula, V.; Rajathy, R. An agile optimization algorithm for vitality management along with fusion of sustainable renewable resources in microgrid. *Energy Sources Part A Recovery Util. Environ. Eff.* **2020**, *42*, 1580–1598. [CrossRef]
149. Moayedi, H.; Kalantar, B.; Foong, L.K.; Tien Bui, D.; Motevalli, A. Application of three metaheuristic techniques in simulation of concrete slump. *Appl. Sci.* **2019**, *9*, 4340. [CrossRef]
150. Heidari, A.A.; Faris, H.; Mirjalili, S.; Aljarah, I.; Mafarja, M. Ant lion optimizer: Theory, literature review, and application in multi-layer perceptron neural networks. In *Nature-Inspired Optimizers*; Springer: Berlin/Heidelberg, Germany, 2020; pp. 23–46.
151. Zhang, S.; Zhou, G.; Zhou, Y.; Luo, Q. Quantum-inspired satin bowerbird algorithm with Bloch spherical search for constrained structural optimization. *J. Ind. Manag. Optim.* **2017**, *13*. [CrossRef]
152. Chintam, J.R.; Daniel, M. Real-power rescheduling of generators for congestion management using a novel satin bowerbird optimization algorithm. *Energies* **2018**, *11*, 183. [CrossRef]
153. Chen, W.; Chen, X.; Peng, J.; Panahi, M.; Lee, S. Landslide susceptibility modeling based on ANFIS with teaching-learning-based optimization and Satin bowerbird optimizer. *Geosci. Front.* **2020**, *12*, 93–107. [CrossRef]
154. Mostafa, M.A.; Abdou, A.F.; Abd El-Gawad, A.F.; El-Kholy, E. SBO-based selective harmonic elimination for nine levels asymmetrical cascaded H-bridge multilevel inverter. *Aust. J. Electr. Electron. Eng.* **2018**, *15*, 131–143. [CrossRef]
155. Zhou, G.; Moayedi, H.; Foong, L.K. Teaching–learning-based metaheuristic scheme for modifying neural computing in appraising energy performance of building. *Eng. Comput.* **2020**, 1–12. [CrossRef]
156. Seyedashraf, O.; Mehrabi, M.; Akhtari, A.A. Novel approach for dam break flow modeling using computational intelligence. *J. Hydrol.* **2018**, *559*, 1028–1038. [CrossRef]
157. Guo, Z.; Moayedi, H.; Foong, L.K.; Bahiraei, M. Optimal modification of heating, ventilation, and air conditioning system performances in residential buildings using the integration of metaheuristic optimization and neural computing. *Energy Build.* **2020**, *214*, 109866. [CrossRef]

sustainability

MDPI

*Review*

# Energy Management System in Microgrids: A Comprehensive Review

**Younes Zahraoui [1], Ibrahim Alhamrouni [1], Saad Mekhilef [2,3,*], M. Reyasudin Basir Khan [4], Mehdi Seyedmahmoudian [2], Alex Stojcevski [2] and Ben Horan [5]**

1 British Malaysian Institute, Universiti Kuala Lumpur, Kuala Lumpur 50250, Malaysia; zahraoui.younes@s.unikl.edu.my (Y.Z.); ibrahim.mohamed@unikl.edu.my (I.A.)
2 School of Science, Computing and Engineering Technologies, Swinburne University of Technology, Hawthorn, VIC 3122, Australia; mseyedmahmoudian@swin.edu.au (M.S.); astojcevski@swin.edu.au (A.S.)
3 Power Electronics and Renewable Energy Research Laboratory, Department of Electrical Engineering, University of Malaya, Kuala Lumpur 50603, Malaysia
4 School of Engineering, Manipal International University, No. 1, Persiaran MIU, Nilai 71800, Malaysia; mohammed.reyasudin@miu.edu.my
5 School of Engineering, Deakin University, Geelong, VIC 3216, Australia; ben.horan@deakin.edu.au
* Correspondence: saad@um.edu.my or smekhilef@swin.edu.au

**Abstract:** As promising solutions to various social and environmental issues, the generation and integration of renewable energy (RE) into microgrids (MGs) has recently increased due to the rapidly growing consumption of electric power. However, such integration can affect the stability and security of power systems due to its complexity and intermittency. Therefore, an optimal control approach is essential to ensure the efficiency, reliability, and quality of the delivered power. In addition, effective planning of policies for integrating MGs can help promote MG operations. However, outages may render these strategies inefficient and place the power system at risk. MGs are considered an ideal candidate for distributed power systems, given their capability to restore these systems rapidly after a physical or cyber-attack and create reliable protection systems. The energy management system (EMS) in an MG can operate controllable distributed energy resources and loads in real-time to generate a suitable short-term schedule for achieving some objectives. This paper presents a comprehensive review of MG elements, the different RE resources that comprise a hybrid system, and the various types of control, operating strategies, and goals in an EMS. A detailed explanation of the primary, secondary, and tertiary levels of MGs is also presented. This paper aims to contribute to the policies and regulations adopted by certain countries, their protection schemes, transactive markets, and load restoration in MGs.

**Keywords:** microgrid; energy management system; restoration; power quality; policy market

**Citation:** Zahraoui, Y.; Alhamrouni, I.; Mekhilef, S.; Basir Khan, M.R.; Seyedmahmoudian, M.; Stojcevski, A.; Horan, B. Energy Management System in Microgrids: A Comprehensive Review. *Sustainability* **2021**, *13*, 10492. https://doi.org/10.3390/su131910492

Academic Editor: Thanikanti Sudhakar Babu

Received: 3 August 2021
Accepted: 29 August 2021
Published: 22 September 2021

**Publisher's Note:** MDPI stays neutral with regard to jurisdictional claims in published maps and institutional affiliations.



## 1. Introduction

Increasing energy demand is a key indicator of economic growth and social development. Such demand has been growing exponentially in various sectors, such as in the transportation, building, and manufacturing industries. However, energy consumption is directly linked to environmental issues due to the frequent use of fuel or coal as the main electricity generation sources that emit greenhouse gases (GHG). Therefore, many global actors, including the World Bank, are encouraging countries to generate clean energy by financially supporting their projects [1].

Renewable energy (RE) is an important energy source with an abundant supply in nature. RE is less carbon-intensive and more sustainable than traditional energy sources, hence explaining its growing popularity. Such green energy resources not only have limited impacts on the environment but also contribute to energy savings and reduce the dependence of industries on fossil fuels. Accordingly, many countries have promoted the use of RE to achieve sustainable development. The electricity produced from RE was estimated to account for 11% of the total energy produced in 2020, as shown in Figure 1 [2].

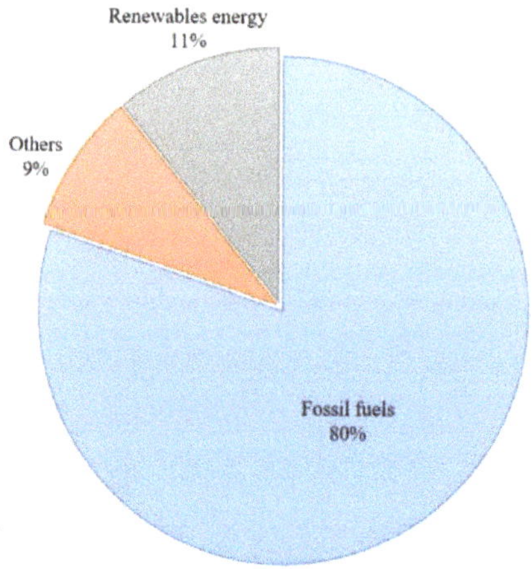


**Figure 1.** The estimated power generation of each power resource in 2020.

RE resources are increasingly being used in distributed generators (DG) to address the shortcomings in centralized energy generation, including its high cost, transmission losses, and environmental effects. However, the efficiency of RE generation is effected by natural environment indicators, such as wind speed, temperature, and solar irradiation, which may introduce challenges in a power network, such as inverse power flow, and voltage deviations and fluctuations. Using a hybrid grid strategy that combines RE with the more efficient and secure microgrid (MG) approach is therefore critical [3].

An MG combines different energy sources (renewable and non-renewable) and energy storage systems (ESS) to fulfill the demand for loads that can be either connected to the main grid at the Point of Common Coupling (PCC) or operated in the islanded mode, where the MG operating system can support green energy. MGs operate autonomously in an isolated mode whenever a fault occurs in linked power systems. MGs provide many benefits, such as reducing GHG, supporting reactive power to increase the voltage profile, decentralizing the energy supply, and offering demand response. The global deployment of MGs reached 1.4 GW in 2015, and is expected to increase to 8.8 GW by 2024. MGs have been deployed in remote areas, communities, and various sectors, including the commercial, industrial, and military sectors, in consideration of their objectives, load types, and geographical and climactic conditions [4].

With its growing popularity, previous studies have examined the application of MGs. For instance, Hirsch et al. [5] discussed some factors that lead to the implementation of an MG in a power system and its contributions to energy security, economics, and clean energy generation. Majumder et al. [6] explored the main features, challenges, and sectors that implement MGs. Cagnano et al. [7] discussed the functions, device configurations, and control topologies of MGs. Dawoud et al. [8] proposed a set of specifications and instructions that can help address the challenges faced in real MG applications. Meng et al. [9] proposed several optimization techniques and tools for improving MG utilization. Muhammad Fahad Zia et al. [4] discussed the development of energy management system (EMS) strategies and solution approaches in MGs. Table 1 summarizes the recent literature on MGs.

**Table 1.** The recent literature on MG.

| Ref | Details |
| --- | --- |
| [10] | Addressed the issues affecting DC MG safety from different aspects, such as fault location detection, and evaluated some protective devices. |
| [11] | Comprehensively reviewed the stability issues being faced by MGs based on extant definitions and classifications of stability and illustrated these issues as modeling examples. |
| [12] | Examined the existing MG architectures in detail, and demonstrated the widely distributed technologies along with their advantages and disadvantages. |
| [13] | Highlighted several issues, challenges, and solutions related to the protection of an AC MG. |
| [14] | Represents features of, and the large-disturbance stability that prevails for, a power-converter-dominated MG, with some stability analysis highlighted. |
| [15] | Comprehensively reviewed the main components, size, and energy management of harbor MGs. |
| This work | Comprehensively reviews the operation strategies and objectives used in EMSs and explains the architecture and elements of an EMS in an MG. |

An EMS ensures the efficiency and economic activity of an MG based on the output power generated from distributed energy resources (DERs), the status of devices, forecasted load and weather, and prices of electricity and fuel. An EMS can correlate and control the output power of DERs, ESSs, and energy exchanges. Consequently, an EMS can be used to achieve single or multiple objectives, such as minimizing daily operational costs, performing real and reactive monitoring of power, reducing losses, and balancing the energy in transmission lines. In this case, an EMS is critical for MGs to operate efficiently, ensure their reliability, and satisfy power balance in both the short and long term [16].

Both the MG and EMS are critical in dealing with the challenges arising from the integration of DER units, such as photovoltaic (PV) systems, wind turbines, microturbines that use the CHP system, and fuel cells and batteries in power systems. However, integrating RE resources into the main grid is unusable given the unpredictable behavior of such integration and the intermittent nature of RE. Therefore, during the intermittent dispatching of RE resources, the reliability index of the power system is reduced. The resiliency of the power system can be improved by using an appropriate protection scheme, improving redundancy, installing isolation systems, and adopting conventional DERs. From this perspective, the necessary policies and regulations should be implemented as benchmarks for interconnecting DERs with traditional electric power systems.

The main objectives of this review are to explore the evolution of the MG and EMS and to review the elements, implementation, classification, objective functions, quality, and protection schemes of the MG. This paper reviews the existing technologies and challenges faced in MGs and EMSs. This article is organized as follows. Section 2 discusses the concept, architecture, and elements of the MG. Section 3 reviews the control schemes of the MG and the objectives of an EMS. Section 4 discusses the transactive energy market and its classification. Sections 5 and 6 present the designs of protection systems and the direction of MG policies in various countries, respectively. Section 7 shares the perspectives of authors toward MGs and concludes the paper. This paper also aims to unlock many possibilities for further research in this area.

The contents of this paper will considerably help researchers mitigate the present shortcomings of MGs and EMSs and formulate new techniques and objective functions for promoting their application.

## 2. MG Architecture and Elements

An MG comprises of DGs, ESSs, balanced and controlled electrical loads, and intelligent devices such as circuit breakers (CBs) and intelligent switches, as shown in Figure 2. DERs and ESSs operate in coordination to reliably supply electricity and to preserve the balance between generated and consumed power. Using MGs in a power system as a model for the massive integration of different DERs will solve the technical problems in traditional centralized distribution. The majority of the DERs that can be connected to an MG cannot be directly integrated into the power system due to the type of power they produce. Therefore, power electronic interfaces, such as inverter controls, are necessary. Moreover, an MG is guaranteed to operate continuously during normal operations and critical cases [17].

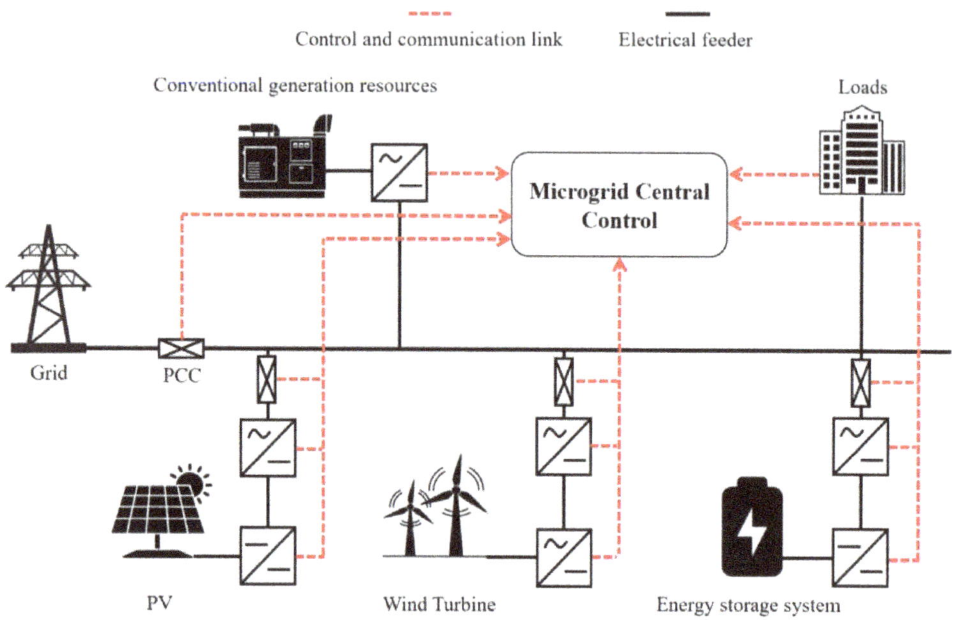


**Figure 2.** Block diagram of an MG model connected with an MGCC.

The MG central controller (MGCC) is considered the brain of an MG, responsible for enhancing its performance, calculating the optimal values, achieving some objective functions in consideration of the constraints, conducting additional operations based on the electricity and gas prices in the market, shouldering extra costs (e.g., for DER startup), and performing weather forecasting to ensure an optimal power generation. The MGCC also controls the loads in an MG by adequately managing the stability of the power system. The following subsections explain the different technologies and architectures that may be integrated into an MG installed in a power system [18].

### *2.1. Microgrid Elements*

Different power technologies, such as DGs and ESSs, are characteristic attributes of MGs. This section discusses some technologies that have been developed to be integrated into MGs.

#### 2.1.1. Distributed Generators

DGs are defined by the Institute of Electrical and Electronics Engineers Inc. (IEEE) as "The generation of electricity by facilities sufficiently smaller than central generating plants as to allow interconnection at nearly any point in a power system. DGs are a subset of distributed resources" [19].

DGs are dispatchable generating units, including fuel cell and diesel generators, or non-dispatchable generators, such as PV plants and wind turbines stations, that are placed at loading sites. The application of these generators is becoming increasingly popular given their role in satisfying the demand of consumers. These units are deployed to improve the efficiency and cleanliness of power generation by using RE resources. Deploying DGs can increase the resilience of a power system by supporting the growth of different resources for the partial distribution of power; for example, a natural disaster may result in large-scale outages, and using diverse DGs will ensure that the power system will not be impacted by such phenomenon [20]. Meanwhile, integrating DERs into the distribution network improves the voltage profile, reduces line loss, and lowers power generation costs. These DGs can provide AC power outputs, such as combined heat and power (CHP), fuel cells, and gas turbines, or DC outputs, such as wind turbines or PV. Therefore, DGs comprise inverters that convert their output to suit the specifications of a power system. The capacity of DGs is also related to space and time [21].

Table 2 presents some DG technologies used in MGs.

**Table 2.** Different DG technologies applied in MGs.

| Ref | CHP [1] | DG [2] | GG [3] | FC [4] | MT [5] | PV [6] | HYD [7] | WT [8] | TI [9] |
|---|---|---|---|---|---|---|---|---|---|
| [22] | ✓ | | | | | | | ✓ | |
| [23] | ✓ | | | ✓ | ✓ | ✓ | | ✓ | |
| [24] | ✓ | | | | | ✓ | | | |
| [25] | | | | ✓ | ✓ | ✓ | | | ✓ |
| [26] | | ✓ | | | | ✓ | | ✓ | |
| [27] | | | | | ✓ | ✓ | | | ✓ |
| [28] | ✓ | ✓ | ✓ | | | ✓ | | ✓ | |
| [29] | ✓ | ✓ | | ✓ | | ✓ | | ✓ | |
| [30] | | ✓ | | | | ✓ | ✓ | | |
| [31] | | | | | | | ✓ | ✓ | |
| [32] | | | | | | | | | ✓ |
| [33] | ✓ | | | | | ✓ | | ✓ | |
| [34] | ✓ | | | ✓ | | | | | |

[1] Combined Heat and Power, [2] Diesel Generator, [3] Gas Generator, [4] Fuel Cell, [5] Microturbine, [6] Photovoltaic, [7] Hydropower, [8] Wind Turbine, [9] Tidal.

### 2.1.2. Energy Storage Devices

An MG has a slight generating capability given that some DGs, such as RE resources, can change the output power and pose technical challenges [35]. Therefore, MGs require an energy storage system (ESS) to solve mismatch problems and suit the power system requirements. ESSs can store and provide surplus energy when needed. These systems can also promote the reliability of the power system, improve the performance of an MG, achieve power balance among end-users, and reduce peak demand. ESS devices also satisfy mismatched energy requirements to ensure a continuous energy supply [36].

ESS technologies have promising uses in MG deployment. Konstantinopoulos et al. [37] used hydrogen for production and storage, given that the power generated from RE resources is assumed to exceed the power demand. Hou et al. [38] integrated hybrid energy storage into MG and used flywheel storage to allow the application of two strategies in the system. Mousavi et al. [39] proposed a novel design for storing surplus energy by using a hydro pump to ensure the efficient performance of MGs in rural areas. Jia et al. [40] used ultracapacitor storage to minimize the total costs and applied the charging/discharging method to manage the power storage. Guo et al. [41] integrated lead-acid batteries into a standalone MG model to minimize the total net present cost and carbon dioxide emissions. Table 3 summarizes the various ESSs that have been studied in the literature.

**Table 3.** Reviews/surveys relating to ESSs.

| Ref | Details |
|---|---|
| [42] | Comprehensively reviewed the challenges, modeling approaches, and estimation of impact on market structures when utilizing energy storage. |
| [43] | Presented an overview of the applications of ESSs, which may introduce challenges to MGs. |
| [44] | Comprehensively reviewed the most recent ESS innovations in MG technologies, including the concepts and optimization techniques, architectures, control techniques, future trends, and challenges in ESSs. |
| [45] | Addressed some factors in sizing of the ESSs in MGs and various applications through the integration with RE. |
| [46] | Presented a comprehensive techno-economic analysis of the battery storage system under various MG system configurations. |

Regardless of their advantages, ESS technologies have not been used in MG applications given some limitations, their cost, and their difficulty to control. Table 4 presents the limitations, advantages, disadvantages, and generation costs of each DER technology during MG operation.

**Table 4.** Comparison of various types of DERs utilized in MGs.

| Type | | Element | Output Type | Capacity | Generation Cost ($/kWh) | Advantages | Disadvantages |
|---|---|---|---|---|---|---|---|
| DG | Dispatchable resources. | CHP | AC | 20 kW–10 MW | – | - Continuous power dispatch.<br>- Startup fast.<br>- Multiple fuel options | - Greenhouse Gas Emissions.<br>- Noise production |
| | | Diesel backup generator | AC | 20 kW–10 MW | 125–300 | | |
| | | Gas generator | AC | 50 kW–5 MW | 250–600 | | |
| | | Fuel cell | AC | 50 kW–1 MW | 1500–3000 | | |
| | | Micro turbine | AC | 25–100 kW | 350–750 | | |
| | Non-Dispatchable resources. | Photovoltaic (PV) | DC | 10 kW–300 MW | – | - Clean energy.<br>- Does not cost power generation. | - Fluctuation in generation.<br>- Comparatively expensive in the installation phase.<br>- Related to geographic locations. |
| | | Hydro | AC | 50 kW–30 MW | – | | |
| | | Wind turbine | AC | 10 kW–300 MW | – | | |
| | | Tidal | AC | 50 kW–200 MW | – | | |
| ESS | | Pumped hydro | AC | 102–107 kWh | 1000–2500 | - Clean<br>- Fast response<br>- High efficiency | - Limited discharge time<br>- Not dispatchable without storage |
| | | Compressed air | | 12,000 kWh–6.42 GWh | 1000–2800 | | |
| | | Thermal storage | | 1000 kWh–1.1 GWh | 1250–1500 | | |
| | | Flywheel | | 2–25 kWh | 250–300 | | |
| | | Li-ion | | 10–120,000 kWh | 250–500 | | |
| | | Lead-acid | | 7–15 kWh | 250–500 | | |
| | | Capacitors | | 3.5–150 kWh | 25–50 | | |

### 2.1.3. Loads

DGs and ESSs can supply either electrical or thermal loads. The defined loads are treated as input parameters in scheduled energy management studies where the load profiles change according to the activities of customers and the weather conditions. The loads in MGs are classified into critical and non-critical loads, of which the latter does not require DERs to be connected to their buses or local generators. These loads should have the potential to disconnect during emergency cases to preserve balance in the power system. Meanwhile, critical loads are a very sensible and high priority; some of these loads, including commercial and industrial loads, need to be supplied continuously, whereas others can be connected individually to DERs [47].

### 2.1.4. Additional Elements

An MG must conduct power management and ensure controllable load sharing. Intelligent circuit breakers are required in an MG to manage and control the interconnection. Most intelligent circuit breakers are located in the point of connection between the MG and the rest of the DERs, and may apply certain techniques, such as power switching, protective relaying, metering, and communication. The interconnection breakers should meet the general standards, such as the IEEE 1547 and UL 1741 in North America, to ensure the safe operation of MGs and to enable the application of DERs or power converters [48].

Power converters, such as intelligent inverters, are used in MGs to ensure their efficient and autonomous operation with limited capacity which is managed via the AC/DC conversion, or vice versa. These converters serve as the interface between the energy generation resources and the end-user, and are utilized to manage, form, and feed the power system. The different technologies described above need to share information with one another before taking the prerequisite actions. Consequently, a robust communication system is needed to ensure a continuous and accurate sharing of information [49]. Figure 3 presents the various communication technologies used in MGs.

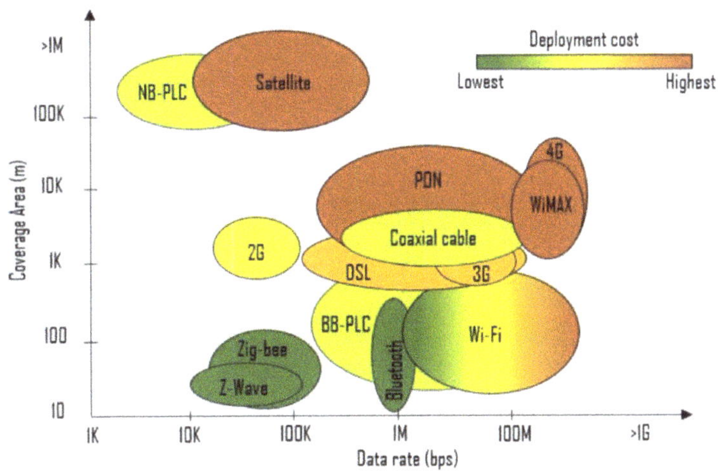


**Figure 3.** Various communication technology is used in MGs. NB-PLC: Narrow band power line communication; BB-PLC: Broad band power line communication; PON: passive optical network; DSL: Digital subscriber line.

## 2.2. Control Scheme of MGs

The control schemes used in MGs can be classified into centralized and decentralized. Centralized control collects all data from a single MGCC unit that can execute the required calculations and define the control procedures and actions. This approach requires comprehensive communication between the MGCC and other substation units [50]. Meanwhile, decentralized control employs the local controller to operate the unit. Figure 4 illustrates the communication and actions that take place between the controller and its unit. Table 5 summarizes previous surveys related to centralized and decentralized control schemes in an MG.

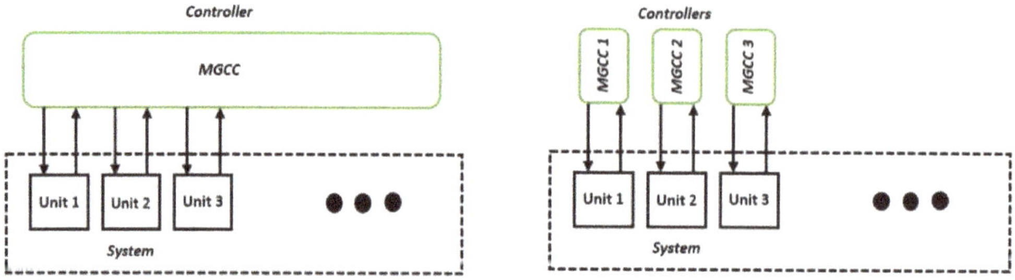


(a) (b)

**Figure 4.** Basic control structures: (**a**) centralized; (**b**) decentralized.

**Table 5.** Previous studies on centralized and decentralized MGs.

| Ref | Type | Remarks |
| --- | --- | --- |
| [51] | Centralized | Proposed an MG control based on a centralized architecture where different DERs are connected to a single bus, and applied a centralized heuristic approach to managing the reliability and economical use of energy. |
| [52] | | Performed a centralized real-time simulation in an MG connected to DERs and found that the optimization model in a centralized control can operate a virtual power of DERs. |
| [53] | | Proposed a centralized control for an intelligent network of greenhouses connected to an MG. The control of stochastic power DERs was based on model predictive control (MPC) to optimize crop production and control indoor climate conditions. |
| [54] | | Managed the active and reactive power in a power system by using centralized control in an MG connected to the primary grid, which can provide an auxiliary to control frequency and voltage. |
| [55] | | Employed an optimal operation approach to schedule energy in multiple MGs and allocated economic benefits. |
| [56] | Decentralized | Developed a multi-agent system relying on an MG cluster (MGC). Performed multi-time scale optimization to control and manage the EMS in the MGC and to schedule the day based on stability and economy. |
| [57] | | Proposed and simulated an adaptive control with DERs, including an ESS, to adjust the power injection by managing the DC voltage bus on an efficiency point. |
| [58] | | Applied the decentralized control of an MG to ensure the robustness and reliability of the power system by considering several objectives, such as economic power dispatch and reduction in power transmission losses. |
| [59] | | Promoted decentralized control by using a near real-time algorithm that operates the elements of an MG at the event of changing conditions. |

The employment of each control scheme is associated with the type of MG, the elements being used, and the geographic area. Although centralized and decentralized control approaches have many advantages (e.g., low-performance complexity), they also have limited reliability, expandability, and flexibility. These approaches typically follow the same hierarchical control structure as illustrated in Figure 5, which shows three levels of control, namely, primary, secondary, and tertiary, with each level having unique features in response, operation, and communication speed [12].

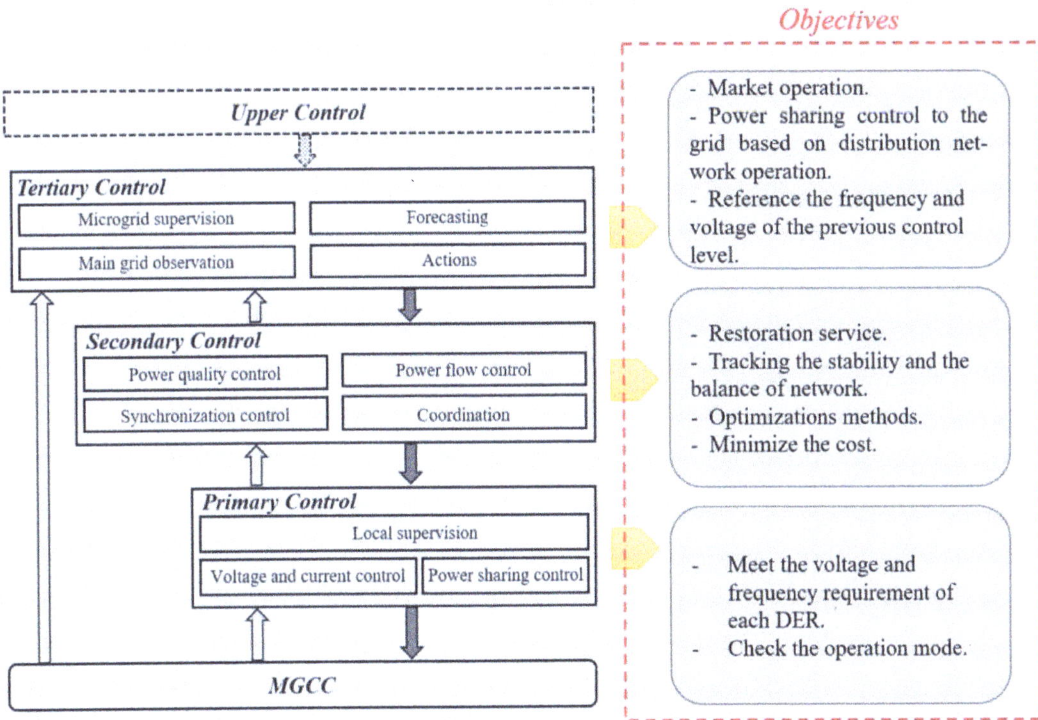


**Figure 5.** Hierarchical control scheme.

2.2.1. Primary Control Level

Primary control, also known as the field control or the first level of control, is completely based on the variables and local measurements (e.g., voltage and frequency) of the system. The different elements and other categories of droop controls at this level do not require communication tools. This level aims to ensure reliability, effective power-sharing, enhanced performance, and plug-and-play capability for DERs. The implementation of active/reactive power mode (PCM) or the voltage control mode (VCM) in DERs allows users to control the active and reactive power output and coordinate power-sharing among DERs as managed by voltage source inverter (VSI) controllers. The PCM and VCM are operated in the grid-connected and island modes in an MG, respectively. To adjust output power-sharing from the VSI, the droop characteristics should be applied to control the active/reactive power or voltage and frequency [60].

Droop control is an autonomous approach for controlling the frequency and voltage amplitude of power dispatch in an MG. Droop is a standard power-sharing method that has been mainly applied in MGs. This method aims to promote power-sharing among DER inverters, given the uncertainty of line impedances and the power delivered from RE resources, which leads to an unbalanced power system. Various approaches for droop control have been designed, such as conventional and non-conventional droop control. The traditional droop control aims to set the steady droop gain. An accurate gain in droop control affects the stability of an MG, the voltage regulation, and the management of power-sharing [61]. Conventional droop control is formulated as:

$$\omega = \omega_0 - k_p * P \quad (1)$$

$$V = V_0 - k_q * Q \quad (2)$$

where $k_p$ and $k_q$ denote active and reactive power gains, respectively, in the droop control; $\omega_0$ and $V_0$ denote the DER output angular frequency and voltage values; $\omega$ and $V$ denote the adjusted frequency and voltage; and $P_g$ and $Q_g$ denote the injected active and reactive powers, where:

$$P = P_0 - P^* \quad (3)$$

$$Q = Q_0 - Q^* \quad (4)$$

$P_0$ and $Q_0$ represent the active and reactive power delivered from DERs, respectively, and $P^*$ and $Q^*$ denote the reference active and reactive power values. Figure 6 presents a block diagram of the conventional droop control strategy.

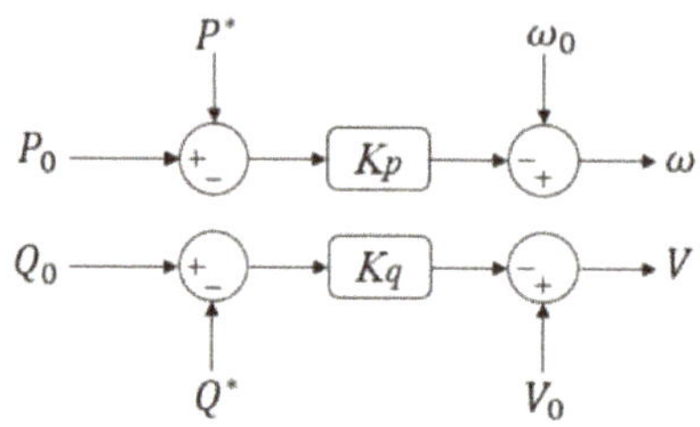


**Figure 6.** Block diagram of conventional droop control.

Many methods have been proposed to further improve the accuracy of droop gain. For instance, Datta et al. [62] proposed a conventional droop control that adjusts the droop gain in two stages, namely, the lower and higher gains for the low and high frequencies, respectively. Datta, Kalam, et al. [63] adjusted a multi-gain using the droop control approach to manage power-sharing in a wind farm and then integrated this method into two types of angle controls in a proportional comparative study. Joung et al. [64] studied the droop gain in traditional droop control for decoupling the frequency and voltage control of DERs, which can preserve the frequency and voltage constants in a grid. Although conventional droop control lacks complexity in its implementation and application, such method has some drawbacks when applied in MGs, such as reducing the voltage due to the current equality, its inability to handle non-linear loads, and the ingrained trade-off between voltage and power-sharing [65]. Therefore, non-conventional droop control has been used to address these shortcomings. Several techniques have also been applied to improve droop control in MGs, such as load sharing [66], voltage-active power droop (VPD) and frequency-reactive power boost (FQB) [67], virtual output impedance [68], and adaptive voltage droop [69], as described in Table 6.

**Table 6.** Non-conventional droop control techniques used in MGs.

| Method | Description | Advantages | Disadvantages |
| --- | --- | --- | --- |
| VPD & FQB method [67] | This approach solves many shortcomings in MG applications. VPD and FQB can support those DERs with power factor impedance and help control the low voltage of highly resistive transmission lines where the common bus voltage Vbus is adjusted to manage a reference voltage Vref for a specific bus. | - Simple implementation. | - May disturb the connection to non-linear loads. |

**Table 6.** *Cont.*

| Method | Description | Advantages | Disadvantages |
|---|---|---|---|
| Adaptive voltage droop control [69] | The voltage response coefficient is changed adaptively, based on the operating state of the converter station in DERs. | - Improves power sharing, reliability, and flexibility in MGs.<br>- More accurate voltage in MGs across various conditions. | - May fail to provide the appropriate voltage and power-sharing in a large DC MG. |
| virtual output impedance [68] | A virtual impedance is used to cancel out the negative impedance by simplifying the coupling relationship of active and reactive power. | - Functions with linear and non-linear loads.<br>- Improves the balance of DER output voltages. | - Voltage regulation is not guaranteed in a large-scale implementation. |
| Virtual transformation method [66] | This method uses an instantaneous power calculation unit, a coordinate rotation transformation unit, and an adaptive inverse control unit, the last of which can adjust and modify the active power frequency droop control module by using a different optimization technique. | - Simple implementation. | - Requires prior knowledge about<br>- the physical parameters that can affect voltage and frequency. |
| Angle droop control [70] | The angle of the reference voltage in the inverters is used to control the active power and the frequency produced from DERs. | - Controls the output voltage of converters with low THD.<br>- Control the voltage at the PCC. | - May fail to handle non-linear loads. |
| Synchronized reactive power compensation [71] | To recognize the errors in power sharing, the system injects a real-reactive power transient coupling term and then compensates for the errors by using a slow integral term for regulating the DG voltage magnitude. | - Effective in slow inner communications.<br>- Does not require configurations in MGs. | - May fail to handle non-linear loads.<br>- Complex implementation. |
| Self-Adjusting control [72] | The control method uses a multi-droop controller whose parameters are adjusted based on the power consumption from the local loads. Virtual inductive impedance is used to improve the control of voltage and transient responses of the power sharing. | - Does not require any communication connection.<br>- Reduces the reactive power flow in medium-voltage lines. | - Can disturb the connection to non-linear loads. |

### 2.2.2. Secondary Control Level

The secondary control level aims to address the shortcomings in the primary control level of MGs, including voltage deviations. This level is also known as the EMS level, which enhances power quality, restores the power system, ensures economical operations, and eliminates the frequency and voltage deviations and fluctuations caused by the droop control at the primary level [73]. Figure 7 presents the problems encountered in MGs that are solved using an EMS. This system can determine the optimal output power for each DER, the optimal network design for the restoration, and the stability of an MG by solving a single objective or multiple objectives in the grid depending on user preferences, geographical areas, available equipment, and their capacity, government regulations, types of tariffs, and DER/battery energy storage system (BESS) constraints. The secondary control level is considered a challenge in MGs. Given that the variations in employment and the power dispatched from DERs, the command and update between the loads and DERs must be in high communication and speed to ensure a subsequent power generation in MGs. The subsections below present the goals and objectives of EMSs in MGs. The following subsections present the goals and objectives of EMSs in MGs.

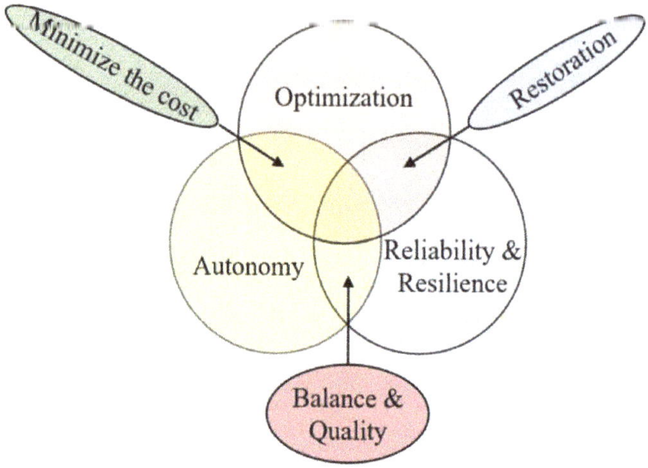


**Figure 7.** Types of problems solved by the EMS.

Minimize the Cost

Minimizing costs has different objectives, as shown in Figure 8. The objectives may be expressed as mathematical models, as shown in Table 7. The literature review reveals that the cost-minimization problems in the EMS are solved using many approaches.

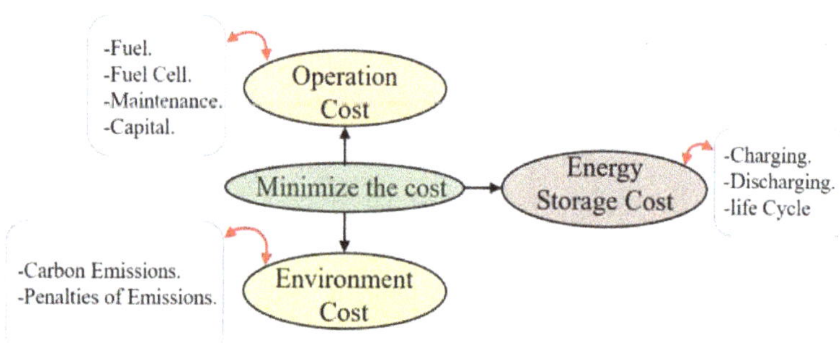


**Figure 8.** Objective function in minimizing costs.

Some studies have applied classical approaches to solve EMS problems. For instance, Lu et al. [74] proposed quadratic programing to solve and analyze the economic power dispatch of DERs in MGs. Economic costs involve the costs of shutting down, starting up, and generating power from DERs. Luna et al. [75] analyzed an EMS in an MG integrated with a grid-connected BESS that applies mixed-integer linear programming (MILP) to minimize the operating costs and improve the self-consumption strategy. Jabarnejad et al. [76] developed an MILP approach to ensure an optimal power flow and to reduce the electricity generation costs and GHG emissions. Sarabi et al. [77] proposed linear programming (LP) to minimize the annual energy invoice of railway station parking using plug-in electric vehicles (EV). Riffonneau et al. [78] proposed an optimal power management strategy using dynamic programming for connected grids, PVs, and BESSs. The proposed control maximizes the economic benefits and minimizes the degradation in BESSs. Maroufmashat et al. [79] used LP to solve EMS problems, such as minimizing the capital, operation, and maintenance costs in a hydrogen refueling station. Dong et al. [80] presented an MG based on CHP and RE while taking economy, environment, and flexibility into consideration to reduce the operation costs and $CO_2$ emissions. Garcia et al. [81] proposed a novel MPC model that minimizes power loss in an ESS in real-time. Sultana et al. [82] developed an EMS controller that reduces the voltage drop and improves the life cycle of Li-ion batteries. Chiang et al. [83] created an EMS controller to reduce the voltage drop in an aim to improve the life cycle of lithium-ion batteries. Ju et al. [84] applied mathematical optimization to prevent shortages in various DERs via a day-ahead output prediction. Zhao et al. [85] developed an optimization strategy for MGs that uses day-head market operations to minimize the demand response costs. Zhen Wang et al. [86] proposed a risk-based method that enhances the overall transient stability of power systems by using LP to minimize shedding costs. Y. Cao et al. [87] proposed an intelligent approach based on a mathematical model to reduce the charging costs in an EV in response to the time-of-use price in the electricity market. In [88] developed a model for reducing the charging and discharging costs in an EV connected to a PV station and grid. Giraldo et al. [89] proposed a comprehensive MG framework that operates in grid-connected and isolated modes, where the objective function is solved using the convex mixed-integer technique.

Several metaheuristic optimization approaches have also been used to solve EMS problems in MGs. For example, Mohammadi-ivatloo et al. [90] used the imperialist competitive algorithm to reduce the operation cost of fuel units, whose objective function involves the dynamic economic dispatch problem. Elsied et al. [91] used an advanced real-time EMS that applies the genetic algorithm (GA) to minimize the energy cost and carbon emissions and to maximize the power penetrating from REs simultaneously. Grisales-Noreña et al. [92] applied particle swarm optimization (PSO) to reduce the cost of energy purchased from utility grids. The MG implemented in this work comprised various DERs, such as PV, WT, and BESS. Marzband et al. [93] used an artificial bee colony (ABC) to obtain the optimal production cost and increase the power penetrating from RE in MGs. K. Roy et al. [94] explored an EMS using an ant-lion optimizer, which parameterizes the uncertainty in solar and wind energy generation in an MG. This optimizer meets the load demand at an optimal cost and takes the constraints into account. Other metaheuristic optimization techniques have also been applied, including the Jaya algorithm (JAYA) [95], teaching-learning based optimization (TLBO) [96], differential evolution (DE) [97], gray wolf optimizer (GWO) [98], firefly algorithm (FA) [99], moth-flame optimization [100], and the crow search algorithm (CSA) [101].

Some papers have used hybrid or modified optimization techniques, such as a hybrid of the bacterial forging optimization algorithm and GA, to reduce the electricity costs and to curtail the peak-to-average ratio [102]; and the hybrid ABC-PSO to analyze the techno-economic MG and reduce the total cost [103]. Other hybrid metaheuristic techniques have also been proposed, including the optimal stopping rule (OSR) and GA (OSR-GA), OSR-TLBO, and OSR-FA in [104]; DE and sequential quadratic programming (DE-SQP) in [105]; the GA and whale optimization algorithm in [106]; Jaya-TLBO in [107]; the genetic harmony search algorithm in [108]; and the wind-driven bacterial foraging algorithm in [109]. Some of these algorithms are developed based on other approaches, such as artificial neural networks (ANN) [110], multi-agent systems [111], and fuzzy control [112].

Other researchers have considered additional objective functions to solve EMS problems. Some of them treat such problems as multi-objective, which may involve minimizing the costs (for operation, maintenance, fuel, and battery charging/discharging), emissions, and penalties. For instance, Swain et al. [113] proposed DE to solve the multiple objectives of the economic emission load dispatch problem. Xiong et al. [114] analyzed the effect of adding EMS to a grid connected to REs. Three objective functions were considered to minimize electricity bills, reduce the cost of power purchased from the main grid, and optimize power quality, and a novel multi-agent system was developed to optimize these objectives. Teo et al. [115] presented a fuzzy logic-based energy management system integrated with a grid-connected integrated with EMS. The model incorporates multi-objective optimization into NSGA-II to reduce the average peak load and operating costs by controlling the BESS status and the electricity market. Ren et al. [116] designed an optimal design for fast EV charging stations using WT, PV, and a BESS and for minimizing electricity costs and pollution emissions. This model is solved by using a hybrid optimization algorithm that combines the multi-objective particle swarm optimization algorithm with TOPSIS.

**Table 7.** Objectives in cost minimization.

| Objective | Equation | Details |
| --- | --- | --- |
| Operation Cost [74] | $\sum_{t=1}^{T}\sum_{i}[F(P_g(i,t)\ I(i,t)) + SU(i,t) + SD(i,t)]$ | $i,\ t$ : number of DERs and time of operation respectively. $P_g$ : $t$ Thermal unit dispatch at hour $t$. $I$ : solar cell terminal current. $SU$ : start-up cost of the thermal unit $i$ at time $t$. $SD$ : shutdown of thermal unit $i$ at time $t$. |
| Operating Cost [75] | $\sum_{t=1}^{T}\sum_{i}E_g^d(i,t) * C(i_d,t)] + \sum_{t=1}^{T}\sum_{i}E_g^{nd}(i,t) * C(i_{nd},t)]$ | $E_g^d$: energy delivered from dispatchable resources. $E_g^{nd}$: energy delivered from non-dispatchable resources. $C(i_{nd})$ and $C(i_d)$: unitary cost of each non-dispatchable and dispatchable generator $i$ at time $t$. |
| Operating Cost [78] | $[P_{grid}\ (\Delta t) * fit\ (\Delta t) * \Delta t\ ] + [P_{grid}\ (\Delta t) * E_gP\ (\Delta t) * \Delta t$ $+BrC\ (\Delta t)]$<br>Where:<br>$BrC = \sum_{t=0}^{T} BrC\ (t) * \left(\frac{-\Delta SoH(xi,xj,t)}{1-SoH_{min}}\right).$ | $P_{grid}$ : power delivered from the grid. $fit$: feed-in tariff.<br>$E_gP$ : electricity grid price. $BrC$ : battery replacement cost. $SoH$ : state of charge at time $t$. |
| Operation Cost [117] | $\sum_{t=1}^{T} a_i\ P_{it}^2 + b_iP_{it} + c_i$ | $a_i,\ b_i,\ c_i$ : coefficients of the appropriate measurement units that depend on DERs. $P$ : generated power. |
| Total Operation [91] | $\sum_{t=1}^{T}\left(\sum_{i=1}^{N} D_{DERi}(t)P_{DERi}(t)\ C_{DERi}(t) + D_{DERi}(t)\ SUC_{DERi}(t) + \sum_{j=1}^{J} D_{ESSj}(t)P_{ESSj}(t)\ C_{ESSj}(t) + P_{gr}(t)\ C_{gr}(t) - P_{gs}(t)\ C_{gs}(t)\right)$ | $C_{DERi}, C_{ESSj}$ : costs of the output power of the $i, j$ generator. $C_{gr},\ C_{gs}$ : cost of buying and selling power to the main grid. $P_{gr},\ P_{gs}$ : power received from and sold to the main grid. $D_{DERi},\ D_{ESSj}$ : state vectors that may be either 0 or 1. $SUC_{DERi}$ : startup cost of each generator $i$. $P_{DERi},\ P_{ESSj}$ : power delivered from DERs and ESS, respectively |
| Economic Emission Dispatch [76] | $\sum_{i}\sum_{k}\frac{C_k^d\ x_{ik}}{(1+i)^{t-1}} + \sum_{t}\sum_{l}\sum_{n}\frac{H_l\ C_n^g\ g_{t\ l\ n}}{(1+i)^{t-1}}$ | $l$: load block. $t$ : time period. $n$ : generator. $C_k^d$: investment for line $k$. $x_{ik}$ : investment state of line $k$ at time $t$. $k$ : number of transmission lines. $N$: number of generators. $H_l$ : a number of hours at load block $l$. $C_n^g$ : operation cost of generator $n$. $g_{t\ l\ n}$ : power generated at time $t$. |
| Grid Cost [81] | $\sum_{t}^{T} P_{grid,t} * C_{grid,t}$ | $P_{grid,t}$ : power consumption from the main grid at time $t$, where $P_{grid} = P_{Load} - P_{pv} - P_{BESS}$. $C_{grid,t}$ : cost of power consumption at time $t$. |
| The production Cost [90] | $\sum_{t} a_i\ P_{it}^2 + b_iP_{it} + c_i + \lvert e_i\ sin(f_i(P_i^{min} - P_{it}))\rvert$ | $a_i, b_i, c_i$ : fuel coefficients of unit $i$. $P_{it}$ : power generated from unit $i$ at time $t$. $e_i$ and $f_i$ : valve-point coefficients of each $i$ unit. $P_i^{min}$ : minimum capacity limit of the $i$ unit. |
| The production Cost [93] | $\sum_{1}^{T}\left(C_t^g + C_t^{g'} + C_t^{ES-} + C_t^l + C_t^{ES+} + \Omega_t\right) * \Delta t$ | $C_t^g,\ C_t^{g'}$ : cost of energy generated by non-dispatchable and dispatchable resources, respectively. $C_t^{ES-}, C_t^{ES+}$ : cost of energy from the charging and discharging of BESS, respectively. $C_t^l$: cost of power from the responsive load demand. $\Omega_t$ : penalty cost. |

**Table 7.** *Cont.*

| Objective | Equation | Details |
| --- | --- | --- |
| Total Operational and Maintenance Costs [79] | $\sum_{s}\left(C_{op}^{s}+C_{fuel}^{s}-income^{s}\right)+\left(\frac{CEPCI_{year}}{CEPCI_{base\ year}}\right) * C_{cap}^{ref\ station} * CRF+C_{pipe}$ | $C_{op}^{s}$ : operation cost of stations. $C_{fuel}^{s}$ : fuel cos t at the station. $CRF$ : capital recovery factor. $C_{pipe}$ : installation of district heating pipelines. $C_{cap}^{ref\ station}$ : total cost related to the hydrogen refueling stations. $CEPCI$ : Chemical Engineering Plant Cost Index, which allows the conversion of costs from their base year to the study year. |
| Carbon Dioxide Emission Cost [80] | $C_{CO_2}=P_{CO_2}\left[C_{gt}\left(P_{GT}^{t}+H_{GT}^{t}\right)+C_{gb}H_{GB}^{t}+C_{grid}Pm_{1}^{t}\right].\Delta t$ | $P_{CO_2}$ : carbon tax. $C_{gt}$, $C_{gb}$, $C_{grid}$ : carbon dioxide emissions per unit. $Pm_1^t$ : electricity purchased from the main grid. $H_{GB}^t$ : power output of the gas boiler. $H_{GT}^t$ : heat produced by the micro gas turbine (GT). |
| Annual Power Loss [118] | $\sum_{g=1}^{N} P_{lossg} * P\{C_g\} * J$ | $P_{lossg}$ : power loss in state g. $P\{C_g\}$ : probability of any combination of load and wind-based DG output. $J$ : takes a value of either 90 or 8760. $N$ : number of discrete states. |
| Power Loss [95] | $\sum_{i=1}^{N}\sum_{j=1}^{N} P_i\, B_{ij}\, P_j+\sum_{i=1}^{N} B_{i0}\, P_i+B_{00}$ | $B_{ij}$, $B_{i0}$, $B_{00}$ : B-matrix coefficients. $P_i$, $P_j$ : power outputs from the generators $i$ and $j$, respectively. |
| Battery Cost [81] | $\sum_{t=1}^{24}\left(\frac{CC_{bat}}{2\cdot Cycles_{bat}} P_{bat,ch}(t).T_s\cdot \eta_{bat,ch}+Cost_{degr,ch}\cdot P_{bat,ch}^{2}(t)+\frac{CC_{bat}\cdot P_{bat,dis}(t)\cdot T_s}{2\cdot Cycles_{bat}\cdot \eta_{bat,dis}}+Cost_{degr,dis}\cdot P_{bat,dis}^{2}(t)\right)$ | $CC_{bat}$ : capital cost. $Cycles_{bat}$ : number of life cycles. $P_{bat,ch}$, $P_{bat,dis}$ : power delivered from the battery during charging and discharging, respectively. $Cost_{degr,ch}$, $Cost_{degr,dis}$ : hourly economic costs during charging and discharging, respectively. $\eta_{bat,ch}$, $\eta_{bat,dis}$ : performance of the battery during charging and discharging, respectively. |
| Charging Cost [87] | $\sum_{t=0}^{T} M(t) * P(t)\, \Delta t$ | $M(t)$ : unit price at time $t$. $P(t)$ : charging power at time $t$. |
| Degradation Cost [119] | $C_d=\frac{c_{bat}}{L_{ET}}$ | $c_{bat}$ : capital cost of the battery. $L_{ET}$ : battery life. |
| Charging and discharging Cost [88] | $\sum_{t=1}^{T}\left(\sum_{n=1}^{N_f} P_{n,d}^{f}(t)e_{n,d}^{f}(t)+P_{g,out}(t)e_{g,out}(t)-\sum_{n=1}^{N_f} P_{n,c}^{f}(t)e_{n,c}^{f}(t)-\sum_{n=1}^{N_f} P_{n,c}^{s}(t)e_{n,c}^{s}(t)+\sum_{n=1}^{N_c} P_{n,c}^{cn}(t)e_{n,c}^{cn}(t)-P_{g,in}(t)e_{g,in}(t)\right)$ | $P_{n,d}^f$ : discharging price per unit of energy for EV. $e_{n,d}^f$ : discharging rate for EV. $P_{g,out}$ : selling price of electricity sold by the grid to the charging station. $e_{g,out}$ : Amount of electricity that the charging station buys from the grid. $P_{n,c}^f$ : charging price per unit of energy for EV. $e_{n,c}^f$ : charging rate for conservative EV. $P_{n,c}^s$ : charging price per unit of energy for premium EV. $e_{n,c}^s$ : charging rate for premium EV. $P_{n,c}^{cn}$ : charging price per unit of energy for conservative EV. $e_{n,c}^{cn}$ : charging rate for conservative EVs. $P_{g,in}$ : price of electricity purchased by the grid from the charging station. $e_{g,in}$ : amount of electricity sold to the grid. |
| Purchase Cost [120] | $\sum_{t=1}^{T}\left(C_{pur}(t)\, P_{pur}(t)\, T_s-C_{sold}(t)\, P_{sold}(t)\, T_s\right)$ | $C_{pur}$ , $C_{sold}$ : prices of the sold and purchased energy at time $t$. $P_{pur}$, $P_{sold}$: purchased and sold power from the grid at time $t$. |
| Start-up Cost [120] | $C_{SU}^{DG}=\gamma_{SU}\cdot \sum_{t=1}^{T}\sigma_{DG}^{ON}$ | $\gamma_{SU}$ : startup cost. $\sigma_{DG}^{ON}$ : ON-OFF binary variable. |
| Maintenance Cost [120] | $C_M=\gamma_M\cdot \sum_{t=1}^{T}\delta_{DG}(t)\, T_s$ | $\gamma_M$ : maintenance cost. $T_s$ : sampling time, set to 0.25 h. $\delta_{DG}$ : ON-OFF binary variable. |
| Shortage Cost [121] | $\sum_{i=1}^{n}\sum_{j=1}^{m}\left(K_{pi}P_{ij}+K_{pi}P_{ij}\right)$ | $K_{pi}$, $K_{pi}$ : loss factors of nodes $i$ and $j$.$P_{ij}$ : power shortage between nodes $i$ and $j$. |

**Table 7.** *Cont.*

| Objective | Equation | Details |
|---|---|---|
| Shortage Cost [84] | $\sum_{t=1}^{T} \rho_{sp,t}\, g_{sp,\, t}$ | $\rho_{sp}$ : penalty price for power shortage. $g_{sp}$ : electricity of power shortages. |
| Operation Cost of Battery [122] | $\frac{C^{max}_{ch,ESS}}{P^{max}_{ch,ESS}} \cdot P^2_{ch,ESS}(t) + \frac{C^{max}_{dis,ESS}}{P^{max}_{dis,ESS}} \cdot P^2_{dis,ESS}(t)$ | $C^{max}_{ch,ESS}$, $C^{max}_{dis,ESS}$ : maximum operation cost of charging and discharging, respectively. $P^{max}_{ch,ESS}$, $P^{max}_{dis,ESS}$ : maximum power dispatched from the ESS during charging and discharging, respectively. $P_{ch,ESS}$, $P_{dis,ESS}$ : power dispatched from the ESS during charging and discharging, respectively. |
| Daily Operation Cost [85] | $\sum_{t=1}^{24} \sum_{s} w_s (C_{grid} + C_{wpc} + C_{ess} + C_{mtg} + C_{dr})$ | $w_s$ : probability of scenario s. $C_{grid}$ : transaction cost in the electricity market. $C_{wpc}$ : cost of wind power curtailment. $C_{ess}$ : cost of the energy storage operation. $C_{mtg}$ : cost of the micro-gas turbine resource. $C_{dr}$ : cost of the electrical demand. |
| Electrical demand response [85] | $C_{dr} = P_{mtg}(P_{down}(t,s) + P_{up}(t,s))\Delta t$ | $P_{down}(t,s)$ : demand response program at time $t$ and sc nario $s$. $P_{up}(t,s)$ : shifted up electrical power by demand response program at time $t$ and scenario $s$.$P_{mtg}$ : unit cost of the electrical demand response. |
| Load Shading Cost [86] | $\sum_{i=1}^{N_D} c_{Di} * \Delta P_{Di}$ | $N_D$ : number of loads. $\Delta P_{Di}$ : active power shedding of the $i$ load. $c_{Di}$ : cost coefficient of $i$ load. |
| Investment Cost [98] | $\sum_{i} \sum_{k} O_{ik}\, \varphi_i^{Equ}\, C_i^{inv}$ | $O_{ik}$ : a variable with a value of either 0 or 1. $\varphi_i^{Equ}$ : capital recovery rate of class $i$ energy conversion and storage equipment cycle. $C_i^{inv}$ : initial investment cost. |

Restoration

Blackout events in power systems have dramatically increased due to weather events, natural disasters, or vandalism. These power outages greatly affect the economic, social, and industrial sectors. Any outage in a network will result in supply interruption for customers of the defective section. To reduce the gravity of the consequences, the scale of different power system damages needs to be evaluated, and system restoration actions need to be taken. Resiliency describes the ability of a power system to persevere in the face of high-impact, low-probability events (HILP) and to quickly restore its operations, either completely or partially, by using all the available resources within a short timeframe with constricted costs [123]. Previous studies have referred to resilience using various terms, such as resourcefulness, self-healing, adaptability, and flexibility. Figure 9 shows how the performance level of a power system changes during HILP events. To improve the resilience of power systems, some measures must be implemented in plant management, restoration service programs, and hardware designs [124]. Figure 10 presents a comprehensive classification of power system resilience.

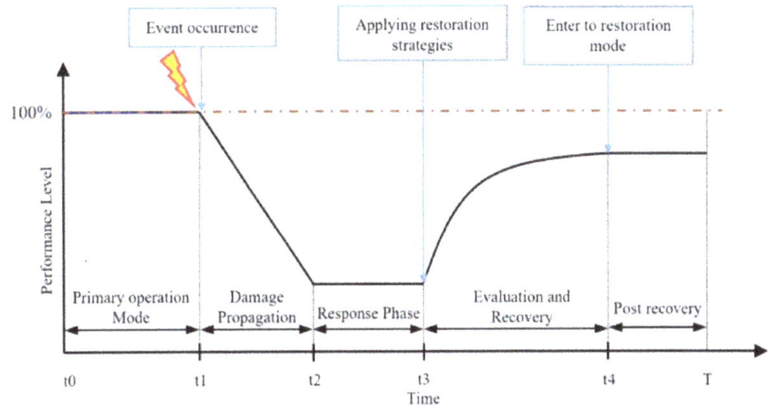


**Figure 9.** Changes in the resilience of a power system during certain events.

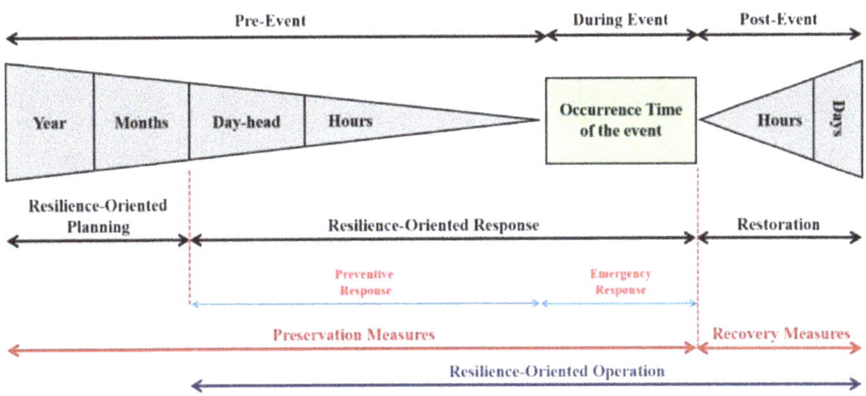


**Figure 10.** Comprehensive classification of power system resilience [124].

The EMS has various computational tools that control and operate the resilience of a power system by introducing advanced computational algorithms. Mathematically, the restoration problem in MGs is viewed as an objective of maximizing the supply for as many customers as possible, minimizing the switching costs, changing the status of line loading, and addressing the radial network constraints as shown in Table 8. Different reconfiguration techniques have been proposed in the literature to solve this problem.

As an important feature of an EMS, service restoration has also received much research attention. Poudel et al. [125] restored the services in a power distribution system by monitoring status switches to isolate the outage area and to maximize the number of restored loads; they modeled this problem in MILP to ensure large-scale flexibility. Gholami et al. [126] proposed two heuristic approaches for solving the restoration problem that involves maximizing the total and priority of loads restored and minimizing the number of switching operations; which are graph-based to optimize the objectives function proposed. Alowaifeer et al. [127] improved the resilience of a power system by relying on a dynamic prioritization of customers. The priority of loads is influenced by many factors, including the criticality of the load and the cost of interruption. Shen et al. [128] proposed a comprehensive framework that involves theoretical and quantitative indicators of self-healing in a power system during its restoration process, including the credibility, rate, speed, and benefits of MGs. Jiao et al. [129] proposed the wide-area measurement/information (WAM/WAI) control to handle the restoration speed problem in MGs. WAM/WAI was applied based on the unified power flow controller, which allows the redistribution of power flow in areas affected by the outage. Yang et al. [130] improved the resilience of a distribution network in three stages. First, the emergency system restored the critical load by applying the tree restoration method. Second, EV was used as an emergency power supply. Third, the restoration model restored the non-critical loads during faults. Zidan et al. [131] proposed a multiagent system that determines, and isolates faults based on several objectives by minimizing the number of switches and power loss, and by maximizing the number of restored critical loads. Romero et al. [132] developed a mathematical model by abstracting multiple objectives into a single objective. Lin et al. [133] proposed the term "electrical betweenness" to determine the optimal restoration path during self-healing operations and used complex network theory to restart the non-black-start generators and to identify the priority loads to be restored. Liu et al. [134] used the WAM system to estimate the restorable size of power load after outages and to control the stability of the system during the load restoration operation. Qianqian et al. [135] developed a two-stage mathematical model for centralized self-healing in an electrical distribution system that isolates the damage zone by minimizing the de-energized area, load shedding, and active power losses. Patsakis et al. [136] proposed an optimal allocation of black-start units to restore the power system. However, these units have a high maintenance cost, which can affect the self-healing process. Cao et al. [137] adopted the concept of top-down power system restoration where black-start resources were used to address the shortcomings of the non-black-start units and sub-transmission systems, and to restore the power system after encountering defaults.

Leite et al. [138] applied a multiagent system to collect and update local information and to quickly isolate the nearest damage location. Wang et al. [139] proposed a multi-objective formulation of service restoration and improved the efficiency of this framework in three steps. Gu et al. [140] introduced a two-level self-healing framework for service restoration, a problem which they formulated as a multi-objective function. Afterward, they applied the lexicographic optimization method to solve this problem. Chen et al. [141] used multiple MGs energized by DERs to restore the critical loads after the occurrence of faults. Zhaoyu et al. [142] proposed a comprehensive framework that applies two strategies based on the self-healing concept. The primary mode minimizes the operation cost and maximizes the profit, and the system enters the self-healing mode after the occurrence of a fault. The sectionalization method applies rolling-horizon optimization to isolate the damaged section and to restart the other utilities in the network system. Dimitrijevic et al. [143] reduced the restoration costs by supplying loads with a low switching cost and by applying the minimum spanning trees algorithm to address the proposed objective function. Rodríguez et al. [144] applied systematic measures to integrate algebraic calculations and heuristic rules that help distribution management systems find the optimal switching selection operation for the rapid isolation and restarting of DERs. Wang et al. [145] proposed the stochastic response method for reducing load loss within a minimum switching time, while allowing for a standard design of the network reconfiguration and islanding section. Widiputra et al. [146] developed a novel restoration algorithm that uses discrete PSO to solve the clod load pick up and conservation voltage reduction problems in service restoration. Vieira et al. [147] integrated the protection constraints in service restoration for a distributed power network by using the multi-objective evolutionary algorithm to enhance the protection efficiency of the device. Ma et al. [148] proposed a three-level optimization problem to minimize the investment and load-shedding costs during extreme weather events, and applied the greedy searching algorithm to optimize the formulation proposed in a multi-study of scenarios. Arif et al. [149] modeled the uncertainty of end-user consumption and power dispatch from DERs to facilitate the service restoration of an MG by using a stochastic mixed-integer linear program to maximize the served load. Xu et al. [150] introduced the resilience-oriented method to optimize the restoration problem in secondary network distribution systems, which are directly controlled by the unit commitment in an EMS. Khatib et al. [151] applied the probabilistic operational planning method to achieve distribution automation by placing the ESS in the fault area. They formulated the objective function to reduce the total energy cost and enhance reliability. Abniki et al. [152] applied a BESS as a backup utility to restore the de-energized portion of the system and formulated the objective function as an MINLP to minimize the total cost of interruption.

**Table 8.** Objectives for improving reliability.

| Objective | Equation | Details |
| --- | --- | --- |
| The restored Load [125] | $\sum_{i \in v} \sum_{\varnothing \in \{a,b,c\}} w_i\, s_i\, P_{Li}^{\varnothing}$ | $w_i$ : the weight factors for each load. $s_i$ : the load pick-up. $P_{Li}^{\varnothing}$ : complex power demand at $i$. |
| Number of Switches [125] | $\sum_{(i \to j) \in E_S^R} (1-\delta_{ij}) + \sum_{(i \to j) \in E_T^R} (\delta_{ij}) + \sum_{(i \to j) \in E_v} (\delta_{ij})$ | $\delta_{ij}$ : Line or switch decision. $E_S^R$ : set normally closed sectionalizing switches. $E_T^R$ : set normally open tie switches. $E_v$ : set of virtual edges for DG connection. |
| Number of switches [131] | $\sum_{i=1}^{N_s} \lvert x_i - x_{i0} \rvert$ | $N_s$ : the total number of switches. $x_i$ : status of switch $i$. $x_{i0}$ : status of switch $i$ after fault occurs. |
| The energized Load [126] | $\sum_{K \in N_{TT}} L_K$ | $L_K$ : the energized loads in the network. $N_{TT}$ : The restorable total buses. |
| The number of switches [126] | $N_{SO}$ | $N_{SO}$ : the number of switches operation. |
| Priority of load [127] | $\sum_{i=1}^{N} PL^i\, x^i$ | $PL^i$ : the priority weight of each load $i$. $x^i$ : the status of the switch in the load $i$. |
| The resilience [130] | $\sum_{i_k=1}^{n_{MG}} C_{i_k} P_{i_k} T_0 + \sum_{s \in S} \sum_{i_s=1}^{n_{EPS}} C_{i_s} P_{i_s} (T_0 - t_{sdi}) + \sum_{i=1}^{n-n_{MG}-n_{EPS}} C_i P_i T_i - \alpha \sum_{s \in S} \sum_{i_s=1}^{n_{EPS}} P_{i_s} (T_0 - t_{sdi})$ | $n_{MG}$ : number of loads. $t_{sdi}$ : the travel time. $n_{EPS}$ : the number of restored loads. $P_i$ and $P_{i_s}$ : the active power dispatch from the microgrid and EV, respectively. $C_i$ : the cost utilities. $\alpha$ : the unit capacity consumption cost. |
| The restoration paths [133] | $\sum_{j \in \Omega_{UR}^{T_i}} \frac{\alpha\, P_{Gj} + \beta\, P_{Lj}}{E^{dsj-1}}$ | $\Omega_{UR}^{T_i}$ : the set of nodes of the power grid. $P_{Gj}$ : the power dispatched from DER.$P_{Lj}$ : the power consumed by each node. $\alpha$, $\beta$ : coefficients for measuring the relative importance. $E$ : coefficient of exponential decay. |
| The centralized Self-healing [135] | $\sum_{z \in \Omega_z} C_z^R (1-x_z) + \sum_{i \in \Omega_b} C_i^{ls} P_i^D r_i + \sum_{ij \in \Omega_l} R_{ij} I_{ij}^{sqr} + C^{lss} \sum_{ij \in \Omega_{sw}} (1-w_{ij}) + C^{sw} \sum_{ij \in \Omega_l} w_{ij}$ | $\Omega_z$ : set of loads, $\Omega_b$ : set of nodes, $\Omega_l$ : set of branches, $\Omega_{sw}$ : set of switches, $P_i^D$ : active power requested in node $i$. $R_{ij}$ : resistance branch $ij$. $I_{ij}^{sqr}$ : current in branch $ij$. $C_z^R$ : cost of de-energizing. $C_i^{ls}$, $C^{lss}$ : cost of load-shedding and loss cost, respectively. $C^{sw}$ : cost of switch operation. |
| The total generation capability [137] | $\sum_{i=1}^{n_L} \sum_{j=1}^{n_{p,i}} E_{P,\,ij}$ | $n_L$ : number of loads, $n_{p,i}$ : number of non-black start generators. $E_{P,\,ij}$ : the power capability of the generator. |
| Out-of-service Area [153] | $\sum_{i=1}^{b_1} L_i - \sum_{i \in B} L_i$ | $b_1$ : number of energized bus, $L_i$ : load $i$. $B$ : set of energized buses. |
| Restoration/maintenance switching sequence [154] | $\sum_{z \in \Omega_z} \sum_{s \in \Omega_s} C_{z,s}^U (1-x_{z,s}) + C^{sw} \sum_{ij \in \Omega_{sw}} \sum_{s \in \Omega_s} \left(\Delta y_{ij,s}^{+} + \Delta y_{ij,s}^{-}\right)$ | $\Omega_z$ : set of zones, $\Omega_s$ : set of sequence, $C_{z,s}^U$ : cost od de-energizing, $x_{z,s}$ : binary variable. $C^{sw}$ : operating cost. $\Omega_{sw}$: set of switches. $\Delta y_{ij,s}^{+}$, $\Delta y_{ij,s}^{-}$ : opening and closing of switch operation. |
| The network layer unit restarting [140] | $\sum_{i=1}^{n_G} \sum_{k\Delta_t}^{T_1} \left( \alpha(t)\, c_{i1}\, P_{G,ij}(t)\, dt + \mu \sum_{j=2}^{n_{G_i}} P_{M,ij} \right)$ | $n_G$ : number of DERs, $G_{ij}$ : the unit $j$ in the plan $i$. $\alpha$ : weight factor. $P_{G,ij}$ : the power delivered from $j$. $\mu$ : distributing factor. $P_{M,ij}$ : the maximum output of DERs. |

**Table 8.** *Cont.*

| Objective | Equation | Details |
|---|---|---|
| Restore the outage area [142] | $\sum_{t}^{t+T_p}\left(\sum_{s}\gamma_s\sum_{k}\left(\lvert V_{k,s,t}-V_n\rvert+\sum_{j}x_{kj}+w_k P^D_{k,s,t}(1-y_{k,t})\right)\right)$ | $\gamma_s$ : probability of the scenario. $V_{k,s,t}$: voltage magnitude, $V_n$ : basic voltage. $x_{kj}$ : indicator of boundary line. $w_k$ : priority index of the load. $P^D_{k,s,t}$ : active power, $y_{k,t}$ : the status of the switch. |
| Served Loads [149] | $\sum_{\forall s} Pr(s)\left(\sum_{\forall t}\left(\sum_{\forall i\in I^D} y_i\, w_i\, P^D_{i,t,s}+\sum_{\forall i\in I^{CL}} P^{CL}_{i,t}\right)\right)$ | $P^{CL}_{i,t}$ : controllable loads, $P^D_{i,t,s}$ : non-controlable loads, $w_i$ : weight factor, $y_i$ : connecting status of the loads. $Pr(s)$ : priority of the loads $s$. |

Power Quality

The proliferation of nonlinear, unbalanced loads and loads shedding during the restoration process may compromise the power quality in MGs and distribution systems. Meanwhile, the intermittency and instability of RE sources can result in fluctuations in power quality and stability [155]. The EMS can improve power quality and stability in the power system by monitoring the control equipment using control theories and optimization techniques [156]. Table 9 shows the objective function used to enhance the power quality in MGs. Several control strategies and approaches in EMSs for improving power quality have been proposed in the literature. For instance, Mei et al. [157] proposed the moth-flame optimization technique to minimize the voltage deviation and total system transmission loss, and to improve power stability via reactive compensators sizing. Jian et al. [158] developed a service model for an unbalanced three-phase active process distributed using a multi-terminal soft open point system to realize power flow in DERs and supply the outage area. The formulation was summarized as a combination of objectives, including maximizing the restored load and minimizing the voltage unbalance and power loss. Mousavi et al. [156] proposed a novel control that uses the PV and battery energy storage interfacing inverter to improve power quality while taking several constraints into consideration, such as battery service life and charging/discharging status. Sahoo et al. [159] proposed a novel centralized energy management approach for stabilizing the voltage flow and the flexibility of inverters in a solar-battery hybrid MG. Ravinder et al. [160] used the shunt active power filter integrated with the PV and battery to improve the quality of power in an MG. The shunt active power filter was controlled using ACO to minimize the total harmonic distortion. Aljohani et al. [161] utilized the vector-decouple algorithm to preserve stability and to control the hybrid MG, and proposed a controller that measures efficiency and robustness and improves the quality of the voltage output and frequency. Nasr et al. [162] proposed a multi-objective function that includes minimizing the voltage deviation in an MG to ensure voltage balance and to satisfy the contingency constraint. Han et al. [163] enhanced the power quality in a power system by using two levels of an EMS. In the first level, the control based on MPPT was used to improve the output power penetration from the PV. In the second level, droop control was applied to monitor the inverter in the power system. Agnoletto et al. [164] formulated the EMS as an optimal power flow problem, and considered both the operating cost and power loss in the optimization process by using the $\in$ constraint method. Zhang et al. [165] addressed a multi-objective function optimal reactive power dispatch problem and proposed a model that minimizes active power loss and voltage deviations using multi-objective optimization techniques. Leonori et al. [166] developed an optimal power flow strategy for a grid connected to a BESS, utilized the BESS to improve power stability, and used a fuzzy EMS controller to manage power in real time.

**Table 9.** Objectives in improving power quality.

| Objective | Equation | Details |
|---|---|---|
| Voltage deviation [157] | $\sum_{i=1}^{Nd} \left\| V_i - V_i^{sp} \right\|$ | $V_i$ : the voltage at load bus- $i$. $V_i^{sp}$ : is the specified value (usually set as 1.0 p. u). |
| Voltage deviation [165] | $\sum_{k=1}^{N_{load}} \left\| \frac{V_K - V_K^{ref}}{V^{upper} - V^{lower}} \right\|$ | $V_K^{ref}$ : reference voltage. $V^{lower}$ : is the lower limit of load bus voltage. $V^{upper}$ : the upper limit of load. |
| The voltage unbalance [158] | $\sqrt{\sum_{t \epsilon T} \sum_{i \in \Omega_b} (1/3\, U_{i,A}^t + 1/3e^{4\pi/3} U_{i,B}^t + 1/3^{2\pi/3} U_{i,C}^t)^2}$ | $\Omega_b$ : the set of the distributed system. $U_i^t$ : is the voltage in each phase. |
| Voltage profile [162] | $\sum_{K_t \in T} \sum_{i \in B} V_{i,\, K_t} - V_{i,\, K_t}^{sp}$ | $B$ : all system buses, $V_{i,\, K_t}$ : bus voltage [p.u]. $V_{i,\, K_t}^{sp}$ : rated voltage [p.u]. |

### 2.2.3. Tertiary Control Level

As the top-level control, the tertiary control level preserves the optimality of the operation, specifically the efficiency and cost between the MG and the primary grid, and vice-versa. This level usually has a slow dynamic response to define the optimal active and reactive power references of each DG due to the complexity of the calculation and the prediction model of economic and meteorological data [167]. The prediction model helps classify the weather, network optimization, and uncertainty quantification. Different methods are applied at this level to formulate the forecasting and prediction model, such as machine learning [168], long short-term memory (LSTM) [169], k-nearest-neighbors (KNN) [170], generalized regression neural network [171], neural network ensemble [172] and deep recurrent neural networks [173]. While the secondary level focuses on the power quality and sharing among DERs, the tertiary control focuses on economic aspects, electricity market participation, and power-sharing trends. This level guarantees high power-sharing quality by defining the error between the reference and actual values, whose values are computed as [73]:

$$\omega^* = k_{pP}\,(P_G^* - P_G) + k_{iP} \int (P_G^* - P_G)\, dt \tag{5}$$

$$V^* = k_{pQ}\,(Q_G^* - Q_G) + k_{iQ} \int (Q_G^* - Q_G)\, dt \tag{6}$$

where $P_G^*$ and $Q_G^*$ are the active and reactive power references from the DER to the main grid, respectively; $\omega^*$ and $V^*$ are the frequency and voltage errors; and $k_p$ and $k_i$ are the gains of the PI controller. The tertiary control level is generally endowed with the familiar concept of the tending of the electricity and the financial market, such as mentioned and discussed in the next section.

## 3. Transactive Energy Market in Microgrids

The MG energy market not only allows the trading of local power generation among consumers but also fosters sustainable and efficient power use. MG markets also help reduce the cost of transporting energy while keeping losses at a minimum [174]. This market design schedules the load profile and power generation from DERs in preparation for the dispatching process to reduce the energy costs. Transactive energy management (TEM) is a comprehensive framework that introduces several features for integrating DER utilities and MGs into power systems. TEM also allows small and large energy consumers or producers to trade energy under market rules. TEM promotes the demand-side based on sharing among prosumers, and the economic signals that are in line with optimal operation targets to ensure the suitability and reliability of the system. This framework optimizes system performance by ensuring a dynamic alignment among local objectives and by using different approaches to determine the tariffs, bilateral contracts, penalties, and organized markets [175]. With TEM, customers can trade their surplus energy either in real-time or on a deferred basis. Nevertheless, the application of TEM to MGs requires a design track to manage complex operations in a way that ensures transparency, freedom, and fairness for prosumers. To design a proper TEM structure, several design principles that are related to agent properties, pricing mechanisms, and internal and external markets must be considered [12].

Xue et al. [176] argued that the technology of the power market and the transactive energy in a large grid are not suitable for MGs. Therefore, the power industry proposed energy trading based on blockchain to allow trading in dynamic P2P networks, distributed networks, cryptography, and others such as those discussed in [177]. Janko et al. [178] proposed multi-agent control as a well-known technique for transactive energy trading due to its ability to improve system scalability, flexibility, autonomy, and transparency. Therefore, this market design can reduce the risks of price oscillation for small customers and increase their profit. Other approaches proposed in the literature including direct acyclic graph [179], hash graph [180], flow chain [181], and game-theory [182].

## 4. Protection Systems

The excess generation capacity of DERs in an MG can provide the primary grid or other connected MGs the necessary system recovery resources to enhance their resilience and shorten the outage duration. However, the resilience of an MG is not entirely protected from short-circuit faults, which could increase the rated current by hundreds of times, thereby necessitating the replacement of CBs. The protection system in an MG is aimed at identifying the fault location. After locating the fault, the protective device in the MG isolates and repairs the fault section quickly [10].

Short-circuit fault is the most common type of fault in an MG that can damage consumer appliances. Therefore, MGs require an overcurrent protection protocol and schemes that protect against external and internal faults. To avoid high voltage in MGs during external faults, a protective relay must be installed to automatically detect faults and assist the CBs to isolate such faults. In the connected mode, the protection is usually placed at the PCC level, whereas in the island mode, the protection is placed depending on the inverters. Rapid automatic detection of faults requires a fast communication system. Therefore, automatic detection schemes should be evaluated based on their speed, sensitivity, selectivity, and reliability [36].

Several schemes have also been proposed in the literature to address the most common issues associated with overcurrent protection in MGs as shown in Table 10.

**Table 10.** Protection schemes applied in the MG.

| Protection Scheme | Ref | Advantages | Disadvantages |
|---|---|---|---|
| Undervoltage-based protection schemes. | [183] | - Protects MGs against both internal and external faults relative to any protective zone.<br>- Detects fault locations and types in MGs. | - Ignores HIFs and symmetrical faults and does not enable single-phase tripping.<br>- Difficult to coordinate.<br>- Any voltage drop in MGs may lead to the malfunctioning of protection devices. |
| Voltage-restrained protection schemes. | [184] | - Protects MGs against electric shock without relying on basic insulation.<br>- Detects low current faults. | - Cannot operate at a high impedance rate. |
| Harmonic content-based schemes. | [185] | - Detects and identifies all types of faults. | - Adaptive settings may be required. |

**Table 10.** *Cont.*

| Protection Scheme | Ref | Advantages | Disadvantages |
| --- | --- | --- | --- |
| Distance protection schemes. | [186] | - Disconnect only the faulted line part.<br>- Avoids the unnecessary disconnection of consumers and/or DGs. | - A synchronizing mechanism may be needed for long lines. |
| Adaptive overcurrent protection schemes | [187] | - Protects group of units or DERs.<br>- Reduces the limitation of settings in larger systems. | - Complex design.<br>- Requires communication links. |
| Differential protection schemes. | [188] | - Provides accurate protection by discriminating the high impedance fault from switching events. | - No backup protection for neighboring sections. |

## 5. Policy of Microgrid

Most power consumer countries are exploring alternative energy sources, such as RE, to reduce their dependence on fossil fuels and lower the associated costs. However, RE lacks a widely accepted framework for implementation due to policy reasons and its experimental nature. Therefore, various policies have been implemented to encourage the deployment of RE and DER technologies [189].

MG regulation in the EU faces many challenges related to protection, consumers and power supplier engagements, legalities, limitations, and interconnection with the main grid. To achieve a sustainable and secure energy supply, the EU issued a policy that aimed to reduce its fuel consumption by 20% by 2020. In 2014, the EU launched its 2030 vision, which involves increasing the penetration of RE technologies by up to 27% and reducing its GHG emissions by 40% to 95% by 2050 [190]. In 2016, the EU launched the IEC TS 62257-9-2 standard, followed by IEC TS 62898-1/2/3 in 2018, and PD IEC TS 62898-2 in 2020.

Since the oil and gas crises in the late 1970s, the US has issued several energy policies, including the IEEE standard 1547-Family, which was launched in 2005. This standard has a vital role in ensuring energy security and power quality. These policies issue financial incentives, such as the exemption of transmission and transmission loss charges, as well as climate change levy exemption. Other policies in the US have focused on R&D programs, software and tools, grants, and funding support to incentivize demonstration projects [191].

The tariff policy in China aims to promote the exploration of RE. This tariff policy is able to offer a continuing purchase price to the electricity seller to the grid corporation with a fit market competition by giving privileged prices [192]. Over the years, China has issued several policies and programs to promote the utilization of the RE, such as the national climate change program in 2002, renewable energy law amendments in 2009, and preferential tax policies for renewable energy in 2015. Table 11 summarizes the MG policies implemented in the EU, US, and China.

**Table 11.** Policies for MG design in different countries.

| Region | Standard/Policy | Description |
|---|---|---|
| EU | PD IEC TS 62898-2 | Applies to the operation and control of MGs, including:<br>• Interconnection modes and mode transfer.<br>• EMS and MG operations.<br>• Communication and monitoring procedures; and<br>• Battery energy storage regulations. |
| | IEC TS 62898-1/2/3 | • AC microgrids.<br>• Interconnection among DERs.<br>• Commissioning, maintenance, and testing. |
| | IEC TS 62257-9-2 | • Low AC voltage.<br>• Three- or single-phase.<br>• Changing the voltage range.<br>• rural electrification. |
| US | IEEE Standard 1547-Family | • Guide for monitoring the design operation and the integration of DR island systems.<br>• Interconnection of DERs.<br>• Tariff policies.<br>• Protection schemes. |
| China | Renewable Energy Law amendments | • Supports emerging RE sectors in the country.<br>• Encourages the industrial power grid to purchase RE. |
| | National Climate Change Program | • A global warming policy initiative. |
| | Preferential Tax Policies for Renewable Energy | • Tax reduction or exemption.<br>• Preferential pricing.<br>• Credit guarantees. |

## 6. Perspective and Discussion

Regardless of the yearly changes in power generation, the authors expect the following services from electricity systems:

- Future MGs may rely on a progressive combination of energy resources, including large-scale decentralized resources, to be suitable and variable. Energy storage systems have the potential to alter the nature of production and transmission;
- The deployment of the ESS only targets a few countries, such as Australia, Germany, and Japan. Such deployment is expected to cover 40% more countries every year until 2025 [193];
- A different change will occur in countries determined by market policy and regulatory structures, and the diversity of the resources supplying customers;
- While MGs are considered the best solution to local and general problems, they are essentially a novel architecture paradigm that offers higher flexibility and reliability against outages;
- Future MGs may improve their fault detection and self-healing capabilities to shorten recovery time, maximize loads restored, and identify gaps between research and implementation;
- The Internet of Things facilitates the emergence of real-time platforms and serves as an important link between decentralized and transactive energy markets. Moreover, from their previous research, the authors have determined that bidirectional exchanges of energy between customers and producers are considered the most challenging for the future; however, future techniques are expected to solve this challenge;

- The application of deep learning, including ANN, in MGs instead of classical and mathematical methods warrants exploration to achieve a dynamic adjustment of energy flow, reduction in GHG emissions, and enhanced protection for MGs;
- The use of blockchains and smart contracts in MGs should be promoted to guarantee secure energy transactions and DER operations.
- Integrating quantum computers into the MG may allow the system to restore more loads within a short period, use deep learning and machine learning methods for improving forecasting models, and apply algorithms for quickly directing DER dispatches;
- MG controllers should be evaluated and tested in controlled laboratory environments to minimize risks. Testing various technologies, such as hardware-in-the-loop, is expected to become a practical approach for evaluating controllers before their deployment.

## 7. Summary

MGs are primarily composed of various DERs, EVs, EMSs, loads, and communication devices. The development of MGs has become a requirement for the integration of REs in remote areas and the deployment of smart grids. MGs with an EMS are promising technologies that not only promote system efficiency and economic achievability but also ensure sustainable development and reduce GHG emissions. Many researchers have examined the development of EMSs and their operations, stability, reliability, costs, and utilization in MGs. This review paper presents a comprehensive and critical review of the elements in MGs. The MG has three levels, namely:

- Primary control, which guarantees reliable operation by maintaining voltage and frequency stability;
- Secondary control, which optimizes the power quality of the system; and
- Tertiary control, which achieves economic optimization according to the prices in the electricity market.

At the secondary control level, the EMS aims to optimize operation, energy planning, and system reliability in either the grid-connected or islanded mode. This extensive review addresses the mathematical objectives of minimizing the cost of restoration and improving power quality. The review also indicates that the design of an autonomous, reliable, and flexible EMS is essential to adapt to different configurations. It is compulsory to design optimal controller's schemes that are capable of controlling MG elements smoothly and fitting the changes in the environment without human interference or restructuring the entire controller. In this context, methods empowered with forecasting models, such as metaheuristics and AI techniques, are promising for the MG. Furthermore, EMSs must be capable of handling the fluctuations of the power generation dispatch from the RE resources by considering data forecasts. Several techniques have illustrated their ability to overcome these problems.

This paper also explores the TEM and the protection schemes mostly applied in MGs. Currently, the MG as a technology is still in its infancy stage. However, several countries, such as China and the US, have already started to encourage its adoption. This paper also discusses the perspectives of authors about the future of EMSs and MGs.

**Author Contributions:** Y.Z.: conceptualization, methodology, data curation, investigation, visualization, writing—reviewing and editing. I.A.: writing—reviewing and editing. S.M.: writing—reviewing and editing. M.R.B.K.: writing—reviewing and editing. M.S.: writing—reviewing and editing. A.S.: writing—reviewing and editing. B.H.: writing—reviewing and editing. All authors have read and agreed to the published version of the manuscript.

**Funding:** The authors would like to thank the Ministry of Higher Education, Malaysia for the financial support under the Long Term Research Grant Scheme (LRGS): LRGS/1/2019/UKM-UM/01/6/3.

**Institutional Review Board Statement:** Not Applicable.

**Informed Consent Statement:** Not Applicable.

**Data Availability Statement:** Not Applicable.

**Conflicts of Interest:** The authors declare no conflict of interest.

## Nomenclature

| | |
|---|---|
| RE | Renewable Energy |
| MG | Microgrid |
| EMS | Energy Management System |
| GHG | Greenhouse Gases |
| DG | Distributed generators |
| PCC | Point of Common Coupling |
| GW | Gigawatt |
| KW | kilowatt |
| DERs | Distributed Energy Resources |
| PV | Photovoltaic |
| MGCC | Microgrid central controller |
| CHP | Combined Heat and Power |
| HYD | Hydropower |
| WT | Wind Turbine |
| AC | Alternating Current |
| DC | Direct Current |
| kWh | kilowatt-hour |
| NB-PLC | Narrow Band Power Line Communication |
| BB-PLC | Broad Band Power Line Communication |
| PON | Passive Optical Network |
| DSL | Digital subscriber line |
| MPC | Predictive Control |
| VCM | Voltage Control Mode |
| PCM | Active/Reactive Power Mode |
| VPD | Voltage-Active Power Droop |
| FQB | Frequency-Reactive Power Boost |
| ESS | Energy Storage System |
| BESS | Battery Energy Storage System |
| WAM | Wide Area Measurement |
| TEM | Transactive energy management |
| ML | Machine Learning |
| DL | Deep Learning |
| LSTM | Long Short-term Memory |
| KNN | K-Nearest-Neighbors |
| GRNN | Generalized Regression Neural Network |
| NNE | Neural Network Ensemble |
| DRNN | Deep Recurrent Neural Networks |
| $\omega$ | Angular frequency |
| $V$ | Voltage |
| $P$ | Active Power |
| $Q$ | Reactive Power |
| $k_p$ and $k_i$ | Gains of the PI controller |

## References

1. Zahraoui, Y.; Khan, M.B.; AlHamrouni, I.; Mekhilef, S.; Ahmed, M. Current Status, Scenario, and Prospective of Renewable Energy in Algeria: A Review. *Energies* **2021**, *14*, 2354. [CrossRef]
2. REN21. Renewbles in Cities—2019 Global Status Report. Available online: https://www.ren21.net/reports/cities-global-status-report/ (accessed on 28 August 2021).
3. Krishna, K.S.; Kumar, K.S. A review on hybrid renewable energy systems. *Renew. Sustain. Energy Rev.* **2015**, *52*, 907–916. [CrossRef]
4. Zia, M.F.; Elbouchikhi, E.; Benbouzid, M. Microgrids energy management systems: A critical review on methods, solutions, and prospects. *Appl. Energy* **2018**, *222*, 1033–1055. [CrossRef]

5. Hirsch, A.; Parag, Y.; Guerrero, J. Microgrids: A review of technologies, key drivers, and outstanding issues. *Renew. Sustain. Energy Rev.* **2018**, *90*, 402–411. [CrossRef]
6. Majumder, R. Some Aspects of Stability in Microgrids. *IEEE Trans. Power Syst.* **2013**, *28*, 3243–3252. [CrossRef]
7. Cagnano, A.; De Tuglie, E.; Mancarella, P. Microgrids: Overview and guidelines for practical implementations and operation. *Appl. Energy* **2020**, *258*, 114039. [CrossRef]
8. Dawoud, S.M.; Lin, X.; Okba, M.I. Hybrid renewable microgrid optimization techniques: A review. *Renew. Sustain. Energy Rev.* **2018**, *82*, 2039–2052. [CrossRef]
9. Meng, L.; Sanseverino, E.R.; Luna, A.; Dragicevic, T.; Vasquez, J.C.; Guerrero, J.M. Microgrid supervisory controllers and energy management systems: A literature review. *Renew. Sustain. Energy Rev.* **2016**, *60*, 1263–1273. [CrossRef]
10. Beheshtaein, S.; Cuzner, R.M.; Forouzesh, M.; Savaghebi, M.; Guerrero, J. DC Microgrid Protection: A Comprehensive Review. *IEEE J. Emerg. Sel. Top. Power Electron.* **2019**, 1. [CrossRef]
11. Farrokhabadi, M.; Canizares, C.A.; Simpson-Porco, J.W.; Nasr, E.; Fan, L.; Mendoza-Araya, P.A.; Tonkoski, R.; Tamrakar, U.; Hatziargyriou, N.D.; Lagos, D.; et al. Microgrid Stability Definitions, Analysis, and Examples. *IEEE Trans. Power Syst.* **2020**, *35*, 13–29. [CrossRef]
12. Zia, M.F.; Benbouzid, M.; Elbouchikhi, E.; Muyeen, S.M.; Techato, K.; Guerrero, J.M. Microgrid Transactive Energy: Review, Architectures, Distributed Ledger Technologies, and Market Analysis. *IEEE Access* **2020**, *8*, 19410–19432. [CrossRef]
13. Patnaik, B.; Mishra, M.; Bansal, R.C.; Jena, R.K. AC microgrid protection—A review: Current and future prospective. *Appl. Energy* **2020**, *271*, 115210. [CrossRef]
14. San, G.; Zhang, W.; Guo, X.; Hua, C.; Xin, H.; Blaabjerg, F. Large-disturbance stability for power-converter-dominated microgrid: A review. *Renew. Sustain. Energy Rev.* **2020**, *127*, 109859. [CrossRef]
15. Roy, A.; Auger, F.; Olivier, J.C.; Schaeffer, E.; Auvity, B. Design, Sizing, and Energy Management of Microgrids in Harbor Areas: A Review. *Energies* **2020**, *13*, 5314. [CrossRef]
16. Koussa, D.S.; Koussa, M. A feasibility and cost benefit prospection of grid connected hybrid power system (wind–photovoltaic)—Case study: An Algerian coastal site. *Renew. Sustain. Energy Rev.* **2015**, *50*, 628–642. [CrossRef]
17. Jiayi, H.; Chuanwen, J.; Rong, X. A review on distributed energy resources and MicroGrid. *Renew. Sustain. Energy Rev.* **2008**, *12*, 2472–2483. [CrossRef]
18. Kaur, A.; Kaushal, J.; Basak, P. A review on microgrid central controller. *Renew. Sustain. Energy Rev.* **2016**, *55*, 338–345. [CrossRef]
19. Dondi, P.; Bayoumi, D.; Haederli, C.; Julian, D.; Suter, M. Network integration of distributed power generation. *J. Power Sources* **2002**, *106*, 1–9. [CrossRef]
20. Ustun, T.S.; Ozansoy, C.; Zayegh, A. Recent developments in microgrids and example cases around the world—A review. *Renew. Sustain. Energy Rev.* **2011**, *15*, 4030–4041. [CrossRef]
21. Zamora, R.; Srivastava, A.K. Controls for microgrids with storage: Review, challenges, and research needs. *Renew. Sustain. Energy Rev.* **2010**, *14*, 2009–2018. [CrossRef]
22. Cavraro, G.; Bernstein, A.; Carli, R.; Zampieri, S. Distributed Minimization of the Power Generation Cost in Prosumer-Based Distribution Networks. In Proceedings of the 2020 American Control Conference (ACC), Denver, CO, USA, 1–3 July 2020; pp. 2370–2375. [CrossRef]
23. Motevasel, M.; Seifi, A.R.; Niknam, T. Multi-objective energy management of CHP (combined heat and power)-based micro-grid. *Energy* **2013**, *51*, 123–136. [CrossRef]
24. Ma, L.; Liu, N.; Zhang, J.; Tushar, W.; Yuen, C. Energy Management for Joint Operation of CHP and PV Prosumers inside a Grid-Connected Microgrid: A Game Theoretic Approach. *IEEE Trans. Ind. Inform.* **2016**, *12*, 1930–1942. [CrossRef]
25. Javidsharifi, M.; Niknam, T.; Aghaei, J.; Mokryani, G. Multi-objective short-term scheduling of a renewable-based microgrid in the presence of tidal resources and storage devices. *Appl. Energy* **2018**, *216*, 367–381. [CrossRef]
26. Zhang, J.; Wu, Y.; Guo, Y.; Wang, B.; Wang, H.; Liu, H. A hybrid harmony search algorithm with differential evolution for day-ahead scheduling problem of a microgrid with consideration of power flow constraints. *Appl. Energy* **2016**, *183*, 791–804. [CrossRef]
27. Neto, P.B.L.; Saavedra, O.R.; Oliveira, D. The effect of complementarity between solar, wind and tidal energy in isolated hybrid microgrids. *Renew. Energy* **2020**, *147*, 339–355. [CrossRef]
28. Ramabhotla, S.; Bayne, S.; Giesselmann, M. Economic dispatch optimization of microgrid in islanded mode. In Proceedings of the International Energy and Sustainability Conference 2014, Farmingdale, NY, USA, 23–24 October 2014; Volume 2, pp. 1–5. [CrossRef]
29. Yu, D.; Zhu, H.; Han, W.; Holburn, D. Dynamic multi agent-based management and load frequency control of PV/Fuel cell/wind turbine/CHP in autonomous microgrid system. *Energy* **2019**, *173*, 554–568. [CrossRef]
30. Haidar, A.M.; Fakhar, A.; Helwig, A. Sustainable energy planning for cost minimization of autonomous hybrid microgrid using combined multi-objective optimization algorithm. *Sustain. Cities Soc.* **2020**, *62*, 102391. [CrossRef]
31. Banerjee, S.; Dasgupta, K.; Chanda, C.K. Short term hydro–wind–thermal scheduling based on particle swarm optimization technique. *Int. J. Electr. Power Energy Syst.* **2016**, *81*, 275–288. [CrossRef]
32. Faridnia, N.; Habibi, D.; Lachowicz, S.; Kavousifard, A. Optimal scheduling in a microgrid with a tidal generation. *Energy* **2019**, *171*, 435–443. [CrossRef]

33. Crisostomi, E.; Liu, M.; Raugi, M.; Shorten, R. Plug-and-Play Distributed Algorithms for Optimized Power Generation in a Microgrid. *IEEE Trans. Smart Grid* **2014**, *5*, 2145–2154. [CrossRef]
34. Nouri, A.; Khodaei, H.; Darvishan, A.; Sharifian, S.; Ghadimi, N. *Optimal Performance of Fuel Cell-CHP-Battery Based Micro-Grid Under Real-Time Energy Management: An Epsilon Constraint Method and Fuzzy Satisfying Approach*; Elsevier: Amsterdam, The Netherlands, 2018; Volume 159. [CrossRef]
35. Aghamohammadi, M.R.; Abdolahinia, H. A new approach for optimal sizing of battery energy storage system for primary frequency control of islanded Microgrid. *Int. J. Electr. Power Energy Syst.* **2014**, *54*, 325–333. [CrossRef]
36. Soshinskaya, M.; Crijns-Graus, W.; Guerrero, J.; Vasquez, J.C. Microgrids: Experiences, barriers and success factors. *Renew. Sustain. Energy Rev.* **2014**, *40*, 659–672. [CrossRef]
37. Konstantinopoulos, S.A.; Anastasiadis, A.G.; Vokas, G.A.; Kondylis, G.P.; Polyzakis, A. Optimal management of hydrogen storage in stochastic smart microgrid operation. *Int. J. Hydrog. Energy* **2018**, *43*, 490–499. [CrossRef]
38. Hou, J.; Song, Z.; Hofmann, H.F.; Sun, J. Control Strategy for Battery/Flywheel Hybrid Energy Storage in Electric Shipboard Microgrids. *IEEE Trans. Ind. Inform.* **2021**, *17*, 1089–1099. [CrossRef]
39. Mousavi, N.; Kothapalli, G.; Habibi, D.; Das, C.K.; Baniasadi, A. A novel photovoltaic-pumped hydro storage microgrid applicable to rural areas. *Appl. Energy* **2020**, *262*, 114284. [CrossRef]
40. Jia, K.; Chen, Y.; Bi, T.; Lin, Y.; Thomas, D.; Sumner, M. Historical-Data-Based Energy Management in a Microgrid With a Hybrid Energy Storage System. *IEEE Trans. Ind. Inform.* **2017**, *13*, 2597–2605. [CrossRef]
41. Guo, L.; Liu, W.; Jiao, B.; Hong, B.; Wang, C. Multi-objective stochastic optimal planning method for stand-alone microgrid system. *IET Gener. Transm. Distrib.* **2014**, *8*, 1263–1273. [CrossRef]
42. Hittinger, E.; Ciez, R.E. Modeling Costs and Benefits of Energy Storage Systems. *Annu. Rev. Environ. Resour.* **2020**, *45*, 445–469. [CrossRef]
43. Faisal, M.; Hannan, M.A.; Ker, P.J.; Hussain, A.; Bin Mansor, M.; Blaabjerg, F. Review of Energy Storage System Technologies in Microgrid Applications: Issues and Challenges. *IEEE Access* **2018**, *6*, 35143–35164. [CrossRef]
44. Tan, X.; Li, Q.; Wang, H. Advances and trends of energy storage technology in Microgrid. *Int. J. Electr. Power Energy Syst.* **2013**, *44*, 179–191. [CrossRef]
45. Yang, Y.; Bremner, S.; Menictas, C.; Kay, M. Battery energy storage system size determination in renewable energy systems: A review. *Renew. Sustain. Energy Rev.* **2018**, *91*, 109–125. [CrossRef]
46. Dhundhara, S.; Verma, Y.P.; Williams, A. Techno-economic analysis of the lithium-ion and lead-acid battery in microgrid systems. *Energy Convers. Manag.* **2018**, *177*, 122–142. [CrossRef]
47. Wang, J.; Zhao, C.; Pratt, A.; Baggu, M. Design of an advanced energy management system for microgrid control using a state machine. *Appl. Energy* **2018**, *228*, 2407–2421. [CrossRef]
48. Kroposki, B.; Basso, T.; DeBlasio, R. Microgrid standards and technologies. In Proceedings of the 2008 IEEE Power and Energy Society General Meeting—Conversion and Delivery of Electrical Energy in the 21st Century, Pittsburgh, PA, USA, 20–24 July 2008. [CrossRef]
49. Kumar, D.; Zare, F.; Ghosh, A. DC Microgrid Technology: System Architectures, AC Grid Interfaces, Grounding Schemes, Power Quality, Communication Networks, Applications, and Standardizations Aspects. *IEEE Access* **2017**, *5*, 12230–12256. [CrossRef]
50. Meng, L.; Shafiee, Q.; Trecate, G.F.; Karimi, H.; Fulwani, D.; Lu, X.; Guerrero, J.M. Review on Control of DC Microgrids and Multiple Microgrid Clusters. *IEEE J. Emerg. Sel. Top. Power Electron.* **2017**, *5*, 1. [CrossRef]
51. Almada, J.; Leão, R.; Sampaio, R.; Barroso, G. A centralized and heuristic approach for energy management of an AC microgrid. *Renew. Sustain. Energy Rev.* **2016**, *60*, 1396–1404. [CrossRef]
52. Abrishambaf, O.; Faria, P.; Gomes, L.; Spínola, J.; Vale, Z.; Corchado, J.M. Implementation of a Real-Time Microgrid Simulation Platform Based on Centralized and Distributed Management. *Energies* **2017**, *10*, 806. [CrossRef]
53. Ouammi, A.; Achour, Y.; Zejli, D.; Dagdougui, H. Supervisory Model Predictive Control for Optimal Energy Management of Networked Smart Greenhouses Integrated Microgrid. *IEEE Trans. Autom. Sci. Eng.* **2020**, *17*, 117–128. [CrossRef]
54. Jmii, H.; Abbes, M.; Meddeb, A.; Chebbi, S. Centralized VSM control of an AC meshed microgrid for ancillary services provision. *Int. J. Electr. Power Energy Syst.* **2020**, *115*, 105450. [CrossRef]
55. Li, Y.; Zhao, T.; Wang, P.; Gooi, H.B.; Wu, L.; Liu, Y.; Ye, J. Optimal Operation of Multimicrogrids via Cooperative Energy and Reserve Scheduling. *IEEE Trans. Ind. Inform.* **2018**, *14*, 3459–3468. [CrossRef]
56. Jayachandran, M.; Ravi, G. Predictive power management strategy for PV/battery hybrid unit based islanded AC microgrid. *Int. J. Electr. Power Energy Syst.* **2019**, *110*, 487–496. [CrossRef]
57. Zhang, Y.; Wei, W. Decentralized coordination control of PV generators, storage battery, hydrogen production unit and fuel cell in islanded DC microgrid. *Int. J. Hydrog. Energy* **2020**, *45*, 8243–8256. [CrossRef]
58. Rahim, S.; Javaid, N.; Khan, R.D.; Nawaz, N.; Iqbal, M. A convex optimization based decentralized real-time energy management model with the optimal integration of microgrid in smart grid. *J. Clean. Prod.* **2019**, *236*, 117688. [CrossRef]
59. Mallick, M.; Srikantha, P. Optimal Decentralized Microgrid Coordination via the Schur's Complement and S-Procedure. *IEEE Trans. Smart Grid* **2020**, *11*, 379–390. [CrossRef]
60. Borazjani, P.; Wahab, N.I.A.; Hizam, H.B.; Soh, A.B.C. A review on microgrid control techniques. In Proceedings of the 2014 IEEE Innovative Smart Grid Technologies—Asia (ISGT ASIA), Kuala Lumpur, Malaysia, 20–23 May 2014; pp. 749–753. [CrossRef]

61. Mohammadi, J.; Ajaei, F.B. Improved Mode-Adaptive Droop Control Strategy for the DC Microgrid. *IEEE Access* **2019**, *7*, 86421–86435. [CrossRef]
62. Datta, U.; Shi, J.; Kalam, A. Primary frequency control of a microgrid with integrated dynamic sectional droop and fuzzy based pitch angle control. *Int. J. Electr. Power Energy Syst.* **2019**, *111*, 248–259. [CrossRef]
63. Datta, U.; Kalam, A.; Shi, J. Frequency performance analysis of multi-gain droop controlled DFIG in an isolated microgrid using real-time digital simulator. *Eng. Sci. Technol. Int. J.* **2020**, *23*, 1028–1041. [CrossRef]
64. Joung, K.W.; Kim, T.; Park, J.-W. Decoupled Frequency and Voltage Control for Stand-Alone Microgrid with High Renewable Penetration. *IEEE Trans. Ind. Appl.* **2018**, *55*, 122–133. [CrossRef]
65. Kumar, R.; Pathak, M.K. Distributed droop control of dc microgrid for improved voltage regulation and current sharing. *IET Renew. Power Gener.* **2020**, *14*, 2499–2506. [CrossRef]
66. Azizi, A.; Peyghami, S.; Mokhtari, H.; Blaabjerg, F. Autonomous and decentralized load sharing and energy management approach for DC microgrids. *Electr. Power Syst. Res.* **2019**, *177*, 106009. [CrossRef]
67. Sao, C.K.; Lehn, P.W. Control and Power Management of Converter Fed Microgrids. *IEEE Trans. Power Syst.* **2008**, *23*, 1088–1098. [CrossRef]
68. Razi, R.; Iman-Eini, H.; Hamzeh, M. An Impedance-Power Droop Method for Accurate Power Sharing in Islanded Resistive Microgrids. *IEEE J. Emerg. Sel. Top. Power Electron.* **2020**, *8*, 3763–3771. [CrossRef]
69. Wang, Y.; Wen, W.; Wang, C.; Liu, H.; Zhan, X.; Xiao, X. Adaptive Voltage Droop Control of Multiterminal VSC-HVDC Systems for DC Voltage Deviation and Power Sharing. *IEEE Trans. Power Deliv.* **2018**, *34*, 1. [CrossRef]
70. Moussa, H.; Shahin, A.; Martin, J.-P.; Pierfederici, S.; Moubayed, N. Optimal Angle Droop for Power Sharing Enhancement With Stability Improvement in Islanded Microgrids. *IEEE Trans. Smart Grid* **2018**, *9*, 5014–5026. [CrossRef]
71. He, J.; Li, Y.W. An Enhanced Microgrid Load Demand Sharing Strategy. *IEEE Trans. Power Electron.* **2012**, *27*, 3984–3995. [CrossRef]
72. Hamzeh, M.; Mokhtari, H.; Karimi, H. A decentralized self-adjusting control strategy for reactive power management in an islanded multi-bus MV microgrid. *Can. J. Electr. Comput. Eng.* **2013**, *36*, 18–25. [CrossRef]
73. Sahoo, S.K.; Sinha, A.K.; Kishore, N.K. Control Techniques in AC, DC, and Hybrid AC–DC Microgrid: A Review. *IEEE J. Emerg. Sel. Top. Power Electron.* **2018**, *6*, 738–759. [CrossRef]
74. Lu, B.; Shahidehpour, M. Short-Term Scheduling of Battery in a Grid-Connected PV/Battery System. *IEEE Trans. Power Syst.* **2005**, *20*, 1053–1061. [CrossRef]
75. Luna, A.C.; Diaz, N.L.; Graells, M.; Vasquez, J.C.; Guerrero, J.M. Mixed-Integer-Linear-Programming-Based Energy Management System for Hybrid PV-Wind-Battery Microgrids: Modeling, Design, and Experimental Verification. *IEEE Trans. Power Electron.* **2017**, *32*, 2769–2783. [CrossRef]
76. Jabarnejad, M. Facilitating emission reduction using the dynamic line switching and rating. *Electr. Power Syst. Res.* **2020**, *189*, 106600. [CrossRef]
77. Sarabi, S.; Davigny, A.; Riffonneau, Y.; Robyns, B. V2G electric vehicle charging scheduling for railway station parking lots based on binary linear programming. In Proceedings of the 2016 IEEE International Energy Conference (ENERGYCON), Leuven, Belgium, 4–8 April 2016; pp. 1–6.
78. Riffonneau, Y.; Bacha, S.; Barruel, F.; Ploix, S. Optimal Power Flow Management for Grid Connected PV Systems with Batteries. *IEEE Trans. Sustain. Energy* **2011**, *2*, 309–320. [CrossRef]
79. Maroufmashat, A.; Fowler, M.; Khavas, S.S.; Elkamel, A.; Roshandel, R.; Hajimiragha, A. Mixed integer linear programing based approach for optimal planning and operation of a smart urban energy network to support the hydrogen economy. *Int. J. Hydrog. Energy* **2016**, *41*, 7700–7716. [CrossRef]
80. Dong, J.; Nie, S.; Huang, H.; Yang, P.; Fu, A.; Lin, J. Research on Economic Operation Strategy of CHP Microgrid Considering Renewable Energy Sources and Integrated Energy Demand Response. *Sustainability* **2019**, *11*, 4825. [CrossRef]
81. Garcia-Torres, F.; Bordons, C. Optimal Economical Schedule of Hydrogen-Based Microgrids with Hybrid Storage Using Model Predictive Control. *IEEE Trans. Ind. Electron.* **2015**, *62*, 5195–5207. [CrossRef]
82. Sultana, U.; Khairuddin, A.B.; Aman, M.; Mokhtar, A.; Zareen, N. A review of optimum DG placement based on minimization of power losses and voltage stability enhancement of distribution system. *Renew. Sustain. Energy Rev.* **2016**, *63*, 363–378. [CrossRef]
83. Chiang, Y.-H.; Sean, W.-Y.; Wu, C.-H.; Huang, C.-Y. Development of a converterless energy management system for reusing automotive lithium-ion battery applied in smart-grid balancing. *J. Clean. Prod.* **2017**, *156*, 750–756. [CrossRef]
84. Ju, L.; Tan, Z.; Yuan, J.; Tan, Q.; Li, H.; Dong, F. A bi-level stochastic scheduling optimization model for a virtual power plant connected to a wind–photovoltaic–energy storage system considering the uncertainty and demand response. *Appl. Energy* **2016**, *171*, 184–199. [CrossRef]
85. Zhao, H.; Lu, H.; Li, B.; Wang, X.; Zhang, S.; Wang, Y. Stochastic Optimization of Microgrid Participating Day-Ahead Market Operation Strategy with Consideration of Energy Storage System and Demand Response. *Energies* **2020**, *13*, 1255. [CrossRef]
86. Wang, Z.; Song, X.; Xin, H.; Gan, D.; Wong, K.P. Risk-Based Coordination of Generation Rescheduling and Load Shedding for Transient Stability Enhancement. *IEEE Trans. Power Syst.* **2013**, *28*, 4674–4682. [CrossRef]
87. Cao, Y.; Tang, S.; Li, C.; Zhang, P.; Tan, Y.; Zhang, Z.; Li, J. An Optimized EV Charging Model Considering TOU Price and SOC Curve. *IEEE Trans. Smart Grid* **2011**, *3*, 388–393. [CrossRef]
88. Tushar, W.; Yuen, C.; Huang, S.; Smith, D.B.; Poor, H.V. Cost Minimization of Charging Stations with Photovoltaics: An Approach. *IEEE Trans. Intell. Transp. Syst.* **2015**, *17*, 156–169. [CrossRef]

89. Giraldo, J.S.; Castrillon, J.A.; Lopez, J.C.; Rider, M.J.; Castro, C.A. Microgrids Energy Management Using Robust Convex Programming. *IEEE Trans. Smart Grid* **2018**, *10*, 4520–4530. [CrossRef]
90. Mohammadi-Ivatloo, B.; Rabiee, A.; Soroudi, A.; Ehsan, M. Imperialist competitive algorithm for solving non-convex dynamic economic power dispatch. *Energy* **2012**, *44*, 228–240. [CrossRef]
91. Elsied, M.; Oukaour, A.; Youssef, T.; Gualous, H.; Mohammed, O. An advanced real time energy management system for microgrids. *Energy* **2016**, *114*, 742–752. [CrossRef]
92. Grisales-Noreña, L.; Montoya, O.D.; Ramos-Paja, C.A. An energy management system for optimal operation of BSS in DC distributed generation environments based on a parallel PSO algorithm. *J. Energy Storage* **2020**, *29*, 101488. [CrossRef]
93. Marzband, M.; Azarinejadian, F.; Savaghebi, M.; Guerrero, J.M. An Optimal Energy Management System for Islanded Microgrids Based on Multiperiod Artificial Bee Colony Combined With Markov Chain. *IEEE Syst. J.* **2017**, *11*, 1712–1722. [CrossRef]
94. Roy, K.; Mandal, K.K.; Mandal, A.C. Ant-Lion Optimizer algorithm and recurrent neural network for energy management of micro grid connected system. *Energy* **2019**, *167*, 402–416. [CrossRef]
95. Yu, J.-T.; Kim, C.-H.; Wadood, A.; Khurshaid, T.; Rhee, S.-B. Jaya Algorithm with Self-Adaptive Multi-Population and Lévy Flights for Solving Economic Load Dispatch Problems. *IEEE Access* **2019**, *7*, 21372–21384. [CrossRef]
96. Barisal, A.K.; Pusty, R.C.; Suna, K.; Panigrahi, T.K. D ouble Teaching Optimization for Short Term Hydro Thermal Scheduling Problem. *Int. J. Electr. Eng. Inform.* **2017**, *9*, 322–341. [CrossRef]
97. Yuan, X.; Cao, B.; Yang, B.; Yuan, Y. Hydrothermal scheduling using chaotic hybrid differential evolution. *Energy Convers. Manag.* **2008**, *49*, 3627–3633. [CrossRef]
98. Qiu, J.; Chen, M.; Wei, Z.; Xu, Z.; Li, Y. Planning and Optimal Scheduling Method of Regional Integrated Energy System Based on Gray Wolf Optimizer Algorithm. *IOP Conf. Ser. Earth Environ. Sci.* **2020**, *546*, 2. [CrossRef]
99. Zhang, H.; Yang, J.; Zhang, J.; Song, P.; Xu, X. A Firefly Algorithm Optimization-Based Equivalent Consumption Minimization Strategy for Fuel Cell Hybrid Light Rail Vehicle. *Energies* **2019**, *12*, 2665. [CrossRef]
100. Mohseni, S.; Brent, A.C.; Burmester, D. A comparison of metaheuristics for the optimal capacity planning of an isolated, battery-less, hydrogen-based micro-grid. *Appl. Energy* **2020**, *259*, 114224. [CrossRef]
101. Mohammadi, F.; Abdi, H. A modified crow search algorithm (MCSA) for solving economic load dispatch problem. *Appl. Soft Comput.* **2018**, *71*, 51–65. [CrossRef]
102. Javaid, N.; Naseem, M.; Rasheed, M.B.; Mahmood, D.; Khan, S.A.; Alrajeh, N.; Iqbal, Z. A new heuristically optimized Home Energy Management controller for smart grid. *Sustain. Cities Soc.* **2017**, *34*, 211–227. [CrossRef]
103. Singh, S.; Chauhan, P.; Singh, N. Capacity optimization of grid connected solar/fuel cell energy system using hybrid ABC-PSO algorithm. *Int. J. Hydrog. Energy* **2020**, *45*, 10070–10088. [CrossRef]
104. Nadeem, Z.; Javaid, N.; Malik, A.W.; Iqbal, S. Scheduling Appliances with GA, TLBO, FA, OSR and Their Hybrids Using Chance Constrained Optimization for Smart Homes. *Energies* **2018**, *11*, 888. [CrossRef]
105. Sivasubramani, S.; Swarup, K.S. Hybrid DE–SQP algorithm for non-convex short term hydrothermal scheduling problem. *Energy Convers. Manag.* **2011**, *52*, 757–761. [CrossRef]
106. Rex, C.R.E.S.; Beno, M.M.; Annrose, J. A Solution for Combined Economic and Emission Dispatch Problem using Hybrid Optimization Techniques. *J. Electr. Eng. Technol.* **2019**, 1–10. [CrossRef]
107. Mokarram, M.J.; Niknam, T.; Aghaei, J.; Shafie-Khah, M.; Catalao, J.P.S. Hybrid Optimization Algorithm to Solve the Nonconvex Multiarea Economic Dispatch Problem. *IEEE Syst. J.* **2019**, *13*, 3400–3409. [CrossRef]
108. Hussain, H.M.; Javaid, N.; Iqbal, S.; Hasan, Q.U.; Aurangzeb, K.; Alhussein, M. An Efficient Demand Side Management System with a New Optimized Home Energy Management Controller in Smart Grid. *Energies* **2018**, *11*, 190. [CrossRef]
109. Hafeez, G.; Wadud, Z.; Khan, I.U.; Khan, I.; Shafiq, Z.; Usman, M.; Khan, M.U.A.; Khan, I. Efficient Energy Management of IoT-Enabled Smart Homes Under Price-Based Demand Response Program in Smart Grid. *Sensors* **2020**, *20*, 3155. [CrossRef]
110. Aljohani, T.; Ebrahim, A.; Mohammed, O. Dynamic Real-Time Pricing Structure for Electric Vehicle Charging Considering Stochastic Microgrids Energy Management System. In Proceedings of the 2020 IEEE International Conference on Environment and Electrical Engineering and 2020 IEEE Industrial and Commercial Power Systems Europe (EEEIC I&CPS Europe), Madrid, Spain, 9–12 June 2020; pp. 1–8.
111. Khan, M.R.B.; Jidin, R.; Pasupuleti, J. Multi-agent based distributed control architecture for microgrid energy management and optimization. *Energy Convers. Manag.* **2016**, *112*, 288–307. [CrossRef]
112. Hosseinnezhad, V.; Shafie-Khah, M.; Siano, P.; Catalao, J.P.S. An Optimal Home Energy Management Paradigm With an Adaptive Neuro-Fuzzy Regulation. *IEEE Access* **2020**, *8*, 19614–19628. [CrossRef]
113. Swain, R.; Sarkar, P.; Meher, K.C.; Chanda, C.K. Population variant differential evolution-based multiobjective economic emission load dispatch. *Int. Trans. Electr. Energy Syst.* **2017**, *27*, e2378. [CrossRef]
114. Xiong, L.; Li, P.; Wang, Z.; Wang, J. Multi-agent based multi objective renewable energy management for diversified community power consumers. *Appl. Energy* **2020**, *259*, 114140. [CrossRef]
115. Teo, T.T.; Logenthiran, T.; Woo, W.L.; Abidi, K.; John, T.; Wade, N.S.; Greenwood, D.M.; Patsios, C.; Taylor, P.C. Optimization of Fuzzy Energy-Management System for Grid-Connected Microgrid Using NSGA-II. *IEEE Trans. Cybern.* **2020**, 1–12. [CrossRef] [PubMed]
116. Sun, B. A multi-objective optimization model for fast electric vehicle charging stations with wind, PV power and energy storage. *J. Clean. Prod.* **2021**, *288*, 125564. [CrossRef]

117. Younes, Z.; Alhamrouni, I.; Mekhilef, S.; Reyasudin, M. A memory-based gravitational search algorithm for solving economic dispatch problem in micro-grid. *Ain. Shams Eng. J.* **2021**, *12*, 1985–1994. [CrossRef]
118. Wang, L.; Collins, E.G.; Li, H. Optimal Design and Real-Time Control for Energy Management in Electric Vehicles. *IEEE Trans. Veh. Technol.* **2011**, *60*, 1419–1429. [CrossRef]
119. Leou, R.-C. Optimal Charging/Discharging Control for Electric Vehicles Considering Power System Constraints and Operation Costs. *IEEE Trans. Power Syst.* **2015**, *31*, 1854–1860. [CrossRef]
120. Taha, M.S.; Abdeltawab, H.; Mohamed, Y.A.-R.I. An Online Energy Management System for a Grid-Connected Hybrid Energy Source. *IEEE J. Emerg. Sel. Top. Power Electron.* **2018**, *6*, 2015–2030. [CrossRef]
121. Yongqiang, Z.; Tianjing, W. Comparison of centralised and distributed energy storage configuration for AC/DC hybrid microgrid. *J. Eng.* **2017**, *2017*, 1838–1842. [CrossRef]
122. Garmabdari, R.; Moghimi, M.; Yang, F.; Gray, E.; Lu, J.-W. Multi Energy System Modelling and Operation Optimisation for University Research Facility. In Proceedings of the 2018 IEEE International Conference on Environment and Electrical Engineering and 2018 IEEE Industrial and Commercial Power Systems Europe (EEEIC/I&CPS Europe), Palermo, Italy, 12–15 June 2018; pp. 1–6.
123. Balasubramaniam, K.; Saraf, P.; Hadidi, R.; Makram, E.B. Energy management system for enhanced resiliency of microgrids during islanded operation. *Electr. Power Syst. Res.* **2016**, *137*, 133–141. [CrossRef]
124. Mahzarnia, M.; Moghaddam, M.P.; Baboli, P.T.; Siano, P. A Review of the Measures to Enhance Power Systems Resilience. *IEEE Syst. J.* **2020**, *14*, 4059–4070. [CrossRef]
125. Poudel, S.; Dubey, A.; Schneider, K.P. A Generalized Framework for Service Restoration in a Resilient Power Distribution System. *IEEE Syst. J.* **2021**, 1–13. [CrossRef]
126. Gholami, M.; Moshtagh, J.; Ghadernejad, N. Service restoration in distribution networks using combination of two heuristic methods considering load shedding. *J. Mod. Power Syst. Clean Energy* **2015**, *3*, 556–564. [CrossRef]
127. AlOwaifeer, M.; Almuhaini, M. Load Priority Modeling for Smart Service Restoration. *Can. J. Electr. Comput. Eng.* **2017**, *40*, 217–228. [CrossRef]
128. Shen, Y.; Chen, Y.; Zhang, J.; Sang, Z.; Zhou, Q. Self-Healing Evaluation of Smart Distribution Network Based on Uncertainty Theory. *IEEE Access* **2019**, *7*, 140022–140029. [CrossRef]
129. Jiao, Z.; Wang, X.; Gong, H. Wide area measurement/wide area information-based control strategy to fast relieve overloads in a self-healing power grid. *IET Gener. Transm. Distrib.* **2014**, *8*, 1168–1176. [CrossRef]
130. Yang, L.-J.; Zhao, Y.; Wang, C.; Gao, P.; Hao, J.-H. Resilience-Oriented Hierarchical Service Restoration in Distribution System Considering Microgrids. *IEEE Access* **2019**, *7*, 152729–152743. [CrossRef]
131. Zidan, A.; El-Saadany, E.F. A Cooperative Multiagent Framework for Self-Healing Mechanisms in Distribution Systems. *IEEE Trans. Smart Grid* **2012**, *3*, 1525–1539. [CrossRef]
132. Romero, R.; Franco, J.F.; Leão, F.B.; Rider, M.J.; De Souza, E.S. A New Mathematical Model for the Restoration Problem in Balanced Radial Distribution Systems. *IEEE Trans. Power Syst.* **2016**, *31*, 1259–1268. [CrossRef]
133. Lin, Z.; Wen, F.; Xue, Y. A Restorative Self-Healing Algorithm for Transmission Systems Based on Complex Network Theory. *IEEE Trans. Smart Grid* **2016**, *7*, 2154–2162. [CrossRef]
134. Liu, W.; Lin, Z.; Wen, F.; Ledwich, G. A wide area monitoring system based load restoration method. *IEEE Trans. Power Syst.* **2013**, *28*, 2025–2034. [CrossRef]
135. Qianqian, L.; Zeng, X.; Xue, M.; Xiang, L. A new smart distribution grid fault self-healing system based on traveling-wave. In Proceedings of the 2013 IEEE Industry Applications Society Annual Meeting, Lake Buena Vista, FL, USA, 6–11 October 2013; pp. 1–6. [CrossRef]
136. Patsakis, G.; Rajan, D.; Aravena, I.; Rios, J.; Oren, S. Optimal Black Start Allocation for Power System Restoration. *IEEE Trans. Power Syst.* **2018**, *33*, 6766–6776. [CrossRef]
137. Cao, X.; Wang, H.; Liu, Y.; Azizipanah-Abarghooee, R.; Terzija, V. Coordinating self-healing control of bulk power transmission system based on a hierarchical top-down strategy. *Int. J. Electr. Power Energy Syst.* **2017**, *90*, 147–157. [CrossRef]
138. Leite, J.B.; Mantovani, J.R.S. Development of a Self-Healing Strategy with Multiagent Systems for Distribution Networks. *IEEE Trans. Smart Grid* **2017**, *8*, 2198–2206. [CrossRef]
139. Wang, S.; Chiang, H.-D. Multi-objective service restoration of distribution systems using user-centered methodology. *Int. J. Electr. Power Energy Syst.* **2016**, *80*, 140–149. [CrossRef]
140. Gu, X.; Zhong, H. Optimisation of network reconfiguration based on a two-layer unit-restarting framework for power system restoration. *IET Gener. Transm. Distrib.* **2012**, *6*, 693–700. [CrossRef]
141. Chen, C.; Wang, J.; Qiu, F.; Zhao, D. Resilient Distribution System by Microgrids Formation after Natural Disasters. *IEEE Trans. Smart Grid* **2016**, *7*, 958–966. [CrossRef]
142. Wang, Z.; Wang, J. Self-Healing Resilient Distribution Systems Based on Sectionalization into Microgrids. *IEEE Trans. Power Syst.* **2015**, *30*, 3139–3149. [CrossRef]
143. Dimitrijevic, S.; Rajaković, N. Service Restoration of Distribution Networks Considering Switching Operation Costs and Actual Status of the Switching Equipment. *IEEE Trans. Smart Grid* **2015**, *6*, 1227–1232. [CrossRef]
144. Rodríguez-Montañés, M.; Macías, J.A.R.; Gómez-Expósito, A. A systematic approach to service restoration in distribution networks. *Electr. Power Syst. Res.* **2020**, *189*, 106539. [CrossRef]

145. Wang, F.; Xiao, X.; Sun, Q.; Chen, C.; Bin, F.; Chen, S.; Fan, J. Service restoration for distribution network with DGs based on stochastic response surface method. *Int. J. Electr. Power Energy Syst.* **2019**, *107*, 557–568. [CrossRef]
146. Widiputra, V.; Jufri, F.H.; Jung, J. Development of service restoration algorithm under cold load pickup condition using conservation voltage reduction and particle swarm optimization. *Int. Trans. Electr. Energy Syst.* **2020**, *30*, 1–16. [CrossRef]
147. Vieira, A.R.; Batista, V.S.; Júnior, G.C.; Vieira, J.P.A. Integrating Protection Constraints to a MEAN-Based Method for Service Restoration in Radial Distribution Systems. *Electr. Power Syst. Res.* **2021**, *191*, 106851. [CrossRef]
148. Ma, S.; Chen, B.; Wang, Z. Resilience Enhancement Strategy for Distribution Systems under Extreme Weather Events. *IEEE Trans. Smart Grid* **2018**, *9*, 1442–1451. [CrossRef]
149. Arif, A.; Wang, Z. Networked microgrids for service restoration in resilient distribution systems. *IET Gener. Transm. Distrib.* **2017**, *11*, 3612–3619. [CrossRef]
150. Xu, Y.; Liu, C.-C.; Wang, Z.; Mo, K.; Schneider, K.P.; Tuffner, F.K.; Ton, D.T. DGs for Service Restoration to Critical Loads in a Secondary Network. *IEEE Trans. Smart Grid* **2019**, *10*, 435–447. [CrossRef]
151. Khatib-Tohidkhaneh, F.; Mohammadi-Hosseininejad, S.M.; Saberi, H. A probabilistic electric power distribution automation operational planning approach considering energy storage incorporation in service restoration. *Int. Trans. Electr. Energy Syst.* **2020**, *30*, 1–13. [CrossRef]
152. Abniki, H.; Taghvaei, S.M.; Mohammadi-Hosseininejad, S.M. Reliability improvement in smart grid through incorporating energy storage systems in service restoration. *Int. Trans. Electr. Energy Syst.* **2019**, *29*, e2661. [CrossRef]
153. Kumar, Y.; Das, B.; Sharma, J. Multiobjective, Multiconstraint Service Restoration of Electric Power Distribution System with Priority Customers. *IEEE Trans. Power Deliv.* **2007**, *23*, 261–270. [CrossRef]
154. Lopez, J.C.; Franco, J.F.; Rider, M.J.; Romero, R. Optimal Restoration/Maintenance Switching Sequence of Unbalanced Three-Phase Distribution Systems. *IEEE Trans. Smart Grid* **2018**, *9*, 6058–6068. [CrossRef]
155. Luo, A.; Xu, Q.; Ma, F.; Chen, Y. Overview of power quality analysis and control technology for the smart grid. *J. Mod. Power Syst. Clean Energy* **2016**, *4*, 1–9. [CrossRef]
156. Mousavi, S.Y.M.; Jalilian, A.; Savaghebi, M.; Guerrero, J.M. Power quality enhancement and power management of a multifunctional interfacing inverter for PV and battery energy storage system. *Int. Trans. Electr. Energy Syst.* **2018**, *28*, e2643. [CrossRef]
157. Mei, R.N.S.; Sulaiman, M.H.; Mustaffa, Z.; Daniyal, H. Optimal reactive power dispatch solution by loss minimization using moth-flame optimization technique. *Appl. Soft Comput.* **2017**, *59*, 210–222. [CrossRef]
158. Wang, J.; Zhou, N.; Wang, Q. Data-driven stochastic service restoration in unbalanced active distribution networks with multi-terminal soft open points. *Int. J. Electr. Power Energy Syst.* **2020**, *121*, 106069. [CrossRef]
159. Sahoo, B.; Routray, S.K.; Rout, P.K. A novel centralized energy management approach for power quality improvement. *Int. Trans. Electr. Energy Syst.* **2020**, e12582. [CrossRef]
160. Kumar, R.; Bansal, H.O.; Kumar, D. Improving power quality and load profile using PV-Battery-SAPF system with metaheuristic tuning and its HIL validation. *Int. Trans. Electr. Energy Syst.* **2020**, *30*, 1–19. [CrossRef]
161. Aljohani, T.M.; Ebrahim, A.F.; Mohammed, O. Hybrid Microgrid Energy Management and Control Based on Metaheuristic-Driven Vector-Decoupled Algorithm Considering Intermittent Renewable Sources and Electric Vehicles Charging Lot. *Energies* **2020**, *13*, 3423. [CrossRef]
162. Nasr, M.-A.; Nikkhah, S.; Gharehpetian, G.B.; Nasr-Azadani, E.; Hosseinian, S.H. A multi-objective voltage stability constrained energy management system for isolated microgrids. *Int. J. Electr. Power Energy Syst.* **2020**, *117*, 105646. [CrossRef]
163. Han, Y.; Chen, W.; Li, Q.; Yang, H.; Zare, F.; Zheng, Y. Two-level energy management strategy for PV-Fuel cell-battery-based DC microgrid. *Int. J. Hydrog. Energy* **2019**, *44*, 19395–19404. [CrossRef]
164. Agnoletto, E.J.; de Castro, D.S.; Neves, R.; Machado, R.Q.; Oliveira, V.A. An Optimal Energy Management Technique Using the ε-Constraint Method for Grid-Tied and Stand-Alone Battery-Based Microgrids. *IEEE Access* **2019**, *7*, 165928–165942. [CrossRef]
165. Zhang, M.; Li, Y. Multi-Objective Optimal Reactive Power Dispatch of Power Systems by Combining Classification-Based Multi-Objective Evolutionary Algorithm and Integrated Decision Making. *IEEE Access* **2020**, *8*, 38198–38209. [CrossRef]
166. Leonori, S.; De Santis, E.; Rizzi, A.; Mascioli, F.F. Optimization of a microgrid energy management system based on a Fuzzy Logic Controller. In Proceedings of the IECON 2016—42nd Annual Conference of the IEEE Industrial Electronics Society, Florence, Italy, 23–26 October 2016; pp. 6615–6620.
167. Yamashita, D.Y.; Vechiu, I.; Gaubert, J.-P. A review of hierarchical control for building microgrids. *Renew. Sustain. Energy Rev.* **2020**, *118*, 109523. [CrossRef]
168. Chaouachi, A.; Kamel, R.M.; Andoulsi, R.; Member, K.N. Multiobjective Intelligent Energy Management for a Microgrid. *IEEE Trans. Ind. Electron.* **2011**, *60*, 1688–1699. [CrossRef]
169. Qing, X.; Niu, Y. Hourly day-ahead solar irradiance prediction using weather forecasts by LSTM. *Energy* **2018**, *148*, 461–468. [CrossRef]
170. Pedro, H.T.; Coimbra, C.F. Short-term irradiance forecastability for various solar micro-climates. *Sol. Energy* **2015**, *122*, 587–602. [CrossRef]
171. Ramsami, P.; Oree, V. A hybrid method for forecasting the energy output of photovoltaic systems. *Energy Convers. Manag.* **2015**, *95*, 406–413. [CrossRef]

172. Raza, M.Q.; Mithulananthan, N.; Li, J.; Lee, K.Y.; Gooi, H.B.; Nadarajah, M. An Ensemble Framework for Day-Ahead Forecast of PV Output Power in Smart Grids. *IEEE Trans. Ind. Inform.* **2019**, *15*, 4624–4634. [CrossRef]
173. Alzahrani, A.; Shamsi, P.; Dagli, C.; Ferdowsi, M. Solar Irradiance Forecasting Using Deep Neural Networks. *Procedia Comput. Sci.* **2017**, *114*, 304–313. [CrossRef]
174. Mengelkamp, E.; Gärttner, J.; Rock, K.; Kessler, S.; Orsini, L.; Weinhardt, C. Designing microgrid energy markets: A case study: The Brooklyn Microgrid. *Appl. Energy* **2018**, *210*, 870–880. [CrossRef]
175. Khorasany, M.; Azuatalam, D.; Glasgow, R.; Liebman, A.; Razzaghi, R. Transactive Energy Market for Energy Management in Microgrids: The Monash Microgrid Case Study. *Energies* **2020**, *13*, 2010. [CrossRef]
176. Xue, L.; Teng, Y.; Zhang, Z.; Li, J.; Wang, K.; Huang, Q. Blockchain technology for electricity market in microgrid. In Proceedings of the 2017 2nd International Conference on Power and Renewable Energy (ICPRE), Chengdu, China, 20–23 September 2017; pp. 704–708. [CrossRef]
177. Di Silvestre, M.L.; Gallo, P.; Sanseverino, E.R.; Sciume, G.; Zizzo, G. Aggregation and remuneration in Demand-Response with a blockchain-based framework. *IEEE Trans. Ind. Appl.* **2020**, *56*, 1. [CrossRef]
178. Janko, S.A.; Johnson, N.G. Scalable multi-agent microgrid negotiations for a transactive energy market. *Appl. Energy* **2018**, *229*, 715–727. [CrossRef]
179. Wang, B.; Dabbaghjamanesh, M.; Kavousi-Fard, A.; Mehraeen, S. Cybersecurity Enhancement of Power Trading Within the Networked Microgrids Based on Blockchain and Directed Acyclic Graph Approach. *IEEE Trans. Ind. Appl.* **2019**, *55*, 7300–7309. [CrossRef]
180. Bansal, G.; Bhatia, A. A Fast, Secure and Distributed Consensus Mechanism for Energy Trading among Vehicles using Hashgraph. In Proceedings of the 2020 International Conference on Information Networking (ICOIN), Barcelona, Spain, 7–10 January 2020; pp. 772–777. [CrossRef]
181. Chen, J. Flowchain: A Distributed Ledger Designed for Peer-to-Peer IoT Networks and Real-time Data Transactions. In Proceedings of the 2nd International Workshop on Linked Data and Distributed Ledgers, Portorož, Slovenia, 29 May 2017.
182. Zhang, N.; Yan, Y.; Xu, S.; Su, W. Game-theory-based electricity market clearing mechanisms for an open and transactive distribution grid. In Proceedings of the 2015 IEEE Power & Energy Society General Meeting, Boston, MA, USA, 26–30 July 2015; pp. 1–5. [CrossRef]
183. Saleh, K.A.; Hooshyar, A.; El-Saadany, E.F.; Zeineldin, H.H. Voltage-Based Protection Scheme for Faults within Utility-Scale Photovoltaic Arrays. *IEEE Trans. Smart Grid* **2017**, *9*, 4367–4382. [CrossRef]
184. Chilukuri, S.; Alla, M.; Johnson, B.K. Enhancing backup protection for thermal power generating stations using sampled values. In Proceedings of the 2017 North American Power Symposium (NAPS), Morgantown, WV, USA, 17–19 September 2017; pp. 1–6. [CrossRef]
185. Bouhouras, A.S.; Gkaidatzis, P.A.; Panagiotou, E.; Poulakis, N.; Christoforidis, G. A NILM algorithm with enhanced disaggregation scheme under harmonic current vectors. *Energy Build.* **2019**, *183*, 392–407. [CrossRef]
186. Tsimtsios, A.M.; Korres, G.N.; Nikolaidis, V.C. A pilot-based distance protection scheme for meshed distribution systems with distributed generation. *Int. J. Electr. Power Energy Syst.* **2019**, *105*, 454–469. [CrossRef]
187. Coffele, F.; Booth, C.; Dysko, A. An Adaptive Overcurrent Protection Scheme for Distribution Networks. *IEEE Trans. Power Deliv.* **2015**, *30*, 561–568. [CrossRef]
188. Chaitanya, B.K.; Yadav, A.; Pazoki, M. An improved differential protection scheme for micro-grid using time-frequency transform. *Int. J. Electr. Power Energy Syst.* **2019**, *111*, 132–143. [CrossRef]
189. Azimian, M.; Amir, V.; Javadi, S. Economic and Environmental Policy Analysis for Emission-Neutral Multi-Carrier Microgrid Deployment. *Appl. Energy* **2020**, *277*, 115609. [CrossRef]
190. Ali, A.; Li, W.; Hussain, R.; He, X.; Williams, B.W.; Memon, A.H. Overview of Current Microgrid Policies, Incentives and Barriers in the European Union, United States and China. *Sustainability* **2017**, *9*, 1146. [CrossRef]
191. Feng, W.; Jin, M.; Liu, X.; Bao, Y.; Marnay, C.; Yao, C.; Yu, J. A review of microgrid development in the United States — A decade of progress on policies, demonstrations, controls, and software tools. *Appl. Energy* **2018**, *228*, 1656–1668. [CrossRef]
192. Chen, W.; Wei, P. Socially optimal deployment strategy and incentive policy for solar photovoltaic community microgrid: A case of China. *Energy Policy* **2018**, *116*, 86–94. [CrossRef]
193. International Renewable Energy Agency, "IRENA". 2019. Available online: https://www.irena.org (accessed on 28 August 2021).

MDPI
St. Alban-Anlage 66
4052 Basel
Switzerland
Tel. +41 61 683 77 34
Fax +41 61 302 89 18
www.mdpi.com

*Sustainability* Editorial Office
E-mail: sustainability@mdpi.com
www.mdpi.com/journal/sustainability

www.ingramcontent.com/pod-product-compliance
Lightning Source LLC
Chambersburg PA
CBHW041947260926
51850CB00018B/1946
* 9 7 8 3 0 3 6 5 8 4 1 6 4 *